GW00373954

PICmicro®
Microcontroller
Pocket
Reference

Myke Predko

McGraw-Hill

New York   San Francisco   Washington, D.C.   Auckland   Bogotá
Caracas   Lisbon   London   Madrid   Mexico City   Milan
Montreal   New Delhi   San Juan   Singapore
Sydney   Tokyo   Toronto

McGraw-Hill

A Division of The McGraw-Hill Companies

1 2 3 4 5 6 7 8 9 0   DOC/DOC   0 6 5 4 3 2 1 0

ISBN 0-07-136175-8

The sponsoring editor for this book was Scott Grillo and the production supervisor was Sherri Souffrance. It was set in Century Schoolbook by The PRD Group, Inc.

Printed and bound by R. R. Donnelley & Sons Company.

McGraw-Hill books are available at special quantity discounts to use as premiums and sales promotions, or for use in corporate training programs. For more information, please write to the Director of Special Sales, Professional Publishing, McGraw-Hill, Two Penn Plaza, New York, NY 10121-2298. Or contact your local bookstore.

This book is printed on acid-free paper.

Library of Congress Cataloging-in-Publication Data
Predko, Michael.
    Picmicro microcontroller pocket reference / Myke Predko.
        p.   cm.
    ISBN 0-07-136175-8 (alk. paper)
    1. Programmable controllers.   I. Title.
TJ223.P76P735   2000
629.8'95—dc21   00-046063

Contents

Chapter 9. PICmicro® MCU Programming   379

x   Contents

1

Conventions Used in This Book

Hz	Hertz (Cycles per Second)
kHz	Kilohertz (Thousands of Cycles per Second)
MHz	Megahertz (Millions of Cycles per Second)
GHz	Gigahertz (Billions of Cycles per Second)
bps	Bits per Second
kbps	Thousands of Bits per Second
mbps	Millions of Bits per Second
KBytes	1,024 Bytes
MBytes	1,048,576 Bytes
GBytes	1,073,741,824 Bytes

K	1,000 ohms
uF	microfarads
ms/msecs	milliseconds
us/usecs	microseconds
0x0nn, $nn, 0nnh, and H'nn'	Hex Numbers
0b0nnn, %nnn, 0nnnb, and B'nnn'	Binary Number
nnn, 0nnnd, and .nnn	Decimal Number
AND and &	Bitwise "AND"
OR and |	Bitwise "OR"
XOR and ^	Bitwise "XOR"
_Label	Negative Active Pin. In some manufacturer's data sheets this is represented with a leading "!" character or with a bar over the entire label.
[parameter]	The parameter is optional.
parameter | parameter	One or another parameter can be used.

2

PICmicro® MCU
Part Number
Feature Comparison

Feature to Part Number Table

The following table lists the different PICmicro® MCU families with the features that are specific to them.

Part Number	Features
PIC12C5xx	8-Pin PICmicro® MCU. 12-Bit (Low-End) Processor. Internal Reset & Oscillator.
PIC12C6xx	8-Pin PICmicro® MCU. 14-Bit (Mid-Range) Processor. 8-Bit ADC/Internal Reset & Oscillator/Optional EEPROM Data Memory.
PIC14C000	28-Pin PICmicro® MCU. 14-Bit (Mid-Range) Processor. Advanced ADC/Internal Voltage Reference/On Chip Temperature Sensor.
PIC16C5x	18- to 28-Pin PICmicro® MCU. 12-Bit (Low-End) Processor.
PIC16C505	14-Pin PICmicro® MCU. 12-Bit (Low-End) Processor. Internal Reset & Oscillator.
PIC16HV540	18-Pin PICmicro® MCU. 12-Bit (Low-End) Processor. Extended Vdd Capabilities with Built-In Regulator.
PIC16C55x	18-Pin PICmicro® MCU. 14-Bit (Mid-Range) Processor.
PIC16C6x	18- to 40-Pin PICmicro® MCU. 14-Bit (Mid-Range) Processor. Optional TMR1 & TMR2/Optional SPI/Optional USART/Optional PSP.
PIC16C62x	18-Pin PICmicro® MCU. 14-Bit (Mid-Range) Processor. Voltage Comparators Built-In with Voltage Reference/Optional EEPROM Data Memory.
PIC16F62x	18-Pin PICmicro® MCU. 14-Bit (Mid-Range) Processor. Flash Program Memory/Voltage Comparators Built-In with Voltage Reference/Internal Reset & Oscillator.
PIC16C642	28-Pin PICmicro® MCU. 14-Bit (Mid-Range) Processor. Voltage Comparators Built-In with Voltage Reference.

PIC16C662	40-Pin PICmicro® MCU. 14-Bit (Mid-Range) Processor. Voltage Comparators Built-In with Voltage Reference.
PIC16C71x	18-Pin PICmicro® MCU. 14-Bit (Mid-Range) Processor. 8-Bit ADC.
PIC16C7x	18- to 40-Pin PICmicro® MCU. 14-Bit (Mid-Range) Processor. 8-Bit ADC/Optional TMR1 & TMR2/Optional USART/Optional SPI/Optional PSP.
PIC16C77x	28- to 40-Pin PICmicro® MCU. 14-Bit (Mid-Range) Processor. 12-Bit ADC/TMR1 & TMR2/USART/I2C/SPI/Optional PSP.
PIC16F8x	18-Pin PICmicro® MCU. 14-Bit (Mid-Range) Processor. Flash Data and Program Memory.
PIC16F87x	28- to 40-Pin PICmicro® MCU. 14-Bit (Mid-Range) Processor. 10-Bit ADC/TMR1 & TMR2/USART/I2C/SPI/Optional PSP.
PIC16C92x	64-Pin PICmicro® MCU. 14-Bit (Mid-Range) Processor. Optional 8-Bit ADC/TMR1 & TMR2/LCD Controller.
PIC17C4x	40-Pin PICmicro® MCU. 16-Bit (High-End) Processor. USART/Multiply.
PIC17C5x	68-Pin PICmicro® MCU. 16-Bit (High-End) Processor. USART/I2C/Multiply/10-Bit ADC.
PIC17C6x	84-Pin PICmicro® MCU. 16-Bit (High-End) Processor. USART/I2C/SPI/Multiply/12-Bit ADC.
PIC18Cxxx	28- to 40-Pin PICmicro® MCU. 16-Bit Advanced (18Cxx) Processor. USART/I2C/SPI/10-Bit ADC.

Mid-range PICmicro® MCU Part Number to Feature Breakout

Part Number	Comments
16Cx1	18-Pin PICmicro® MCU. 1K Program Memory. No USART/SPI/I2C. PIC16C61 and PIC16C71 are currently "obsoleted."
16Cx2	28-Pin PICmicro® MCU. 2K Program Memory. SPI/TMR1 & TMR2.
16Cx3	28-Pin PICmicro® MCU. 4K Program Memory. USART/SPI/TMR1 & TMR2.
16Cx4	40-Pin PICmicro® MCU. 4K Program Memory. USART/SPI/PSP/TMR1 & TMR2.
16Cx5	40-Pin PICmicro® MCU. 4K Program Memory. USART/SPI/PSP/TMR1 & TMR2.
16Cx6	28-Pin PICmicro® MCU. 8K Program Memory. USART/SPI/I2C/TMR1 & TMR2.
16Cx7	40-Pin PICmicro® MCU. 8K Program Memory. USART/SPI/I2C/PSP/TMR1 & TMR2.

3

Device Pinouts

As a rule of thumb, Pin-Through-Hole ("PTH") parts ("P" and "JW") are standard 0.300″ and 0.600″ widths with pins 0.100″ apart in dual in-line packages. The height of the device is dependent on the package used. Surface Mount Technology ("SMT") parts are either in dual in-line packages ("SO") or in quad plastic chip carriers ("PT", "PQ", and "L").

For actual device dimensions, check the datasheets (on the CD-ROM or from the Microchip Web site) for the PICmicro® MCU that you are planning on using. Different packages for different PICmicro® MCUs have different via pad and clearance specifications.

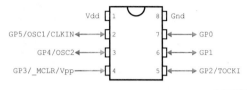

"JW"/"P"/"SO"
Packages
0.300" PTH Package Width

Figure 3.1 "PIC12C508"/"PIC12C509" Pinout

Low-End

There are no PLCC or QFP packages used for the low-end devices and the pinouts remain the same whether or not the PICmicro® MCU is in a surface mount technology or pin-through-hole package (Figs. 3.1–3.4).

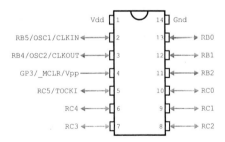

"JW"/"P"/"SO"
Packages
0.300" PTH Package Width

Figure 3.2 "PIC16C505" Pinout

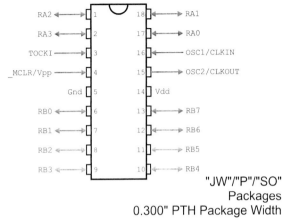

Figure 3.3 "PIC16C54"/"PIC16C56" Pinout

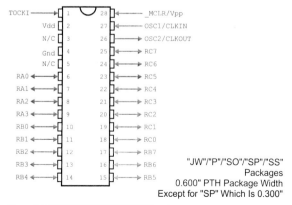

Figure 3.4 "PIC16C55"/"PIC16C57" Pinout

Mid-Range

The mid-range devices have the widest range of pinouts of any of the PICmicro® MCU families (Figs. 3.5–3.10). For many of the devices, the pinout is similar, but the pin functions may be different. In these cases, the pins marked with "*" show that these pins have other, optional purposes. Actual part number functions can be confirmed with Microchip Datasheets.

The PIC14000, which is designed for "Mixed Signals" uses the 28-pin packaging of the standard devices, but the pinouts are different as shown in Fig. 3.11.

The PIC16C92x LCD Driver microcontrollers are fairly high pin count devices. Figure 3.12 shows the 64-pin "DIP" ("Dual In-line Package") part. There is also a "PLCC" and "TQFP" package for the parts as well.

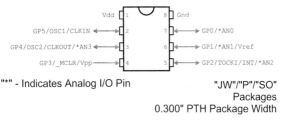

"*" - Indicates Analog I/O Pin

"JW"/"P"/"SO"
Packages
0.300" PTH Package Width

Figure 3.5 "PIC12C67x" Pinout

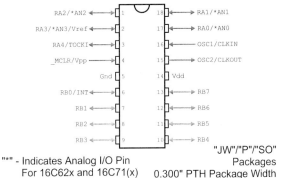

Figure 3.6 Mid-Range 18-Pin PICmicro® MCU Pinout

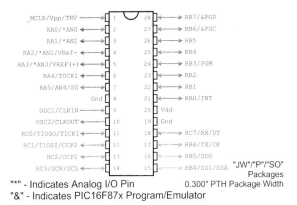

Figure 3.7 Mid-Range PICmicro® MCU 28-Pin Device Pinout

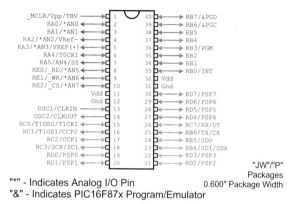

Figure 3.8 Mid-Range PICmicro® MCU 40-Pin Device Pinout

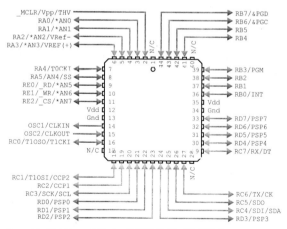

Figure 3.9 Mid-Range PICmicro® MCU 44-Pin "PLCC" Pinout

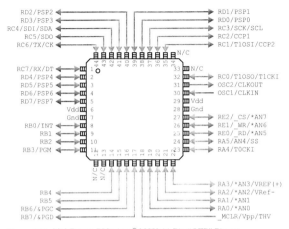

Figure 3.10   Mid-Range PICmicro® MCU 44-Pin "QFP" Pinout

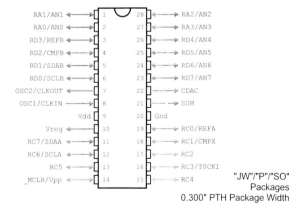

Figure 3.11   PIC14000 28-Pin Device Pinout

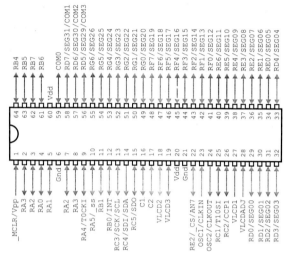

Figure 3.12   PIC16C92x 64-Pin Device Pinout

PIC17Cxx

The PIC17Cxx PICmicro® MCUs are available in 40- or 64-pin DIP packages as shown in Figs. 3.13 and 3.14. "PLCC" and "TQFP" surface mount packages as well for the 40-pin parts are displayed in the following graphics (Figs. 3.13–3.16).

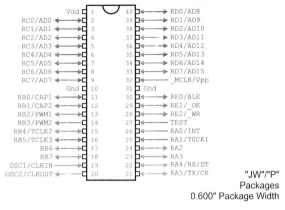

Figure 3.13 PIC17C4x 40-Pin Device Pinout

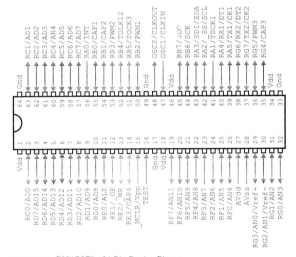

Figure 3.14 PIC17C75x 64-Pin Device Pinout

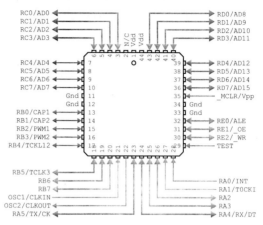

Figure 3.15 PIC17C4x 44-Pin "PLCC" Pinout

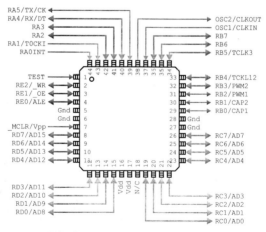

Figure 3.16 PIC17C4x 44-Pin "QFP" Pinout

PIC18Cxx

There is a lot of similarity between the mid-range
PICmicro® MCU's pinouts and the PIC18Cxx parts,
as will be seen in the following pinouts (Figs. 3.17–
3.20). Note that several pins that are optional in one
PICmicro® MCU family are not optional in others.

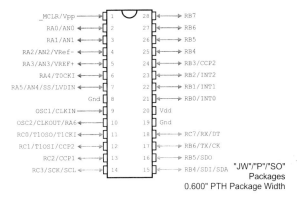

Figure 3.17   PIC18C2X2 28-Pin Device Pinout

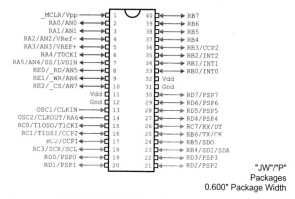

Figure 3.18   PIC18C4X2 40-Pin Device Pinout

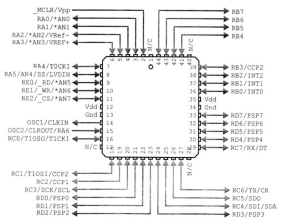

Figure 3.19   PIC18C4X2 44-Pin "PLCC" Pinout

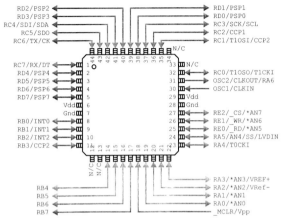

Figure 3.20   PIC18C4X2 44-Pin "QFP" Pinout

PICmicro® MCU Instruction Sets

Unless otherwise noted, all instructions execute in one instruction cycle.

Parameters

There are a number of parameters that are used with the instructions. The parameters are defined as:

Parameter	Symbol	Op code letter	Value range
Don't Care	N/A	x	
Byte Constant	k	k	0 to 0x0FF
Register Address	Reg	f	PICmicro® MCU Architecture Specific
Destination	d	d	0 or 1
Selection Bit	Bit	b	0 to 7
Destination Address	Address	a	0 to 0x07FF
Destination Port	Port	p	PORTA (5) to PORTC (7)

Low-End Instruction Set

Register Banks are 32 bytes in size in the low-end devices. This makes "Reg" in the range of 0x00 to 0x01F.

Description	Instruction	Operation	Op code

Add Register Contents to "w" and optionally store result in "w"	addwf Reg, d	if (d == 1)
		  Reg = Reg + w
		else
		  w = Reg + w
		endif
		C = (Reg + w)
		> 0xOFF
		Z = ((Reg + w) &
		  0xOFF) == 0
		DC = ((Reg & 0xOF)
		  + (w & 0xOF))
		> 0xOF
			0001 11df ffff

AND Immediate with "w"	andlw k	w = w & k Z = (w & k) == 0	1110 kkkk kkkk

AND Register Contents with "w" and Optionally store result in "w"	andwf Reg, d	if (d == 1)
		  Reg = Reg & w
		else
		  w = Reg & w
		endif
		Z = (Reg & w) == 0
			0001 01df ffff

Clear the Specified Bit in the Register	bcf Reg, bit	Reg = Reg &
		  (0xOFF ^
		  (1 << Bit))
			0100 bbbf ffff

Description	Instruction	Operation	Op code
Set the Specified Bit in the Register	bcf Reg, bit	Reg = Reg l (1 << Bit)	0101 bbbf ffff
Skip if the Specified Bit In the Register is Clear. One Instruction Cycle if Skip not Executed, two if it is	btfsc Reg, bit	if ((Reg & (1 << Bit))) == 0) PC = PC + 1 endif	0110 bbbf ffff
Skip if the Specified Bit In the Register is Set. One Instruction Cycle if Skip not Executed, two if it is	btfsc Reg, bit	if ((Reg & (1 << Bit)) != 0) PC = PC + 1 endif	0111 bbbf ffff
Save the Stack Pointer and jump to the Specified Address (two Instruction cycles)	call Address	[SP] = PC SP = SP + 1 PC = ((STATUS & 0xE0) << 4) + Address	1001 aaaa aaaa
Clear the Specified Register	clrf Reg	Reg = 0 Z = 1	0000 011f ffff
Clear the "w" Register	clrw	w = 0 Z = 1	0000 0100 0000
Clear the Watchdog Timer's Counter	clrwdt	WDT = 0 _TO = 1 _PD = 1	0000 0000 0100

Description	Instruction	Operation	Encoding

Complement the Contents of the Specified Register and optionally store the results in "w"	`comf Reg, d`	if (d == 1)
		  Reg = Reg ^ 0xOFF
		else
		  w = Reg ^ 0xOFF
		endif
		Z = (Reg ^ 0xOFF) == 0
			`0010 01df ffff`

Decrement the Contents of the Register and Optionally store the results in "w"	`decf Reg, d`	if (d == 1)
		  Reg = Reg − 1
		else
		  w = Reg − 1
		endif
		Z = (Reg − 1) == 0
			`0011 11df ffff`

Decrement the Contents of the Register and Optionally store the results in "w" and Skip the next instruction if the results are equal to Zero. Two Instruction Cycles taken if Skip Executed	`decfsz Reg, d`	if (d == 1)
		  Reg = Reg − 1
		else
		  w = Reg − 1
		endif
		PC = PC + 1
		if ((Reg − 1) == 0)
		  PC = PC + 1
		endif
			`0010 11df ffff`

Jump to the Specified Address (two Instruction cycles)	`goto Address`	PC = ((STATUS & 0xOE0) << 4) + Address	`101a aaaa aaaa`

25

Description	Instruction	Operation	Op code

Increment the Contents of the Register and Optionally store the results in "w"	incf Reg, d	if (d == 1)
		    Reg = Reg + 1
		else
		    w = Reg + 1
		endif
		Z = (Reg + 1) == 0
			0010 10df ffff

Increment the Contents of the Register and Optionally store the results in "w" and Skip the next instruction if the results are equal to Zero. Two Instruction Cycles taken if Skip Executed	incfsz Reg, d	if (d == 1)
		    Reg = Reg + 1
		else
		    w = Reg + 1
		endif
		PC = PC + 1
		if ((Reg + 1) == 0)
		    PC = PC + 1
		endif
			0011 11df ffff

OR Immediate with "w"	iorlw k	w = w | k Z = (w | k) == 0	1101 kkkk kkkk

OR Register Contents with "w" and Optionally store result in "w"	iorwf Reg, d	if (d == 1)
		    Reg = Reg | w
		else
		    w = Reg | w
		endif
		Z = (Reg | w) == 0
			0001 00df ffff

Description	Instruction	Operation	Opcode

Check Register Contents equal to zero and Optionally store result in "w"	`movf Reg, d`	if (d == 0)
		  w = Reg
		endif
		Z = Reg == 0
			0010 00df ffff

Load "w" with an Immediate value	`movlw k`	w = k	1100 kkkk kkkk
Store the value in "w"	`movwf Reg`	Reg = w	0000 001f ffff
Waste one Instruction	`nop`		0000 0000 0000
Move the contents of "w" into the OPTION Register	`option`	TRIS(Port) = w	0000 0000 0010
Resume Execution after Subroutine and Place a constant value in "w" (Two Cycles used)	`retlw k`	w = k SP = SP − 1 PC = [SP]	1000 kkkk kkkk
Resume Execution after Subroutine and Place Zero in "w" (Two Cycles used). This is actually a "retlw 0" instruction that MPLAB provides	`return`	w = 0 SP = SP − 1 PC = [SP]	1000 0000 0000

Description	Instruction	Operation	Op code

Rotate the Register left through carry and optionally Save the Result in "w"	rlf Reg, d	Temp = C
		C = (Reg >> 7) & 1
		if (d === 1)
		  Reg = (Reg << 1)
		    + Temp
		else
		  w = (Reg << 1)
		    + Temp
		endif
			0011 01df ffff

Rotate the Register right through carry and optionally Save the Result in "w"	rrf Reg, d	Temp = C
		C = Reg & 1
		if (d === 1)
		  Reg = (Reg >> 1)
		    + (Temp << 7)
		else
		  w = (Reg >> 1)
		    + (Temp << 7)
		endif
			0011 00df ffff

Go into "Standby" Mode (Indeterminate number of cycles used)	sleep	_TO = 1 _PD = 0	0000 0000 0011

Subtract "w" Register Contents from Register and Optionally store Result in "w"

`subwf Reg, d`

`0000 10df ffff`

```
if (d == 1)
    Reg = Reg +
        (w ^ 0x0FF) + 1
else
    w = Reg +
        (w ^ 0x0FF) + 1
endif
C = (Reg +
    (w ^ 0x0FF) + 1)
    > 0x0FF
Z = ((Reg +
    (w ^ 0x0FF) + 1)
    & 0x0FF) == 0
DC = ((Reg & 0x0F)
    + ((w ^ 0x0F)
    & 0x0F) + 1)
    > 0x0F
```

Swap the Upper and lower nybbles of a Register and Optionally store result in "w"

`swapf Reg, d`

`0011 10df ffff`

```
if (d == 1)
    Reg =
        ((Reg & 0x0F0)
        >> 4) +
        ((Reg & 0x00F)
        << 4)
else
    w = ((Reg & 0x0F0)
        >> 4) +
        ((Reg & 0x00F)
        << 4)
endif
```

Description	Instruction	Operation	Op code
Move the contents of "w" into the Tri-state control Register of the Port	`tris Port`	TRIS(Port) = w	0000 0000 0ppp
XOR Immediate with "w"	`xorlw k`	w = w ^ k Z = (w ^ k) == 0	1111 kkkk kkkk

XOR Register Contents with "w" and Optionally store result in "w"	`xorwf Reg, d`	if (d == 1)
		    Reg = Reg ^ w
		else
		    w = Reg ^ w
		endif
		Z = (Reg ^ w) == 0
			0001 10df ffff

Mid-Range Instruction Set

Register Banks are 128-bytes in size in the low-end devices. This makes "Reg" in the range of 0 to 0x07F.

Description	Instruction	Operation	Op code
Add Immediate to "w"	addlw k	$w = w + k$ $C = (w + k) > 0xFF$ $Z = ((w + k)\ \&\ 0xFF)$ $== 0$ $DC = ((w\ \&\ 0x0F) +$ $(k\ \&\ 0x0F)) > 0x0F$	11 111x kkkk kkkk

Add Register Contents to "w" and optionally store result in "w"	addwf Reg, d	if (d == 1)
		$\quad Reg = Reg + w$
		else
		$\quad w = Reg + w$
		endif
		$C = (Reg + w)$
		$> 0xFF$
		$Z = ((Reg + w)\ \&$
		$0xFF) == 0$
		$DC = ((Reg\ \&\ 0x0F)$
		$+ (w\ \&\ 0x0F))$
		$> 0x0F$
			00 0111 dfff ffff

AND Immediate with "w"	andlw k	$w = w\ \&\ k$ $Z = (w\ \&\ k) == 0$	11 1001 kkkk kkkk

31

Description	Instruction	Operation	Op code
AND Register Contents with "w" and Optionally store result in "w"	andwf Reg, d	if (d == 1) Reg = Reg & w else w = Reg & w endif Z = (Reg & w) == 0	00 0101 dfff ffff
Clear the Specified Bit in the Register	bcf Reg, bit	Reg = Reg & (0xOFF^ (1 << Bit))	01 00bb bfff ffff
Set the Specified Bit in the Register	bcf Reg, bit	Reg = Reg l (1 << Bit)	01 01bb bfff ffff
Skip if the Specified Bit in the Register is Clear. One Instruction Cycle if Skip not Executed, two if Skip Executed	btfsc Reg, bit	if ((Reg & (1 << Bit))) == 0) PC = PC + 1 endif	01 10bb bfff ffff

Skip if the Specified Bit in the Register is Set. One Instruction Cycle if Skip not Executed, two if it is	btfsc Reg, bit	if ((Reg & (1 << Bit)) != 0) PC = PC + 1 endif	01 11bb bfff ffff
Save the Stack Pointer and jump to the Specified Address (two Instruction cycles)	call Address	[SP] = PC SP = SP + 1 PC = ((PCLATH << 8) & 0x01800) + Address	10 0aaa aaaa aaaa
Clear the Specified Register	clrf Reg	Reg = 0 Z = 1	00 0001 1fff ffff
Clear the "w" Register	clrw	w = 0 Z = 1	00 0001 0xxx xxxx
Clear the Watchdog Timer's Counter	clrwdt	WDT = 0 _TO = 1 _PD = 1	00 0000 0110 0100
Complement the Contents of the Specified Register and Optionally store the results in "w"	comf Reg, d	if (d == 1) Reg = Reg ^ 0xFF else w = Reg ^ 0xFF endif Z = (Reg ^ 0xFF) == 0	00 1001 dfff ffff

Description	Instruction	Operation	Op code

Decrement the Contents of the Register and Optionally store the results in "w"	decf Reg, d	if (d == 1)
		  Reg = Reg − 1
		else
		  w = Reg − 1
		endif
		Z = (Reg − 1) == 0
			00 0011 dfff ffff

Decrement the Contents of the Register and Optionally store the results in "w" and Skip the next instruction if the results are equal to Zero. Two Instruction Cycles taken if skip executed	decfsz Reg, d	if (d == 1)
		  Reg = Reg − 1
		else
		  w = Reg − 1
		endif
		if ((Reg − 1) == 0)
		  PC = PC + 1
		endif
			00 1011 dfff ffff

Jump to the Specified Address (two Instruction cycles)	goto Address	PC = ((PCLATH << 8) & 0x01800) + Address	10 1aaa aaaa aaaa

Increment the Contents of the Register and Optionally store the results in "w"	incf Reg, d	if (d == 1)
		  Reg = Reg + 1
		else
		  w = Reg + 1
		endif
		Z = (Reg + 1) == 0
			00 1010 dfff ffff

Description	Instruction	Operation	Opcode

Increment the Contents of the Register and Optionally store the results in "w" and Skip the next instruction if the results are equal to Zero. Two Instruction Cycles taken if Skip Executed	incfsz Reg, d	if (d == 1)
		  Reg = Reg + 1
		else
		  w = Reg + 1
		endif
		
		if ((Reg + 1) == 0)
		  PC = PC + 1
		endif
			00 1111 dfff ffff

OR Immediate with "w"	iorlw k	w = w | k Z = (w | k) == 0	11 1000 kkkk kkkk

OR Register Contents with "w" and Optionally store result in "w"	iorwf Reg, d	if (d == 1)
		  Reg = Reg | w
		else
		  w = Reg | w
		endif
		Z = (Reg | w) == 0
			00 0100 dfff ffff

Check Register Contents equal to zero and Optionally store Register contents in "w"	movf Reg, d	if (d == 0)
		  w = Reg
		endif
		Z = Reg == 0
			00 1000 dfff ffff

Description	Instruction	Operation	Op code
Load "w" with an immediate value	movlw k	w = k	11 00xx kkkk kkkk
Store the value In "w"	movwf Reg	Reg = w	00 0000 1fff ffff
Waste one Instruction	nop		00 0000 0xx0 0000
Move the contents of "w" into the OPTION Register. Use of this instruction is not recommended	option	TRIS(Port) = w	00 0000 0110 0010
Resume Execution after Interrupt (Two Cycles used)	retfie	GIE = 1 SP = SP − 1 PC = [SP]	00 0000 0000 1001
Resume Execution after Subroutine and Place a constant Value in "w" (Two Cycles used)	retlw k	w = k SP = SP − 1 PC = [SP]	11 01xx kkkk kkkk

Description	Mnemonic	Operation	Encoding
Resume Execute after Subroutine (Two Cycles used)	`return`	SP = SP - 1 PC = [SP]	00 0000 0000 1000

Rotate the Register left through carry and Optionally Save the Result in "w"	`rlf Reg, d`	Temp = C
		C = (Reg >> 7) & 1
		if (d == 1)
		    Reg = (Reg << 1)
		        + Temp
		else
		    w = (Reg << 1)
		        + Temp
		endif
			00 1101 dfff ffff

Rotate the Register right through carry and Optionally Save the Result in "w"	`rrf Reg, d`	Temp = C
		C = Reg & 1
		if (d == 1)
		    Reg = (Reg >> 1)
		        + (Temp << 7)
		else
		    w = (Reg >> 1)
		        + (Temp << 7)
		endif
			00 1100 dfff ffff

Go into "Standby" Mode (Indeterminate number of cycles used)	`sleep`	_TO = 1 _PD = 0	00 0000 0110 0011

Description	Instruction	Operation	Op code
Subtract "w" Contents from Immediate and Store the Result In "w"	sublw k	w = k + (w ^ 0xFF) + 1 C = (k + (w ^ 0xFF) + 1) > 0xFF Z = ((k + (w ^ 0xFF) + 1) & 0xFF) == 0 DC = ((k & 0x0F) + ((w ^ 0xFF) & 0x0F) + 1) > 0x0F	11 110x kkkk kkkk

Subtract "w" Register Contents from Register and Optionally store Result in "w"	subwf Reg, d	if (d == 1)
		  Reg = Reg + (w ^ 0xFF) + 1
		else
		  w = Reg + (w ^ 0xFF) + 1
		endif
		C = (Reg + (w ^ 0xFF) + 1) > 0xFF
		Z = ((Reg + (w ^ 0xFF) + 1) & 0xFF) == 0
		DC = ((Reg & 0x0F) + ((w ^ 0xFF) & 0x0F) + 1) > 0x0F
			00 0010 dfff ffff

Description	Instruction	Operation	Encoding

Swap the Upper and lower Nybbles of a Register and Optionally store result in "w"	`swapf Reg, d`	if (d == 1)
		  Reg =
		    ((Reg & 0x0F0)
		    >> 4) +
		    ((Reg & 0x00F)
		    << 4)
		  else
		  w = ((Reg & 0x0F0)
		    >> 4) +
		    ((Reg & 0x00F)
		    << 4)
		  endif
			00 1110 dfff ffff

Move the contents of "w" into the Tri-state control Register of the Port. Use of this Instruction is not recommended	`tris Port`	TRIS(Port) = w	00 0000 0110 0ppp
XOR Immediate with "w"	`xorlw k`	w = w ^ k Z = (w ^ k) == 0	11 1010 kkkk kkkk

XOR Register Contents with "w" and Optionally store result in "w"	`xorwf Reg, d`	if (d == 1)
		  Reg = Reg ^ w
		  else
		  w = Reg ^ w
		  endif
		Z = (Reg ^ w) == 0
			00 0110 dfff ffff

PIC17Cxx Instruction Set

The PIC17Cxx's instruction set is very similar to both the low-end and the mid-range instruction sets except for the basic "move" instructions. These instructions are quite a bit different because of the operation of the PIC17Cxx's "primary" register subset of the total 256 possible addresses.

Instruction	Format	Operation	Bit pattern

Add a Constant to the "wreg" and store the Result in "wreg"	`addlw Constant`	wreg = wreg + Constant
		if (((wreg > 0)
		& (Constant > 0))
		& ((wreg + Constant > 0x07F))
		OV = 1
		else
		OV = 0
		if ((wreg + Constant) > 0x0FF)
		C = 1
		else
		C = 0
		if (((wreg & 0x0F) + (Constant & 0x0F)) > 0x0F)
		DC = 1
		else
		DC = 0
		if ((wreg + Constant) & 0x0FF) == 0x000)
		Z = 1
		else
		Z = 0
			1011 0001 kkkk kkkk

Instruction	Format	Operation	Bit pattern

Add "wreg" to the Contents of "Reg" and store the Result According to "d" Result in "wreg"	addwf Reg, d	if ("d" == 1)
		 wreg = wreg + Reg
		 else
		 Reg = wreg + Reg
		 if (((wreg > 0) & (Reg > 0)) & ((wreg + Reg) > 0x07F))
		 OV = 1
		 else
		 OV = 0
		 if ((wreg + Reg) > 0x0FF)
		 C = 1
		 else
		 C = 0
		 if (((wreg & 0x0F) + (Reg & 0x0F)) > 0x0F)
		 DC = 1
		 else
		 DC = 0
			0000 111d ffff ffff

Add "wreg" to the Contents
of "Reg" and "C", store the
Result According to "d"
Result in "wreg"

`addwfc Reg, d`

```
0001 000d ffff ffff

if (("d" == 1)
    wreg = wreg
        + Reg + C
else
    Reg = wreg
        + Reg + C
if (((wreg > 0)
    & (Reg > 0))
    & ((wreg
        + Reg + C
        > 0x07F))
    OV = 1
else
    OV = 0
if ((wreg +
    Reg + C) > 0xFF)
    C = 1
else
    C = 0
if (((wreg & 0x0F) +
    (Reg & 0x0F) + C
    > 0x0F)
    DC = 1
if (((wreg +
    Reg) & 0x0FF)
    == 0x000)
    Z = 1
else
    Z = 0
```

Instruction	Format	Operation	Bit pattern

		else
		DC = 0
		if (((wreg + Reg + C) & 0xFF) == 0x000)
		Z = 1
		else
		Z = 0	

AND a Constant to the "wreg" and store the Result in "wreg"	andlw Constant	wreg = wreg & Constant
		if ((wreg & Constant) == 0x000)
		Z = 1
		else
		Z = 0
			1011 0101 kkkk kkkk

AND "wreg" to the Contents of "Reg" and store the Result According to "d" Result in "wreg"	andwf Reg, d	if ("d" == 1)
		wreg = wreg & Reg
		else
		Reg = wreg & Reg
		if ((wreg & Reg) == 0x000)
		Z = 1
		else
		Z = 0
			0000 101d ffff ffff

Description	Instruction	Operation	Opcode
Clear the Specified Bit in "Reg"	bcf Reg, Bit	Reg = Reg & (0xOFF ^ (1 << Bit))	1000 1bbb ffff ffff
Set the Specified Bit in "Reg"	bsf Reg, Bit	Reg = Reg | (1 << Bit)	1000 0bbb ffff ffff
Test the Specified Bit in "Reg" and skip if Clear. One Instruction Cycle if Skip not Executed, two if Skip Executed	btfsc Reg, Bit	if ((Reg & (1 << Bit) == 0) PC = PC + 1	1001 1bbb ffff ffff
Test the Specified Bit in "Reg" and skip if Set. One Instruction Cycle if Skip not Executed, two if Skip Executed	btfss Reg, Bit	if ((Reg & (1 << Bit)) != 0) PC = PC + 1	1001 0bbb ffff ffff
Toggle the Specified Bit in "Reg"	btg Reg, Bit	Reg = Reg ^ (1 << Bit)	0011 1bbb ffff ffff
Call the "Label" Address. Two Instruction Cycles are Required	call Label	PUSH (PC) PCLATH = PC (15:13) + Label (12:8) PCL = Label (7:0)	111k kkkk kkkk kkkk
Clear the Specified Register and Optionally "wreg"	clrf Reg, s	Reg = 0 if (s == 0) wreg = 0	0010 100s ffff ffff

Instruction	Format	Operation	Bit pattern
Clear the Watchdog Register and STATUS flags	`clrwdt`	WDT = 0 WDT Postscaler = 0 _TO = 1 _PD = 1	0000 0000 0000 0100

Complement the Contents of the Specified Register	`comf Reg, d`	if ("d" == 0)
		  wreg = Reg ^ 0xOFF
		else
		  Reg = Reg ^
		if ((Reg ^ 0xOFF)
		  == 0x000)
		  Z = 1
		else
		  Z = 0
			0001 001d ffff ffff

Compare the Specified Register with wreg and skip if the Register == Wreg. If Skip Executed,	`cpfseq Reg`	if ((Reg − wreg) == 0) PC = PC + 1	0011 0001 ffff ffff

two Instruction Cycles
Executed Else one Cycle

Description	Mnemonic	Operation	Encoding
Compare the Specified Register with wreg and skip if the Register > wreg. If Skip Executed, two Instruction Cycles Executed Else one Cycle	cpfsgt Reg	if ((Reg − wreg) > 0) PC = PC + 1	0011 0010 ffff ffff
Compare the Specified Register with wreg and skip if the Register < wreg. If Skip Executed, two Instruction Cycles Executed Else one Cycle	cpfslt Reg	if ((Reg − wreg) < 0) PC = PC + 1	0011 0000 ffff ffff
Do a Decimal Adjust after Addition of Two BCD Values	daw Reg, s	if ((wreg & 0x0F) > 9) if (s == 0) wreg = (wreg & 0x0F) + 0x010 else Reg = (wreg & 0x0F) + 0x010	0010 111s ffff ffff

Instruction	Format	Operation	Bit pattern

Decrement the Contents of the Specified Register	decf Reg, d	if ("d" == 0)
		  wreg = Reg − 1
		else
		  Reg = Reg − 1
		if ((Reg − 1)
		  == 0x000)
		    Z = 1
		else
		    Z = 0
		if (((Reg > 0)
		  & ((Reg − 1)
		  < 0x080))
		    OV = 1
		else
		    OV = 0
		if (((Reg & 0x00F)
		  & 0x080) != 0)
		    DC = 0
		else
		    DC = 1
		if ((Reg − 1)
		  < 0)
		    C = 0
		else
		    C = 1
			0000 011d ffff ffff

Description	Mnemonic	Operation	Encoding
Decrement the Contents of the Specified Register and skip the Next Instruction if Result == 0. If Skip Executed, two Instruction Cycles Executed Else one Cycle	decfsz Reg, d	if ("d" == 0) wreg = Reg − 1 else Reg = Reg − 1 if ((Reg − 1) == 0x000) PC = PC + 1	0001 011d ffff ffff
Decrement the Contents of the Specified Register and skip the Next Instruction if Result != 0. If Skip Executed, two Instruction Cycles Executed Else one Cycle	dcfsnz Reg, d	if ("d" == 0) wreg = Reg − 1 else Reg = Reg − 1 if ((Reg − 1) != 0x000) PC = PC + 1	0010 011d ffff ffff
Goto the "Label" Address. Two Instruction Cycles	goto Label	PCLATH = PC (15:13) + Label (12:8) PCL = Label (7:0)	1101 kkkk kkkk kkkk

Instruction	Format	Operation	Bit pattern

Increment the Contents of the Specified Register	`incf Reg, d`	if ("d" == 0)
		   wreg = Reg + 1
		 else
		   Reg = Reg + 1
		 if ((Reg + 1) == 0x000)
		   Z = 1
		 else
		   Z = 0
		 if ((Reg > 0)
		   & ((Reg + 1)
		   > 0x07F))
		   OV = 1
		 else
		   OV = 0
		 if (((Reg & 0x00F)
		   & 0x010) != 0)
		   DC = 1
		 else
		   DC = 0
		 if ((Reg + 1)
		   == 0x0100)
		   C = 1
		 else
		   C = 0
			0001 010d ffff ffff

Description	Instruction	Operation	Encoding

Increment the Contents of the Specified Register and skip the Next Instruction if Result == 0. If Skip Executed, two Instruction Cycles Executed Else one Cycle	incfsz Reg, d	if ("d" == 0)
		  wreg = Reg + 1
		else
		  Reg = Reg + 1
		if ((Reg + 1)
		  == 0x000)
		  PC = PC + 1
			0001 111d ffff ffff

Increment the Contents of the Specified Register and skip the Next Instruction if Result != 0. If Skip Executed, two Instruction Cycles Executed Else one Cycle	infsnz Reg, d	if ("d" == 0)
		  wreg = Reg + 1
		else
		  Reg = Reg + 1
		if ((Reg + 1)
		  != 0x000)
		  PC = PC + 1
			0010 010d ffff ffff

OR a Constant to the "wreg" and store the Result in "wreg"	iorlw Constant	wreg = wreg
		
			Constant
			if ((wreg |
			  Constant)
			  == 0x000)
			  Z = 1
			else
			  Z = 0
				1011 0011 kkkk kkkk

Description	Format	Operation	Bit pattern
...ording to "d" ...sult in "wreg"	iorwf Reg, d	if ("d" == 1) wreg = wreg | Reg else Reg = wreg | Reg if ((wreg | Reg) == 0x000) Z = 1 else Z = 0	0000 100d ffff ffff
Call the "Label" Address Using PCLATH and the Least Significant Eight bits of "Label". Two Instruction Cycles	lcall Label	PUSH (PC) PCL = Label (7:0)	1011 0111 kkkk kkkk
Move data from 256 Address Register Data to Primary Register Set	movfp Reg, p	p = Reg	011p pppp ffff ffff
Move Constant into low Nybble of BSR	movlb Constant	BSR (3:0) = Constant	1011 1000 0000 kkkk
Move Constant into high Nybble of BSR	movlr Constant	BSR (7:4) = Constant	1011 1010 kkkk 0000

Move Constant into wreg	`movlw Constant`	wreg = Constant	1011 0000 kkkk kkkk

Move data from Primary Register Data to 256 Address Register Set	`movpf p, Reg`	Reg = p
		if (p == 0)
		Z = 1
		else
		Z = 0
			010p pppp ffff ffff

Move contents of "wreg" into "Reg"	`movwf Reg`	Reg = wreg	0000 0001 ffff ffff
Multiply Constant by "wreg"	`mullw Constant`	PRODH:PROGL = Constant * wreg	1011 1100 kkkk kkkk
Multiply Register by "wreg"	`mullwf Reg`	PRODH : PROGL = Reg * wreg	0011 0100 ffff ffff

Instruction	Format	Operation	Bit pattern

Negate the Contents of "wreg" and Optionally store the Result in a Register	`negw Reg, s`	if (s == 0)
		  Reg = −wreg
		wreg = −wreg
		if ((wreg < 0)
		  & (−wreg < 0x080))
		    OV = 1
		else
		  OV = 0
		if (−wreg > 0x0FF)
		  C = 1
		else
		  C = 0
		if ((−wreg & 0x0F)
		  > 0x0F)
		    DC = 1
		else
		  DC = 0
		if (−wreg == 0x000)
		  Z = 1
		else
		  Z = 0
			0010 110s ffff ffff

Do Nothing for one Instruction Cycle	`nop`		0000 0000 0000 0000
Return from Interrupt Handler. Two Instruction Cycles	`retfie`	PC = POP () GLTIND = 0	0000 0000 0000 0101

Description	Instruction	Operation	Encoding
Return from Subroutine with new value in wreg. Two Instruction Cycles	`retlw Constant`	wreg = Constant PC = POP ( )	1011 0110 kkkk kkkk
Return from Subroutine. Two Instruction Cycles	`return`	PC = POP ( )	0000 0000 0000 0010

Rotate Left Through the Carry Flag	`rlcf Reg, d`	if (d == 0)
		  wreg (7:1) =
		    Reg (6:0)
		  wreg (0) = C
		  C = Reg (7)
		else
		  Reg (7:1) =
		    Reg (6:0)
		  Reg (0) = C
		  C = Reg (7)
			0001 101d ffff ffff

Rotate Left	`rlcnf Reg, d`	if (d == 0)
		  wreg (7:1) =
		    Reg (6:0)
		  wreg (0) =
		    Reg (7)
		else
		  Reg (7:1) =
		    Reg (6:0)
		  Reg (0) =
		    Reg (7)
			0010 001d ffff ffff

Instruction	Format	Operation	Bit pattern

Rotate Right Through the Carry Flag	rrcf Reg, d	if (d == 0)
		wreg (6:0) =
		Reg (7:1)
		wreg (7) = C
		C = Reg (0)
		else
		Reg (6:0) =
		Reg (7:1)
		Reg (7) = C
		C = Reg (0)
			0001 100d ffff ffff

Rotate Right	rrcnf Reg, d	if (d == 0)
		wreg (6:0) =
		Reg (7:1)
		wreg (7) =
		Reg (0)
		else
		Reg (6:0) =
		Reg (7:1)
		Reg (7) =
		Reg (0)
			0010 000d ffff ffff

Set the Specified Register and Optionally "wreg"	setf Reg, s	Reg = 0xOFF if (s == 0) wreg = 0xOFF	0010 101s ffff ffff

Put the PICmicro® MCU in a "Power Down" State	`sleep`	WDT = 0 WDT Postscaler = 0 _TO = 1 _PD = 0 PICmicro MCU Power Down	0000 0000 0000 0011

Subtract "wreg" from a Constant and store the Result in "wreg"	`sublw Constant`	wreg = Constant
		    - wreg
		if (((wreg < 0)
		& (Constant < 0))
		& ((Constant
		    - wreg)
		    < 0x080))
		    OV = 1
		else
		    OV = 0
		if ((Constant -
		    wreg) > 0x0FF)
		    C = 1
		else
		    C = 0
		if (((Constant & 0x0F) -
		    wreg & 0x0F))
		    > 0x0F)
		    DC = 1
		else
		    DC = 0
		if (((Constant -
		    wreg) & 0x0FF)
		    == 0x000)
			1011 0010 kkkk kkkk

Instruction	Format	Operation	Bit pattern

Subtract "wreg" from the Contents of "Reg" and store the Result According to "d" Result in "wreg"	subwf Reg, d	if ("d" == 1)
		  wreg = Reg −
		    wreg
		else
		  Reg = Reg
		    − wreg
		if (((wreg < 0)
		  & (Reg < 0))
		  & ((Reg
		  − wreg)
		  < 0x080))
		    OV = 1
		else
		  OV = 0
		if ((Reg −
		  wreg) > 0x0FF)
		  C = 1
		else
		  C = 0
		if (((Reg & 0x0F) −
		  (wreg & 0x0F))
		  > 0x0F)
		    DC = 1
			0000 010d ffff ffff

$Z = 1$
else
$Z = 0$

Subtract "wreg" from the Contents of "Reg" and "C", store the Result According to "d" Result in "wreg"

`subwfb Reg, d`

`0000 001d ffff ffff`

```
                                    else
                                        DC = 0
                                    if (((Reg –
                                    wreg) & 0x0FF)
                                    == 0x000)
                                        Z = 1
                                    else
                                        Z = 0

if ("d" == 1)
    wreg = Reg
        – wreg – !C
else
    Reg = Reg
        – wreg – !C
if (((Reg > 0)
    & (wreg > 0))
    & ((Reg
    – wreg – !C)
    < 0x080))
        OV = 1
else
        OV = 0
if ((Reg –
    wreg – !C) > 0xFF)
        C = 1
else
        C = 0
if (((Reg & 0x0F) –
```

Instruction	Format	Operation	Bit pattern

		(wreg & 0x0F) − !C)
		> 0x0F)
		DC = 1
		else
		DC = 0
		if (((Reg −
		wreg − !C) & 0x0FF)
		== 0x000)
		Z = 1
		else
		Z = 0	

Swap the Contents of "Reg" and store the Result According to "d" Result in "wreg"	swapf Reg, d	if ("d" == 1)
		wreg =
		((Reg & 0x00F)
		<< 4) +
		((Reg & 0x0F0)
		>> 4)
		else
		Reg =
		((Reg & 0x00F)
		<< 4) +
		((Reg & 0x0F0)
		>> 4)
			0001 110d ffff ffff

Description	Mnemonic	Operation	Encoding

Read the Contents of the Table Pointer or Read Program Memory into the Table Pointer. Two or three Instruction Cycles	tablrd t, i, f	if (t == 1)
		 f = TBLATH
		else
		 f = TBLATH
		TBLAT = ProgMem(TBLPTR)
		if (i == 1)
		 TBLPTR = TBLPTR + 1
			1010 10ti ffff ffff

Write new Contents of the Table Pointer or Write Program Memory from the Table Pointer. If the Destination is Internal EPROM, the Instruction Does not End until an Interrupt. Two Instruction Cycles or Until Timer Interrupt	tablwt t, i, f	if (t == 0)
		 TBLATL = f
		else
		 TBLATH = f
		TBLAT = ProgMem(TBLPTR)
		if (i == 1)
		 TBLPTR = TBLPTR + 1
			1010 11ti ffff ffff

Read the Contents of the Table Pointer	tlrd t, f	if (t == 1)
		 f = TBLATH
		else
		 f = TBLATL
			1010 00t0 ffff ffff

Write the Contents of the Register into the Table Pointer	tlwt t, f	if (t == 1)
		 TBLATH = f
		else
		 TBLATL = f
			1010 01t0 ffff ffff

Instruction	Format	Operation	Bit pattern
Compare the Specified Register zero and skip if the Register == 0. One Instruction Cycle if Skip not Executed, Two if it is	`tstfsz Reg`	if (Reg == 0) PC = PC + 1	0011 0011 ffff ffff
XOR a Constant to the "wreg" and store the Result in "wreg"	`xorlw Constant`	wreg = wreg ^ Constant if ((wreg ^ Constant) == 0x000) Z = 1 else Z = 0	1011 0100 kkkk kkkk
XOR "wreg" to the Contents of "Reg" and store the Result According to "d" Result in "wreg"	`xorwf Reg, d`	if ("d" == 1) wreg = wreg ^ Reg else Reg = wreg ^ Reg if ((wreg ^ Reg) == 0x000) Z = 1 else Z = 0	0000 110d ffff ffff

PIC18Cxx Instruction Set

Instruction	Format	Operation	Bit pattern

Add a Constant to the "wreg" and store the Result in "wreg"	addlw Constant	wreg = wreg + Constant
		if (((wreg > 0) & (Constant > 0)) & ((wreg + Constant) > 0x07F))
		    OV = 1
		else
		    OV = 0
		if ((wreg + Constant) > 0x0FF)
		    C = 1
		else
		    C = 0
		if ((wreg + Constant) & 0x080) != 0)
		    N = 1
		else
		    N = 0
		if (((wreg & 0x0F) + (Constant & 0x0F)) > 0x0F)
		    DC = 1
		else
			0000 1111 kkkk kkkk

Instruction	Format	Operation	Bit pattern

Add "wreg" to the Contents of "Reg" and store the Result According to "d" Result in "wreg". If "a" is set then BSR used for Reg, else Access Bank is used.	addwf Reg, d, a	DC = 0
		if (((wreg + Constant) & 0x0FF) == 0x000)
		Z = 1
		else
		Z = 0
		
		if ("d" == 1)
		wreg = wreg + Reg
		else
		Reg = wreg + Reg
		if (((wreg > 0) & (Reg > 0)) & ((wreg + Reg) > 0x07F))
		OV = 1
		else
		OV = 0
		if ((wreg + Reg) & 0x080) != 0)
		N = 1
		else
		N = 0
			0010 01da ffff ffff

Add "wreg" to
the Contents
of "Reg", store
the
Result
According to
"d" Result in
"wreg".
If "a" is set
then Reg is

addwfc Reg, d, a

```
if ((wreg +
Reg) > 0x0FF)
    C = 1
else
    C = 0
if (((wreg & 0x0F) +
(Reg & 0x0F))
> 0x0F)
    DC = 1
else
    DC = 0
if (((wreg +
Reg) & 0x0FF)
== 0x000)
    Z = 1
else
    Z = 0

if ("d" == 1)
    wreg = wreg
    + Reg + C
else
    Reg = wreg
    + Reg + C
if (((wreg > 0)
& (Reg > 0))
& ((wreg
+ Reg + C)
> 0x07F))
```

0010 00da ffff ffff

Instruction	Format	Operation	Bit pattern

in the BSR Bank else, Reg is in the Access Bank		OV = 1
		else
		OV = 0
		if ((wreg + Reg + C) & 0x080) != 0)
		N = 1
		else
		N = 0
		if ((wreg + Reg + C) > 0x0FF)
		C = 1
		else
		C = 0
		if (((wreg & 0x0F) + (Reg & 0x0F) + C) > 0x0F)
		DC = 1
		else
		DC = 0
		if (((wreg + Reg + C) & 0x0FF) == 0x000)
		Z = 1
		else
		Z = 0	

andlw Constant

AND a Constant to the "wreg" and store the Result in "wreg"

```
wreg = wreg
    & Constant
if ((wreg &
    Constant)
    == 0x000)
    Z = 1
else
    Z = 0
if (((wreg & Constant)
    & 0x080) ! = 0)
    N = 1
else
    N = 0
```

`0000 1011 kkkk kkkk`

andwf Reg, d, a

AND "wreg" to the Contents of "Reg" and store the Result According to "d" Result in "wreg". If "a" is set, then Reg is in the BSR Bank, else it is in the Access Bank

```
if ("d" == 1)
    wreg = wreg
        & Reg
else
    Reg = wreg
        & Reg
if ((wreg & Reg)
    == 0x000)
    Z = 1
else
    Z = 0
if ((wreg & Reg)
    & 0x080) ! = 0)
    N = 1
else
    N = 0
```

`0001 01da ffff ffff`

Instruction	Format	Operation	Bit pattern
Branch if the Carry Flag is Set. Label is a Two's Complement Offset. One Instruction Cycle if Branch Not Executed, Two if Branch Executed.	bc Label	if (C == 1) PC = PC + 2 + Label	1110 0010 kkkk kkkk
Clear the Specified Bit in "Reg". If "a" is set then Reg is in the BSR Bank else, Reg is in the Access Bank	bcf Reg, Bit, a	Reg = Reg & (0x0FF ^ (1 << Bit))	1001 bbba ffff ffff

Branch if Negative Flag is Set. Label is a Two's Complement Offset. One Instruction Cycle if Branch Not Executed, Two if Branch Executed.	bn Label	if (N == 1) PC = PC + 2 + Label	1110 0110 kkkk kkkk
Branch if the Carry Flag is Reset. Label is a Two's Complement Offset. One Instruction Cycle if Branch Not Executed, Two if Branch Executed.	bnc Label	if (C == 0) PC = PC + 2 + Label	1110 0011 kkkk kkkk

Instruction	Format	Operation	Bit pattern
Branch if Negative Flag is Reset. Label is a Two's Complement Offset. One Instruction Cycle if Branch Not Executed, Two if Branch Executed.	bnn Label	if (N == 0) PC = PC + 2 + Label	1110 0111 kkkk kkkk
Branch if Overflow Flag is Reset. Label is a Two's Complement Offset. One Instruction Cycle if Branch Not Executed, Two if Branch Executed.	bnov Label	if (OV == 0) PC = PC + 2 + Label	1110 0101 kkkk kkkk

Branch if the Zero Flag is Reset. Label is a Two's Complement Offset. One Instruction Cycle if Branch Not Executed, Two if Branch Executed.	bnz Label	if (Z == 0) PC = PC + 2 + Label	1110 0001 kkkk kkkk
Branch if Overflow Flag is Set. Label is a Two's Complement Offset. One Instruction Cycle if Branch Not Executed, Two if Branch Executed.	bov Label	if (OV == 1) PC = PC + 2 + Label	1110 0100 kkkk kkkk

Instruction	Format	Operation	Bit pattern
Branch Always. Label is a Two's Complement Offset. Two Instruction Cycles	`bra Label`	PC = PC + 2 + Label	`1110 0kkk kkkk kkkk`
Set the Specified Bit in "Reg". If "a" is set then Reg is in the BSR Bank, else Reg is in the Access Bank	`bsf Reg, Bit, a`	Reg = Reg | (1 << Bit)	`1000 bbba ffff ffff`
Test the Specified Bit in "Reg" and skip if Clear. If "a" is set then the BSR is used for Reg, else the Access Bank is Used. One Instruction Cycle if Skip Not Executed, Two if Skip Executed.	`btfsc Reg, Bit, a`	if ((Reg & (1 << Bit) == 0) PC = NextIns	`1011 bbba ffff ffff`

Description	Syntax	Operation	Encoding
Test the Specified Bit in "Reg" and skip if Set. If "a" is set then the BSR is used for Reg, else the Access Bank is used. One Instruction Cycle if Skip Not Executed, Two if Skip Executed.	btfss Reg,Bit,a	f ((Reg & (1 << Bit)) != 0) PC = NextIns	1010 bbba ffff ffff
Toggle the Specified Bit in "Reg". If "a" is set then the BSR is used for Reg, else the Access Bank is used	btg Reg, Bit, a	Reg = Reg ^ (1 << Bit)	0111 bbba ffff ffff

Instruction	Format	Operation	Bit pattern
Branch if the Zero Flag is Set. Label is a Two's Complement Offset. One Instruction Cycle if Branch Not Executed, Two if Branch Executed.	bz Label	if (Z == 1) PC = PC + 2 + Label	1110 0000 kkkk kkkk
Call the 20-Bit "Label" Address. If "s" is set, Save the Context Registers. Two Instruction Cycles	call Label, s	PUSH(PC) if (s == 1) PUSH (W, STATUS BSR) PC = Label	1110 110s kkkk kkkk 1111 kkkk kkkk kkkk

Description	Instruction	Operation	Encoding
Clear the Specified Register. If "a" is set then the BSR is used for Reg, else the Access Bank is used	`clrf Reg, a`	Reg = 0 Z = 1	0110 101a ffff ffff
Clear the Watchdog Register and STATUS flags	`clrwdt`	WDT = 0 WDT Postscaler = 0 _TO = 1 _PD = 1	0000 0000 0000 0100

Complement the Contents of the Specified Register. If "a" is set then the BSR is used for Reg, else the Access Bank is used	`comf Reg, d, a`	if ("d" == 0)
		  wreg = Reg ^ 0x0FF
		else
		  Reg = Reg ^ 0x0FF
		if ((Reg ^ 0x0FF) == 0x000)
		  Z = 1
		else
		  Z = 0
		if ((Reg ^ 0x0FF) & 0x080) != 0)
		  N = 1
		else
		  N = 0
			0001 11da ffff ffff

Instruction	Format	Operation	Bit pattern
Compare the Specified Register with wreg and skip if the Register == Wreg. If "a" is set then the BSR is used for Reg, else the Access Bank is used. One Instruction Cycle if Skip Not Executed, Two if Skip Executed.	cpfseq Reg, a	if ((Reg − wreg) == 0) PC = NextIns	0110 001a ffff ffff

Compare the Specified Register with wreg and skip if the Register > wreg. If "a" is set then the BSR is used for Reg, else the Access Bank is used. One Instruction Cycle if Skip Not Executed, Two if Skip Executed

cpfsgt Reg, a

if ((Reg − wreg) > 0)
PC = NextIns

`0110 010a ffff ffff`

Instruction	Format	Operation	Bit pattern
Compare the Specified Register with wreg and skip if the Register < wreg. If "a" is set then the BSR is used for Reg, else the Access Bank is used. One Instruction Cycle if Skip Not Executed, Two if Skip Executed	cpfslt Reg, a	if ((Reg − wreg) < 0) PC = NextIns	0110 000a ffff ffff
Do a Decimal Adjust after Addition of Two BCD Values	daw	if ((wreg & 0x0F) > 9) wreg = (wreg & 0x0F) + 0x010	0000 0000 0000 0111

78

Decrement
the Contents
of the
Specified
Register.
If "a" is set
then Reg is
in BSR Bank
else Access
Bank is used

decf Reg, d, a

```
if ("d" == 0)
    wreg = Reg − 1
else
    Reg = Reg − 1
if ((Reg − 1)
    == 0x000)
    Z = 1
else
    Z = 0
if (((Reg > 0)
    & ((Reg − 1)
    < 0x080))
    OV = 1
else
    OV = 0
if ((Reg − 1)
    & 0x080) != 0)
    N = 1
else
    N = 0
if (((Reg & 0x00F) − 1)
    & 0x00F) != 0)
    DC = 0
else
    DC = 1
if ((Reg − 1)
    < 0)
    C = 0
else
    C = 1
```

Instruction	Format	Operation	Bit pattern

Decrement the Contents of the Specified Register and skip the Next Instruction if Result == 0. If "a" is set then the BSR is used for Reg, else the Access Bank is used. One Instruction Cycle if Skip Not Executed, Two if Skip Executed	decfsz Reg,d,a	if ("d" == 0)
		wreg = Reg − 1
		else
		Reg = Reg − 1
		if ((Reg − 1)
		== 0x000)
		PC = NextIns
			0010 11da ffff ffff

Description		Operation	Encoding

Decrement the Contents of the Specified Register and skip the Next Instruction if Result != 0. If "a" is set then the BSR is used for Reg, else the Access Bank is used. One Instruction Cycle if Skip Not Executed, Two if Skip Executed	dcfsnz Reg, d, a	if ("d" == 0)
		  wreg = Reg − 1
		else
		  Reg = Reg − 1
		if ((Reg − 1)
		  != 0x000)
		    PC = NextIns
			0100 11da ffff ffff

Goto the 20-Bit "Label" Address. Two Instruction Cycles	goto Label	PC = Label	1110 1111 kkkk kkkk kkkk 1111 kkkk kkkk kkkk

Instruction	Format	Operation	Bit pattern

Increment the Contents of the Specified Register. If "a" is set then Reg is in the BSR Bank else Access Bank is used.	incf Reg, d, a	if ("d" == 0)
		  wreg = Reg + 1
		else
		  Reg = Reg + 1
		if ((Reg + 1) == 0x000)
		  Z = 1
		else
		  Z = 0
		if (((Reg > 0) & ((Reg + 1) > 0x07F))
		  OV = 1
		else
		  OV = 0
		if ((Reg + 1) & 0x080) != 0)
		  N = 1
		else
		  N = 0
		if (((Reg & 0x00F) + 1) & 0x010) != 0)
		  DC = 1
		else
		  DC = 0
		if ((Reg + 1) == 0x0100)
			0010 10da ffff ffff

Increment the Contents of the Specified Register and skip the Next Instruction if Result == 0. If "a" is set then the BSR is used for Reg, else the Access Bank is used. One Instruction Cycle if Skip Not Executed, Two if Skip Executed

incfsz Reg, d, a

```
                    C = 1
                  else
                    C = 0

if ("d" == 0)
  wreg = Reg + 1
else
  Reg = Reg + 1
if ((Reg + 1)
  == 0x000)
  PC = NextIns
```

0011 11da ffff ffff

Instruction	Format	Operation	Bit pattern

Increment the Contents of the Specified Register and skip the Next Instruction if Result != 0. If "a" is set then the BSR is used for Reg, else the Access Bank is used. One Instruction Cycle if Skip Not Executed, Two if Skip Executed	infsnz Reg, d, a	if ("d" == 0)
		  wreg = Reg + 1
		else
		  Reg = Reg + 1
		if ((Reg + 1)
		  != 0x000)
		  PC = NextIns
			0100 10da ffff ffff

| OR a Constant to the "wreg" and store the Result in "wreg" | `iorlw Constant` | wreg = wreg \| Constant
if ((wreg \| Constant) == 0x000)
  Z = 1
else
  Z = 0
if ((wreg \| Constant) & 0x080) != 0)
  N = 1
else
  N = 0 | 0000 1001 kkkk kkkk |
| OR "wreg" to the Contents of "Reg" and store the Result According to "d". If "d" Result in "wreg". If "a" is set then Reg is in the BSR Bank, else it is in the Access Bank | `iorwf Reg,d,a` | if ("d" == 1)
  wreg = wreg
    \| Reg
else
  Reg = wreg
    \| Reg
if ((wreg \| Reg) == 0x000)
  Z = 1
else
  Z = 0
if ((wreg \| Reg) & 0x080) != 0)
  N = 1
else
  N = 0 | 0001 00da ffff ffff |

Instruction	Format	Operation	Bit pattern
Load the Specified FSR Register with the Constant. Two Instruction Cycles	`lfsr f, Const`	FSR(f) = Const	`1110 1110 00ff kkkk` `1111 0000 kkkk kkkk`

Move data from 256 Address Register Data to Primary Register Set. If "a" is set then the BSR is used for Reg, else the Access Bank is used	`movf Reg, d, a`	if (d == 0)
		  wreg = Reg
		if (Reg == 0)
		  Z = 1
		else
		  Z = 0
		if ((Reg & 0x080)
		  != 0)
		  N = 1
		else
		  N = 0
			`0101 00da ffff ffff`

Move Contents of the Source Register into the Destination Register. The Full 12-Bit Addresses are Specified. Two Instruction Cycles	`movff Regs, Regd`	Regd = Regs	`1100 ffff ffff ffff` `1111 ffff ffff fffd`

Description	Mnemonic	Operation	Opcode
Move Constant into low Nybble of BSR	movlb Constant	BSR(3:0) = Constant	0000 0001 kkkk kkkk
Move Constant into wreg	movlw Constant	wreg = Constant	0000 1110 kkkk kkkk
Move contents of "wreg" into "Reg". If "a" is set then the BSR is used for Reg, else the Access Bank is used	movwf Reg, a	Reg = wreg	0110 111a ffff ffff
Multiply Constant by "wreg"	mullw Constant	PRODH:PROGL = Constant * wreg	0000 1101 kkkk kkkk
Multiply Register by "wreg". If "a" is set then the BSR is used for Reg, else the Access Bank is used	mullwf Reg	PRODH:PROGL = Reg * wreg	0000 0010a ffff ffff

Instruction	Format	Operation	Bit pattern

Negate the Contents of "Reg" and store the result back in "Reg". If "a" is set, then Reg is in the BSR Bank, else Reg is in the Access Bank	negw Reg, a	Reg = −Reg
		if (−Reg < 0x080)
		  OV = 1
		else
		  OV = 0
		if ((−Reg & 0x080)
		!= 0)
		  N = 1
		else
		  N = 0
		if (−Reg > 0x0FF)
		  C = 1
		else
		  C = 0
		if ((−Reg & 0x0F)
		> 0x0F)
		  DC = 1
		else
		  DC = 0
		if (−Reg == 0x000)
		  Z = 1
		else
		  Z = 0
			0110 110a ffff ffff

Do Nothing for one Instruction Cycle. Note Two Different Op Codes.	nop	0000 0000 0000 0000 1111 1111 1111 1111
Pop the top of the Instruction Pointer Stack and Discard the Result.	pop	POP ( )	0000 0000 0000 0110
Push the top of the Instruction Pointer Stack.	push	PUSH (PC + 2)	0000 0000 0000 0101
Call the 11-Bit 2's Complement "Offset". Two Instruction Cycles	rcall Label	PUSH (PC) PC = PC + 2 + Label	1101 1kkk kkkk kkkk

89

Instruction	Format	Operation	Bit pattern
Reset the PICmicro® MCU Processor and all the Registers Affected by _MCLR Reset	Reset	_MCLR = 0 _MCLR = 1	0000 0000 1111 1111
Return from Interrupt Handler. If "s" is set, Restore the wreg, STATUS and BSR Registers. Two Instruction Cycles	retfie, s	PC = POP ( ) GIE = 0 if (s === 1) wreg = POP ( ) STATUS = POP ( ) BSR = POP ( )	0000 0000 0001 000s
Return from Subroutine with new value in wreg. Two Instruction Cycles	retlw Constant	wreg = Constant PC = POP ( )	0000 1100 kkkk kkkk

Description	Instruction	Operation	Encoding

Return from Subroutine. If "s" is set, Restore the Wreg, STATUS and BSR Registers. Two Instruction Cycles	`return, s`	PC = POP ()
		if (s === 1)
		  wreg = POP ()
		  STATUS = POP ()
		  BSR = POP ()
			0000 0000 0001 001s

Rotate Left Through the Carry Flag. If "a" is set then Reg is in BSR Bank else Reg is in the Access Bank	`rlcf Reg, d, a`	if (d == 0)
		  wreg(7:1) =
		    Reg(6:0)
		  wreg(0) = C
		  C = Reg (7)
		else
		  Reg (7:1) =
		    Reg (6:0)
		  Reg (0) = C
		  C = Reg (7)
		if (Reg (6)!= 0)
		  N = 1
		else
		  N = 0
			0011 01da ffff ffff

Instruction	Format	Operation	Bit pattern

Rotate Left. If "a" is set then Reg is in the BSR Bank else Reg is in the Access Bank	`rlcnf Reg, d, a`	if (d == 0)
		$\quad$ wreg (7:1) = Reg (6:0)
		$\quad$ wreg (0) = Reg (7)
		else
		$\quad$ Reg (7:1) = Reg (6:0)
		$\quad$ Reg(0) = Reg(7)
		if (Reg(6)!= 0)
		$\quad$ N = 1
		else
		$\quad$ N = 0
			0100 01da ffff ffff

Rotate Right Through the Carry Flag. If "a" is set then Reg is in the BSR Bank else Reg is in the Access Bank	`rrcf Reg, d, a`	if (d == 0)
		$\quad$ wreg (6:0) = Reg (7:1)
		$\quad$ wreg (7) = C
		$\quad$ C = Reg (0)
		else
		$\quad$ Reg (6:0) = Reg (7:1)
		$\quad$ Reg (7) = C
		$\quad$ C = Reg (0)
		if (Reg (0)!= 0)
		$\quad$ N = 1
		else
		$\quad$ N = 0
			0011 00da ffff ffff

Rotate Right.
If "a" is set
then Reg is in
the BSR Bank
else Reg is in
the Access Bank

```
rrcnf Reg, d, a
```

f (d == 0)
   wreg (6:0) =
      Reg (7:1)
   wreg (7) =
      Reg (0)
else
   Reg (6:0) =
      Reg (7:1)
   Reg (7) =
      Reg (0)
if (Reg (0) != 0)
   N = 1
else
   N = 0

`0100 00da ffff ffff`

Set the
Specified
Register and
Optionally
"wreg".
If "a" is set
then the BSR
is used for
Reg, else the
Access Bank
is used

```
setf Reg, s, a
```

Reg = 0x0FF
if (s == 0)
   wreg = 0x0FF

`0110 100a ffff ffff`

93

Instruction	Format	Operation	Bit pattern
Put the PICmicro® MCU in a "Power Down" State	`sleep`	WDT = 0 WDT Postscaler = 0 _TO = 1 _PD = 0 PICmicro MCU Power Down	0000 0000 0000 0011

Subtract the Contents of "Reg" and C from wreg and store the Result According to "d" Result in "wreg". If "a" is set then Reg is in the BSR Bank else it is in the Access Bank	`subwfb Reg, d, a`	if ("d" == 1)
		  wreg = wreg
		    – Reg – !C
		else
		  Reg = wreg
		    – Reg – !C
		if ((((Reg > 0)
		  & (wreg > 0))
		  & ((wreg
		  = Reg – !C)
		  < 0x080))
		  OV = 1
		else
		  OV = 0
		if (((wreg –
		  Reg –C) & 0x080)
		  != 0)
		  N = 1
		else
		  N = 0
			0101 01da ffff ffff

| Subtract "wreg" from a Constant and store the Result in "wreg" | `sublw Constant` | if ((wreg – Reg – !C) > 0x0FF)
  C = 1
else
  C = 0
if (((wreg & 0x0F) – (Reg & 0x0F) – !C) > 0x0F)
  DC = 1
else
  DC = 0
if (((wreg – Reg – !C) & 0x0FF) == 0x000)
  Z = 1
else
  Z = 0

wreg = Constant – wreg
if (((wreg < 0) & (Constant < 0)) & ((Constant – wreg) < 0x080))
  OV = 1
else
  OV = 0
if (((Constant – | `0000 1000 kkkk kkkk` |

Instruction	Format	Operation	Bit pattern

		wreg) & 0x080) != 0)
		 N = 1
		 else
		 N = 0
		 if ((Constant − wreg) < 0x0FF)
		 C = 1
		 else
		 C = 0
		 if ((((Constant & 0x0F) − (wreg & 0x0F)) > 0x0F)
		 DC = 1
		 else
		 DC = 0
		 if ((((Constant − wreg) & 0x0FF) == 0x000)
		 Z = 1
		 else
		 Z = 0	

Subtract "wreg"
from the Contents
of "Reg" and
store the
Result
According to
"d" Result in
"wreg". If
"a" is set then
Reg is in the
BSR Bank else
Reg is in the
Access Bank

```
subwf Reg, d, a
```

```
if ("d" == 1)
    wreg = Reg –
    wreg
else
    Reg = Reg
    – wreg
if (((wreg < 0)
& (Reg < 0))
& ((Reg
– wreg)
< 0x080))
    OV = 1
else
    OV = 0
if (((Reg –
wreg) & 0x080)
!= 0)
    N = 1
else
    N = 0
if ((Reg –
wreg) > 0x0FF)
    C = 1
else
    C = 0
if (((Reg & 0x0F) –
(wreg & 0x0F)
> 0x0F)
    DC = 1
```

Instruction	Format	Operation	Bit pattern

Subtract "wreg" from the Contents of "Reg" and "C", store the Result According to "d" Result in "wreg". If "a" is set, then Reg is in the BSR Bank else Reg is in the Access Bank	subwfb Reg, d, a	else
		  DC = 0
		if (((Reg −
		  wreg) & 0x0FF)
		  == 0x000)
		  Z = 1
		else
		  Z = 0
		
		if ("d" == 1)
		  wreg = Reg
		  − wreg −!C
		else
		  Reg = Reg
		  − wreg −!C
		if (((Reg > 0)
		& (wreg > 0))
		& ((Reg
		− wreg −!C
		< 0x080))
		  OV = 1
		else
		  OV = 0
		if (((Reg −
		wreg −C) & 0x080)
		!= 0)
		  N = 1
			0101 10da ffff ffff

```
        else
          N = 0
        if ((Reg –
          wreg – !C) > 0x0FF)
            C = 1
        else
            C = 0
        if (((Reg & 0x0F) –
          (wreg & 0x0F) – !C)
          > 0x0F)
            DC = 1
        else
            DC = 0
        if (((Reg –
          wreg – !C) & 0x0FF)
          == 0x000)
            Z = 1
        else
            Z = 0
```

Instruction	Format	Operation	Bit pattern

Swap the Contents of "Reg" and store the Result According to "d" Result in "wreg". If "a" is set, then Reg is, in the BSR Bank else Reg is in the Access Bank	swapf Reg, d, a	if ("d" == 1)
		wreg =
		((Reg & 0x00F)
		<< 4) +
		((Reg & 0x0F0)
		>> 4)
		else
		Reg =
		((Reg & 0x00F)
		<< 4) +
		((Reg & 0x0F0)
		>> 4)
			0011 10da ffff ffff

Read the Program Memory Contents at the Table Pointer and Execute as "Option" Specifies. Two Instruction Cycles	tablrd Option	switch(Option) case * TABLAT = ProgMem (TBLPTR) case *+ TABLAT = ProgMem (TBLPTR) TBLPTR = TBLPTR + 1 case *- TABLAT = ProgMem (TBLPTR)	0000 0000 0000 10nn

nn	Option
00	*
01	*+
10	*-
11	+*

tablwt Option

Write the
Contents of
the Table
Latch into
Program
Memory based
on the
"Option"
Specification.
If the
Destination
is Internal
EPROM, the
Instruction
does not
End until an
Interrupt.
Two
Instruction Cycles or
Many if EPROM Write

```
TBLPTR =
TBLPTR – 1
case +*
TBLPTR =
TBLPTR + 1
TABLAT =
ProgMem (TBLPTR)

switch(Option)
case *
  ProgMem (TBLPTR)
  = TABLAT
case *+
  ProgMem(TBLPTR)
  = TABLAT
  TBLPTR =
  TBLPTR + 1
case *–
  ProgMem(TBLPTR)
  = TABLAT
  TBLPTR =
  TBLPTR + 1
case +*
  TBLPTR =
  TBLPTR + 1
  ProgMem (TBLPTR)
  = TABLAT
```

```
0000 0000 0000 11nn
nn      Option
00       *
01       *+
10       *–
11       +*
```

101

Instruction	Format	Operation	Bit pattern
Compare the Specified Register zero and skip if the Register == 0. If "a" is set then the BSR is used for Reg, else the Access Bank is used. One Instruction Cycle if Skip Not Executed, Two if Skip Executed	`tstfsz Reg, a`	if (Reg == 0) PC = NextIns	`0110 011a ffff ffff`

XOR a Constant to the "wreg" and store the Result in "wreg"	`xorlw Constant`	wreg = wreg ^ Constant
		if ((wreg ^ Constant) == 0x000)
		  Z = 1
		else
		  Z = 0
		if ((wreg ^ Constant) & 0x080) != 0)
		  N = 1
		else
		  N = 0
			0000 1010 kkkk kkkk

XOR "wreg" to the Contents of "Reg" and store the Result According to "d" Result in "wreg". If "a" is Set then Reg is in the BSR Bank else Reg is in the Access Bank	`xorwf Reg,d,a`	if ("d" == 1)
		  wreg = wreg ^ Reg
		else
		  Reg = wreg ^ Reg
		if ((wreg ^ Reg) == 0x000)
		  Z = 1
		else
		  Z = 0
		if ((wreg ^ Reg) & 0x080) != 0)
		  N = 1
		else
		  N = 0
			0001 10da ffff ffff

Microchip Special Instruction Mnemonics

The following "special instructions" are macros built into MPASM by Microchip to help make some low-end and mid-range PICmicro® MCU instructions more intuitive. These instructions are built into MPASM and their labels should *never* be used for macros, addresses (code or variable), or defines.

Most of these special instructions are made up of one or more standard low-end or mid-range PICmicro® MCU instructions. Note that some of these special instructions may change the value of the zero flag.

"LCALL" should never be used because the PCLATH bits are not returned to the appropriate value for the code following "LCALL". When a "goto" or "call" is executed after an "LCALL" statement and the PCLATH bits are not set appropriately for the current page, execution will jump into the "LCALL" page.

For the low-end PICmicro® MCUs, "LCALL" should be

```
bcf/bsf         STATUS, PA0
bcf/bsf         STATUS, PA1
bcf/bsf         STATUS, PA2
call            Label
bsf/bcf         STATUS, PA0
bsf/bcf         STATUS, PA1
bsf/bcf         STATUS, PA2
```

and for the mid-range, "LCALL" should be

```
bcf/bsf         PCLATH, 3
bcf/bsf         PCLATH, 4
call            Label
bsf/bcf         PCLATH, 3
bsf/bcf         PCLATH, 4
```

"negf" should never be used unless the destination is back into the file register source. If the destination is "w", note that the contents of the file register source will be changed with the complement of the value. Because of this added complexity, use of this special instruction is not recommended.

Description	Instruction	Actual Instructions	Operation

Add Carry to File Register	addcf Reg, d	btfsc STATUS, C
		incf Reg, d
			if (C == 1)
			  if (d == 1)
			    Reg = Reg + 1;
			  else
			    w = Reg + 1

Add Digit Carry to File Register	adddcf Reg, d	btfsc STATUS, DC
		incf Reg, d
			if (DC == 1)
			  if (d == 1)
			    Reg = Reg + 1;
			  else
			    w = Reg + 1;

Branch to Label	B Label	goto Label	PC = ((PCLATH << 8) & 0x01800) + Label;
Branch on Carry Set	BC Label	btfsc STATUS, C goto Label	if (C == 1) PC = ((PCLATH << 8) & 0x01800) + Label;
Branch on Digit Carry Set	BDC Label	btfsc STATUS, DC goto Label	if (DC == 1) PC = ((PCLATH << 8) & 0x01800) + Label;
Branch on Carry Reset	BNC Label	btfss STATUS, C goto Label	if (C == 0) PC = ((PCLATH << 8) & 0x01800) + Label;

Branch on Digit Carry Reset	BNDC Label	btfss STATUS, DC goto Label	if (DC == 0) PC = ((PCLATH << 8) & 0x01800) + Label;
Branch on Zero Reset	BNZ Label	btfss STATUS, Z goto Label	if (Z == 0) PC = ((PCLATH << 8) & 0x01800)+Label;
Branch on Zero Set	BZ Label	btfsc STATUS, Z goto Label	if (Z == 1) PC = ((PCLATH << 8) & 0x01800)+Label;
Clear Carry	clrc	bcf STATUS, C	C = 0;
Clear Digit Carry	clrdc	bcf STATUS, DC	DC = 0;
Clear Zero	clrz	bcf STATUS, Z	Z = 0;
Long Call — Do NOT use as Described Above	lcall Label	Low-End: bcf/bsf STATUS, PA0 bcf/bsf STATUS, PA1 bcf/bsf STATUS, PA2 call Label Mid-Range: bcf/bsf PCLATH, 3 bcf/bsf PCLATH, 3 call Label	

Description	Instruction	Actual Instructions	Operation

Long Goto	`lgoto Label`	Low-End:
		`bcf/bsf STATUS, PA0`
		`bcf/bsf STATUS, PA1`
		`bcf/bsf STATUS, PA2`
		`goto   Label`
		Mid-Range:
		`bcf/bsf PCLATH, 3`
		`bcf/bsf PCLATH, 3`
		`goto   Label`	

Load "w" with Contents of "Reg"	`movfw Reg`	`movf  Reg, w`	w = Reg
			if (Reg == 0)
			  Z = 1;
			else
			  Z = 0;

Negate a File Register.
— Do NOT use as
Described Above
	`negf Reg, d`	`comf  Reg, f`
		`incf  Reg, d`
			Reg = Reg ^ 0x0FF;
			if (d == 0)
			  w = Reg + 1;
			else
			  Reg = Reg + 1;

Set Carry	`setc`	`bsf   STATUS, C`	C = 0;
Set Digit Carry	`setdc`	`bsf   STATUS, DC`	DC = 0;
Set Zero	`setz`	`bsf   STATUS, Z`	Z = 0;

Description	Mnemonic	Operands	Operation
Skip the next Instruction if the Carry Flag is Set	skpc	STATUS, C	if (C == 1) PC = PC + 1;
Skip the next Instruction if the Digit Carry Flag is Set	skpdc	STATUS, DC	if (DC == 1) PC = PC + 1;
Skip the next Instruction if the Carry Flag is Reset	skpnc	STATUS, C	if (C == 0) PC = PC + 1;
Skip the next Instruction if the Digit Carry Flag is Reset	skpndc	STATUS, DC	if (DC == 0) PC = PC + 1;
Skip the next Instruction if the Zero Flag is Reset	skpnz	STATUS, Z	if (Z == 0) PC = PC + 1;
Skip the next Instruction if the Zero Flag is Set	skpz	STATUS, Z	if (Z == 1) PC = PC + 1;

Subtract Carry from File Register	subcf Reg, d	btfsc STATUS, C
		decf Reg, d
			if (C == 1)
			if (d == 1)
			Reg = Reg − 1;
			else
			w = Reg − 1

Description	Instruction	Actual Instructions		Operation

Subtract Digit Carry To File Register	adddcf Reg, d	btfsc
		incf
			STATUS, DC
			Reg, d
				if (DC == 1)
				if (d == 1)
				  Reg = Reg − 1;
				else
				  w = Reg − 1;

Load "Z" with 1 if Contents of "Reg" equal 0	movfw Reg	movf	Reg, w	if (Reg == 0)
				  Z = 1;
				else
				  Z = 0;

Parallax PICmicro® MCU Instruction Set

Parallax Inc. (manufacturers of the "Basic Stamp") have written a very popular assembler for the Microchip PICmicro® MCUs. "PASM" (as it is known) implements an assembler language that is similar to the Intel 8051 instruction set. The assembler also supports MPASM (standard Microchip) instruction formats as well.

Some of these instructions are designed specifically for the low-end PICmicro® MCUs (they have been noted). If you're working with a mid-range PICmicro® MCU, these instructions MUST NOT be used.

Note that many of these mnemonics result in multiple PICmicro® MCU instructions with unexpected changes to the STATUS and "w" register.

PASM is available from the Parallax web site.

− Literal Instructions

fr − File Register

PASM Data Instructions

Instruction Description		Cycles	Context Resources Affected	Actual PICmicro® MCU Instructions
CLR Parm	**Clear Parameter**			
"W"	w = 0	1	Zero	clrw
fr	fr = 0	1	Zero	clr fr
WDT	WDT = 0	1	_TO, _PD	clrwdt
MOV Parm	**Move Data**			
"W, #"	w = #	1	None	movlw #
"W, fr"	w = fr	1	Zero	movf fr, w
"W, /fr"	w = fr ^ 0x0FF	1	Zero	comf fr, w
"W, fr-W"	w = fr+(w^0x0FF)+1	1	Z, C, DC	subwf fr, w
"W, ++fr"	w = fr + 1	1	Z	incf fr, w
"W, --fr"	w = fr - 1	1	Z	decf fr, w
"W, >>fr"	w = fr >> 1	1	Carry	rrf fr, w
"W, <<fr"	w = fr << 1	1	Carry	rlf fr, w
"W, <>fr"	w = NibSwap fr	1	None	swapf fr, w
"fr, W"	fr = w	1	None	movwf fr
"!Port, W"	TRIS = w	1	None	TRIS Port
"!Port, #"	TRIS = #	2	w	movlw # TRIS Port
"!Port, fr"	TRIS = fr	2	w, Zero	movf fr, w TRIS Port
"OPTION, W"	OPTION = w	1	None	OPTION
"OPTION, #"	OPTION = #	2	w	movlw # OPTION

Parm	Operation	#	Status	Instruction	Operand
"OPTION, fr"	OPTION = fr	2	w, Zero	movf	fr, w
				OPTION	
"fr, #"	fr = #	2	None	movlw	#
				movwf	fr
"fr, fr2"	fr = fr2	2	Zero	movf	fr2, w
				movwf	fr
ADD two Values					
"W, fr"	w = w + fr	1	Z, C, DC	addwf	fr, w
"fr, W"	fr = w + fr	1	Z, C, DC	addwf	fr, f
"fr, #"	fr = fr + #	2	w, Z, C, DC	movlw	#
				addwf	fr, f
"fr, fr2"	fr = fr + fr2	2	w, Z, C, DC	movf	fr2, w
				addwf	fr, f
Subtraction					
"fr, W"	fr = fr+(w^0xFF)+1	1	Z, C, DC	subwf	fr, f
"fr, #"	fr = fr+((#^0xFF)+1)	2	w, Z, C, DC	movlw	#
				subwf	fr, f
"fr, fr2"	fr=fr+(fr2^0xFF)+1	2	w, Z, C, DC	movf	fr2, w
				subwf	fr, f
Bitwise AND					
"W, #"	w = w & #	1	Zero	andlw	#
"W, fr"	w = w & fr	1	Zero	andwf	fr, w
"fr, W"	fr = w & fr	1	Zero	andwf	fr, f
"fr, #"	fr = fr & #	2	w, Zero	movlw	#
				andwf	fr, f
"fr, fr2"	fr = fr & fr2	2	w, Zero	movf	fr2, w
				andwf	fr, f

PASM Data Instructions Instruction Description	Cycles	Context Resources Affected	Actual PICmicro® MCU Instructions
OR Parm Bitwise Inclusive OR			
"W, #" w = w | #	1	Zero	iorlw #
"W, fr" w = w | fr	1	Zero	iorwf fr, w
"fr, W" fr = fr | w	1	Zero	iorwf fr, f
"fr, #" fr = fr | #	2	w, Zero	movlw #
			iorwf fr, f
"fr, fr2" fr = fr | fr2	2	w, Zero	movf fr2, w
			iorwf fr, f
XOR Parm Bitwise Exclusive OR			
"W, #" w = w ^ #	1	Zero	xorlw #
"W, fr" w = w ^ fr	1	Zero	xorwf fr, w
"fr, W" fr = fr ^ w	1	Zero	xorwf fr, f
"fr, #" fr = fr ^ #	2	w, Zero	movlw #
			xorwf fr, f
"fr, fr2" fr = fr ^ fr2	2	w, Zero	movf fr2, w
			xorwf fr, f
DEC Parm Decrement Register			
"fr" fr = fr - 1	1	Zero	decf fr, f
INC Parm Increment Register			
"fr" fr = fr + 1	1	Zero	incf fr, f
NEG Parm Two's Complement Negation			
"fr" fr = 0 - fr	2	Zero	comf fr, f
			incf fr, f

NOT Parm — Bitwise Complement

Instruction	Description	Cycles	Context Resources Affected	Actual PICmicro® MCU Instructions
"W"	w = w ^ 0xFF	1	Zero	xorlw 0xOFF
"fr"	fr = fr ^ 0xOFF	1	Zero	comf fr

TEST Parm — Test Parm Equal to Zero

Instruction	Description	Cycles	Context Resources Affected	Actual PICmicro® MCU Instructions
"W"	Z = (w == 0)	1	Zero	iorlw 0
"fr"	Z = (fr == 0)	1	Zero	movf fr, f

RR Parm — Rotate Register to Right

Instruction	Description	Cycles	Context Resources Affected	Actual PICmicro® MCU Instructions
"fr"	fr = fr >> 1	1	Carry	rrf fr, f

RL Parm — Rotate Register to Left

Instruction	Description	Cycles	Context Resources Affected	Actual PICmicro® MCU Instructions
"fr"	fr = fr << 1	1	Carry	rlf fr, f

SWAP Parm — Swap Nybbles of Register

Instruction	Description	Cycles	Context Resources Affected	Actual PICmicro® MCU Instructions
"fr"	fr = <>fr	1	None	swapf fr, f

PASM Bit Instructions

Instruction	Description	Cycles	Context Resources Affected	Actual PICmicro® MCU Instructions
CLRB fr, bit	fr.bit = 0	1	None	bcf fr, bit
SETB fr, bit	fr.bit = 1	1	None	bsf fr, bit
CLC	Carry = 0	1	None	bcf STATUS, C
STC	Carry = 1	1	None	bsf STATUS, C
CLZ	Zero = 0	1	None	bcf STATUS, Z
STZ	Zero = 0	1	None	bsf STATUS, Z
ADDB fr, bit	fr = fr + Bit	2	Zero	btfsc fr, bit incf fr, f
SUBB fr, bit	fr = fr - bit	2	Zero	btfss fr, bit decf fr, f

PASM Bit Instructions

Instruction Description	Cycles	Context Resources Affected	Actual PICmicro® MCU Instructions
MOVB fr.b, fr2.b2 Move Bit	4	None	btfss fr2, b2 bcf fr, b btfsc fr2, b2 bsf fr, b
MOVB fr.b, /fr2.b2 Move Invert	4	None	btfsc fr2, b2 bcf fr, b btfss fr2, b2 bsf fr, b

PASM PICmicro® MCU Microcontroller Instructions

Instruction Description	Cycles	Context Resources Affected	Actual PICmicro® MCU Instructions
NOP Do Nothing	1	Nothing	nop
SLEEP Put PICmicro® MCU to Sleep	N/A	_TO, _PD	sleep
LSET Addr Jump Setup	0-2	PA0, PA1	bcf/bsf STATUS, PA0

* – Low End Instruction, bcf/bsf of STATUS PAx Bits Address Dependant

PASM PICmicro® MCU Conditional Skip Instructions

Instruction	Description	Cycles	Context Resources Affected	Actual PICmicro® MCU Instructions
MOVSZ Parm	Skip if Result = 0			
"W, ++fr"	w = fr + 1	1/2	w	incfsz fr, w
"W, --fr"	w = fr - 1	1/2	w	decfsz fr, w
INCSZ fr	w=fr+1, if Z Skip	1/2	w	incfsz fr, f
DECSZ fr	w=fr-1, if Z Skip	1/2	w	decfsz fr, f
SB fr, bit	Skip if Bit Set	1/2	None	btfsc fr, bit
SNB fr, bit	Skip if Bit Reset	1/2	None	btfss fr, bit
SC	Skip if Carry Set	1/2	None	btfsc STATUS, C
SNC	Skip if C Reset	1/2	None	btfss STATUS, C
SZ	Skip if Zero Set	1/2	None	btfsc STATUS, Z
SNZ	Skip if Zero Reset	1/2	None	btfss STATUS, Z
CJA fr, #	if fr > # Skip_Next	3/4	w, C, DC, Z	movlw # addwf fr, w btfss STATUS, C

CJA fr, fr2	if fr > fr2
	Skip_Next
		3/4	w, C, DC, Z	movf fr, w
				subwf fr2, w
				btfss STATUS, C

CJAE fr, #	if fr > # Skip_Next	3/4	w, C, DC, Z	movlw # subwf fr, w btfss STATUS, C

CJAE fr, fr2	if fr > fr2
	Skip_Next
		3/4	w, C, DC, Z	movf fr2, w
				subwf fr, w
				btfss STATUS, C

PASM PICmicro® MCU Conditional Skip Instructions Instruction Description	Cycles	Context Resources Affected	Actual PICmicro® MCU Instructions
CSB fr, # if fr < # Skip_Next	3/4	w, C, DC, Z	movlw # subwf fr, w btfsc STATUS, C

CSB fr, fr2 if fr < fr2
Skip_Next
	3/4	w, C, DC, Z	movf fr2, w
			subwf fr, w
			btfsc STATUS, C

CSBE fr, # if fr <= # Skip_Next	3/4	w, C, DC, Z	movlw # addwf fr, w btfsc STATUS, C

CSBE fr, fr2 if fr <= fr2
Skip_Next
	3/4	w, C, DC, Z	movf fr, w
			subwf fr2, w
			btfss STATUS, C

CSE fr, # if fr == # Skip_Next	3/4	w, C, DC, Z	movlw # subwf fr, w btfss STATUS, Z
CSE fr, fr2 if fr == fr2 Skip_Next	3/4	w, C, DC, Z	movf fr2 w subwf fr, w btfss STATUS, Z
CSNE fr, # if fr == # Skip_Next	3/4	w, C, DC, Z	movlw # subwf fr, w btfsc STATUS, Z
CSNE fr, fr2 if fr == fr2 Skip_Next	3/4	w, C, DC, Z	movf fr2 w subwf fr, w btfsc STATUS, Z

PASM PICmicro® MCU Unconditional Branch Instructions

Instruction	Description	Cycles	Context Resources Affected	Actual PICmicro® MCU Instructions
JMP Parm "addr9"	Jump to Address PC = 9 Bit Address	2	None	goto addr9
* – Low End Instruction				
"PC+W"	PC = PC + Offset w	2	Z, C, DC	addwf PCL, f
"W"	PC = w	2	None	movwf PCL
CALL addr8	Call Subroutine	2	None	call addr8
* – Low End Instruction				
RET	Return & w = 0	2	w	retlw 0
SKIP	Skip Over Next Ins	2	None	btfss FSR, 7
* – Low End Instruction, Bit 7 of FSR is always Set				
LJMP Addr	LSET before JMP	2–5	PA0–PA2	bcf/bsf STATUS, PAx ; goto Addr
* – Low End Instruction				
LCALL Addr	LSET before CALL	2–5	PA0–PA2	bcf/bsf STATUS, PAx ; call Addr
* – Low End Instruction				
RETW 'String'	Table Return	2	w	retlw 'S' retlw 't' retlw 'r' retlw 'i' retlw 'n' retlw 'g'

PASM PICmicro® MCU Conditional Branch Instructions

Instruction Description	Cycles	Context Resources Affected	Actual PICmicro® MCU Instructions
IJNZ fr, addr9 Increment/Jump	2/3	None	incfsz fr, f goto addr9
DJNZ fr, addr9 Decrement/Jump	2/3	None	decfsz fr, f goto addr9
JB fr, bit, addr9 Jump on Bit	2/3	None	btfsc fr, bit goto addr9
JNB fr, bit, addr9	2/3	None	btfss fr, bit goto addr9
JC addr9 Jump on Carry	2/3	None	btfsc STATUS, C goto addr9
JNC addr9 Jump on !Carry	2/3	None	btfss STATUS, C goto addr9
JZ addr9 Jump on Zero	2/3	None	btfsc STATUS, Z goto addr9
JNZ addr9 Jump on !Zero	2/3	None	btfss STATUS, Z goto addr9

CJA fr, #, addr9
  if fr > #
    goto addr9
	4/5	w, C, DC, Z	movlw #
			subwf fr, w
			btfss STATUS, C
			goto addr9

CJA fr, fr2, addr9
  if fr > fr2
    goto addr9
	4/5	w, C, DC, Z	movf fr, w
			subwf fr2, w
			btfss STATUS, C
			goto addr9

Macro	Operation	Cycles	Affects	Expansion
CJAE fr, #, addr9	if fr >= # goto addr9	4/5	w, C, DC, Z	movlw # subwf fr, w btfss STATUS, C goto addr9

CJAE fr, fr2, addr9	if fr >= fr2
	goto addr9
		4/5	w, C, DC, Z	movf fr2, w
				subwf fr, w
				btfsc STATUS, C
				goto addr9

CJB fr, #, addr9	if fr < # goto addr9	4/5	w, C, DC, Z	movlw # subwf fr, w btfss STATUS, C goto addr9

CJB fr, fr2, addr9	if fr < fr2
	goto addr9
		4/5	w, C, DC, Z	movf fr2, w
				subwf fr, w
				btfss STATUS, C
				goto addr9

CJBE fr, #, addr9	if fr <= # goto addr9	4/5	w, C, DC, Z	movlw # addwf fr, w btfss STATUS, C goto addr9

CJBE fr, fr2, addr9	if fr <= fr2
	goto addr9
		4/5	w, C, DC, Z	movf fr2, w
				subwf fr, w
				btfsc STATUS, C
				goto addr9

CJE fr, #, addr9	if fr == # goto addr9	4/5	w, C, DC, Z	movlw # subwf fr, w btfsc STATUS, Z goto addr9

PASM PICmicro® MCU Unconditional Branch Instructions

Instruction Description	Cycles	Context Resources Affected	Actual PICmicro® MCU Instructions

CJE fr, fr2, addr9
if fr == fr2
goto addr9
	4/5	w, C, DC, Z	movf fr2, w
			subwf fr, w
			btfsc STATUS, Z
			goto addr9

CJNE fr, #, addr9 if fr == # goto addr9	4/5	w, C, DC, Z	movlw # subwf fr, w btfss STATUS, Z goto addr9

CJNE fr, fr2, addr9
if fr == fr2
goto addr9
	4/5	w, C, DC, Z	movf fr2, w
			subwf fr, w
			btfss STATUS, Z
			goto addr9

(#### - End Table)

PICmicro® MCU Processor Architectures

The PICmicro® MCU's Arithmetic Logic Unit

Standard PICmicro® MCU Processor ALU Operations	
Operation	Equivalent Operation
Move	AND with 0x0FF
Addition	None
Subtraction	Addition to a Negative
Negation	XOR with 0x0FF (Bitwise "Invert") and Increment
Increment	Addition to One
Decrement	Subtract by One/Addition by 0x0FF

Standard PICmicro® MCU Processor ALU Operations (*Continued*)

AND	None
OR	None
XOR	None
Complement	XOR with 0x0FF
Shift Left	None
Shift Right	None

Along with these functions, the PIC17Cxx and PIC18Cxx also have an 8-bit × 8-bit multiplier.

The PICmicro® MCU's "ALU" ("Arithmetic Logic Unit") could be blocked out as shown in Fig. 5.1.

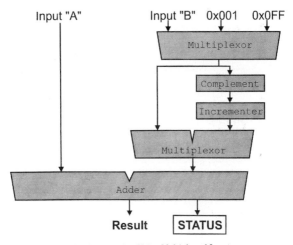

Figure 5.1 ALU Implementation Using Multiplexed Inputs

Low-End PICmicro® MCUs

The "Low-End" PICmicro® MCUs have the part numbers:

PIC12C5xx

PIC16C5x

PIC16C50x

where "x" can be any digit.

A sample low-end PICmicro® MCU processor architecture is shown in Fig. 5.2.

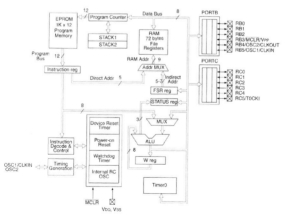

Figure 5.2 Low-end PICmicro® MCU Architecture

Reset addresses are at the last location in program memory. The following table lists the different reset vectors for different device program memory sizes.

Low-end PICmicro® MCU Program Memory Size to Reset Vector	
Program Memory Size	Reset Vector
512	0x01FF
1024	0x03FF
2048	0x07FF

It is recommended that the reset address is ignored in the Low-End PICmicro® MCUs (or used to load the oscillator calibration register ["OSCCAL"] value into "w") before rolling over the Program Counter and starting at address 0x0000, like the other PICmicro® MCU devices.

Register access

The low-end register space is shown in Fig. 5.3.

The first 16 addresses of each bank are common. The 16 bank unique file registers are located in the last 16 addresses of the bank. This limitation of only being able to address data 16 bytes at a time prevents the construction of arrays or other data structures longer than 16 bytes.

Bank 0 can be accessed directly within instructions. Other banks can only be accessed using the FSR (and INDF) index registers. The following table lists bank offsets.

Bank 0	Bank 1	Bank 2	Bank 3	
Addr - Reg	Addr - Reg	Addr - Reg	Addr - Reg	
00 - INDF	20 - INDF	40 - INDF	60 - INDF	Shared
01 - TMR0	21 - TMR0	41 - TMR0	61 - TMR0	Registers
02 - PCL	22 - PCL	42 - PCL	62 - PCL	
03 - STATUS	23 - STATUS	43 - STATUS	63 - STATUS	
04 - FSR	24 - FSR	44 - FSR	64 - FSR	
05 - PORTA*	25 - PORTA*	45 - PORTA*	65 - PORTA*	
06 - PORTB	26 - PORTB	46 - PORTB	66 - PORTB	
07 - PORTC	27 - PORTC	47 - PORTC	67 - PORTC	
08-0F Shared File Regs	28-2F Shared File Regs	28-2F Shared File Regs	68-8F Shared File Regs	
10-1F Bank 0 File Regs	30-3F Bank 1 File Regs	50-4F Bank 2 File Regs	70-7F Bank 3 File Regs	Bank Unique Registers

* - "OSCCAL" may take place of "PORTA" in PICmicro® MCUs
   with Internal Oscillators

OPTION - Accessed via "option" Instruction
TRIS# - Accessed via "TRIS PORT#" Instruction

Figure 5.3 Low-End PICmicro® MCU Register Map

Low-end PICmicro® MCU Unique Bank Address Table		
Bank	FSR	Start of 16 Unique Registers
0	0x000	0x010
1	0x020	0x030
2	0x040	0x050
3	0x060	0x070

Note that the PICmicro® MCU's FSR ("index") register can never equal zero. The table below lists which bits will be set in the low-end's FSR depending on how many bank registers the PICmicro® MCU has.

Low-end PICmicro® MCU Minimum FSR Value to Number of Banks		
Number of Banks	Set FSR bits	Minimum FSR value
1	7, 6, 5	0x0E0
2	7,6	0x0C0
4	7	0x080

STATUS register

Low-end PICmicro® MCU Register Definitions		
Address 0x003	Bits 7	Bit Function GPWUF - in PIC12C5xx and PIC16C505: when Set, Reset from Sleep on Pin Change. When Set, power up or _MCLR reset. In other Devices the Bit 7 is Unused.
	6-5	PA1-PA0 - Select the Page to execute out of: 00 - Page 0 (0x0000 to 0x01FF) 01 - Page 1 (0x0200 to 0x03FF) 10 - Page 2 (0x0400 to 0x05FF) 11 - Page 3 (0x0600 to 0x07FF)
	4	_TO - Set after Power Up, clrwdt and sleep instructions
	3	_PD - Set after Power Up, clrwdt instruction. Reset after sleep instruction
	2	Z - Set if the eight bit result is equal to zero
	1	DC - Set for low order Nybble carry after addition or subtraction instruction
	0	C - Set for Carry after addition or subtraction instruction

Program counter

The low-end PICmicro® MCU's program counter block
diagram is given in Fig. 5.4.

The "PA0" and "PA1" bits of the STATUS register (bits
five and six) perform the same function as the "PCLATH"
register of the other PICmicro® MCUs. Bit PA0 is used to
provide bit nine of the destination address to jump to
during a "goto" or "call" instruction or when "PCL" is

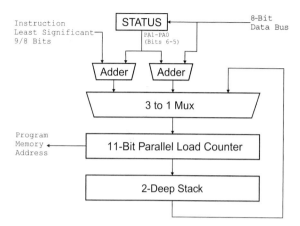

Figure 5.4 Low-End Program Counter Block Diagram

written to. Bit PA1 is address bit ten. In some low-end PICmicro® MCUs, bit seven of the STATUS register is referred to as "PA2". This bit is not used by any of the current PICmicro® MCUs.

To jump to a new "page" address, the following instruction sequence is used:

Low-end PICmicro® MCU Program Counter Update Operation

```
STATUS = (STATUS & 0x01F) + ((HIGH new_address &
  0x0FE) << 4);
PCL    = LOW new_address;
```

Note that subroutines and tables at addresses 0x0100 to 0x01FF, 0x0300 to 0x03FF, 0x0500 to 0x05FF, and

00x700 to 0x07FF cannot be accessed directly. Instead, redirection using a "goto" instruction is required.

The "call stack" is two elements deep.

Mid-Range PICmicro® MCUs

The mid-range PICmicro® MCUs have the part numbers:

PIC12C6xx

PIC14000

PIC16C55x

PIC16C6x (x)

PIC16C7x (x)

PIC16C8x

PIC16F8x (x)

PIC16C9xx

The mid-range PICmicro® MCU's have the block diagram shown in Fig. 5.5.

Upon Reset, execution starts at address 0x00000. Interrupts are handled at address 0x00004. The configuration registers are located at address 0x02007.

Register access

The Mid-Range PICmicro® MCUs can have up to four register "banks" of 0x080 (128) registers. Each register is accessed using the "RPx" bits of the STATUS register. For the different Register Banks and Register Addresses, the following table is used to set the RPx bits.

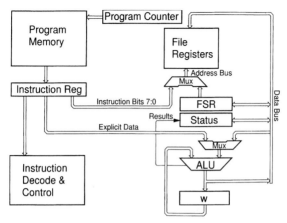

Figure 5.5 Mid-Range PICmicro® MCU Architecture

PICmicro® MCU PORT and TRIS Bit Access			
Register Address Range	RP1	RP0	XOR Value
0x000-0x07F	0	0	0
0x080-0x0FF	0	1	0x080
0x0100-0x017F	1	0	0x0100
0x0180-0x01FF	1	1	0x0180

The "XOR Value" is the value that is XORed with the Register Address to ensure it is within the bank address range of 0 to 0x07F.

When the FSR (Index) register is used to access data in Banks 1 through 3, the "IRP" bit of the STATUS register will be set appropriately and the least significant 8 bits of the address are loaded into the FSR register.

The Register Address Map looks like the following:

Mid-Range Bank0/Bank1 Register Definitions			
Offset	Bank 0	Bank 1	Comments
0x000	INDF	INDF	
0x001	TMR0	OPTION	
0x002	PCL	PCL	
0x003	STATUS	STATUS	
0x004	FSR	FSR	
0x005	PORTA	TRISA	
0x006	PORTB	TRISB	
0x007	PORTC	TRISC	Available in 28/40 Pin Parts
0x008	PORTD	TRISD	Available in 40 Pin Parts
0x009	PORTE	TRISE	Available in 40 Pin Parts
0x00A	PCLATH	PCLATH	
0x00B	INTCON	INTCON	

The File Registers (Variable registers) start at either 0x00C or 0x020 of the bank depending on the "Hardware I/O" or "Special Function Registers" ("SFRs") built into the device. It is recommended to start all variable declarations at 0x020 to avoid issues porting between a PICmicro® MCU that has file registers starting at 0x00C and one that has file registers starting at 0x020.

STATUS register

Mid-Range STATUS Register Definition	
Bit	Function
7	IRP - FSR Select Between the High and Low Register Banks
6-5	RP1:RP0 - Direct Addressing Select Banks (0 through 3)
4	_TO - Time Out Bit. Reset after a Watchdog Timer Reset

Mid-Range STATUS Register Definition (*Continued*)
3   _PD - Power-down Active Bit. Reset after sleep instruction
2   Z - Set when the eight bit result is equal to zero
1   DC - Set when the low Nybble of addition/subtraction result carries to the high Nybble
0   C - Set when the addition/subtraction result carries to the next byte. Also used with the Rotate Instructions

Program counter

The mid-range PICmicro® MCU's program counter can be represented by the block diagram shown in Fig. 5.6.

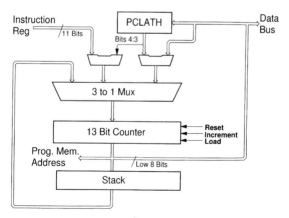

Figure 5.6   Mid-Range PICmicro® MCU Program Counter Block Diagram

To jump to another bank, the following instruction sequence is used:

Mid-range PICmicro® MCU Program Counter Update Operation

```
PCLATH = (HIGH new_address;
PCL    = LOW new_address;
```

All addresses within the mid-range PICmicro® MCU can be accessed with the PCLATH and PCL registers.

The "call stack" is eight elements deep.

Interrupt operation

Interrupts are controlled by the state of the INTCON Register and optionally the "PIE" and "PIR" registers. Interrupt handlers always start executing at address 0x004.

Mid-Range INTCON Register Definition

Bit	Function
7	GIE - Global Interrupt Enable. For any Interrupt Requests to be acknowledged, this bit must be set
6	Device Specific Interrupt Enable
5	T0IE - TMR0 Interrupt Overflow Request Enable
4	INTE - RB0/INT Pin Interrupt Request Enable
3	RBIE - PORTB Change Interrupt Request Enable
2	T0IF - TMR0 Interrupt Overflow Request
1	INTF - RB0/INT Pin Interrupt Request
0	RBIF - PORTB Change Interrupt Request

For an Interrupt Request (which sets the bit ending in "F"), the corresponding "Enable" bit (which is the bit ending in "E") has to be set along with the "GIE" bit.

Some Enable and Interrupt Request bits may be in auxiliary registers or "PIR" or "PIE".

Interrupt handler skeleton

The mid-range PICmicro® MCU has an interrupt skeleton of:

```
org     4
movwf   _w              ; Save Context Registers
movf    STATUS, w
bcf     STATUS, RP1     ; Make Bank 0 Active
bcf     STATUS, RP0
movwf   _status
movf    FSR, w
movwf   _fsr
movf    PCLATH, w
movwf   _pclath
clrf    PCLATH          ; Make sure Execution in Page 0

   :                    ; Execute Interrupt Handler

movf    _pclath, w      ; Restore the Context
                              Registers
movwf   PCLATH
movf    _fsr, w
movwf   FSR
movf    _status, w
movwf   STATUS
swapf   _w, f
swapf   _w, w
retfie
```

To enable the TMR0 Interrupt Request, the following code is used:

Mid-Range Timer Interrupt Enable Code
``clrf TMR0   ; Reset TMR0
bcf   INTCON, T0IF ; Reset TMR0 Interrupt Request
bsf   INTCON, T0IE ; Enable TMR0 Interrupt Request
bsf   INTCON, GIE ; Enable PICmicro® MCU Interrupts```

PIC17Cxx

The PIC17Cxx architecture encompasses parts with the part numbers:

PIC17Cxx(x)

The unique features of the PIC17Cxx, as compared to the other PICmicro® MCU's, include:

1. The ability to access external, parallel memory.

2. Up to seven I/O ports.

3. A built-in 8×8 multiplier.

4. Up to 902 file registers in up to 16 banks.

5. Up to 64K address space.

6. The ability to read and write program memory.

7. Multiple interrupt vectors.

The PIC17Cxx processor has a block diagram as shown in Fig. 5.7.

The important differences between the PIC17Cxx architecture and the low-end and mid-range PICmicro® MCU architectures are as follows:

1. The accumulator, "WREG," can be addressed in the register space.

2. The STATUS and OPTION register functions are spread across different registers.

3. The program counter works slightly differently from the other architectures.

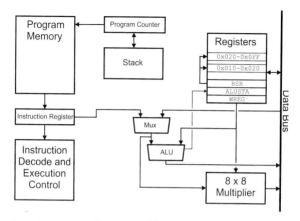

Figure 5.7 PIC17Cxx Processor Architecture

4. The "registers" are accessed differently and accesses can bypass the "WREG".

The reset address of the PIC17Cxx is 0x00000.

Register access

The PIC17Cxx has a single 256 Register address space. Addresses 0x010 to 0x017 are banked and addressed using the lower nybble of the "BSR" register and contain the register selection information of the "Special Function Registers" ("SFRs") or Hardware I/O Registers. The high nybble of BSR is used to select the File Register Bank at addresses 0x020 to 0x0FF.

PIC17Cxx Register Configuration

Addr	Register	Function/Bit	Definition
0x000		INDF0	Register Pointed to by FSR0
0x001		FSR0	Index Register 0
0x002		PCL	Low Byte of the Program Counter
0x003		PCLATH	Latched High Byte of the Program Counter
0x004		ALUSTA	Processor Status and Control Register
0x005		T0STA	TMR0 Status and Control Register
0x006		CPUSTA	Processor Operating Status Register
0x007		INTSTA	Interrupt Status and Control Register
0x008		INDF1	Register Pointed to by FSR1
0x009		FSR1	Index Register 1
0x00A		WREG	Processor Accumulator
0x00B		TMR0L	Low Byte of TMR0
0x00C		TMR0H	High Byte of TMR0
0x00D		TBLPTRL	Low Byte of the Table Pointer
0x00E		TBLPTRH	High Byte of the Table Pointer
0x00F		BSR	Bank Select Register
0x010-0x017		Special Function Registers	
0x018		PRODL	Low Byte of Multiplication Product
0x019		PRODH	High Byte of Multiplication Product
0x01A-0x01F		Unbanked File Registers	
0x020-0x0FF		Banked File Registers	

STATUS register

The PIC17Cxx has two registers that provide the same functions as the single "STATUS" register of the other three PICmicro® MCU architectures. The PIC17Cxx Bank Selection is made by the "BSR" Register.

PIC17Cxx ALUSTA Register Definition

```
Bit     Function
7-6     FSR1 Mode Select
        1x - FSR1 Does not Change after Access
        01 - Post Increment FSR1
        00 - Post Decrement FSR1
5-4     FSR0 Mode Select
        1x - FSR0 Does not Change after Access
        01 - Post Increment FSR0
        00 - Post Decrement FSR0
3       OV - Set when there is a two's complement
        overflow after addition/subtraction
2       Z - Set when the eight bit result is equal to
        Zero
1       DC - Set for low order Nybble carry after
        addition or subtraction instruction
0       C - Set for Carry after addition or
        subtraction instruction
```

PIC17Cxx CPUSTA Register Definition

```
Bit     Function
7-6     Unused
5       STKAV - When Set, there is Program Counter
        Stack Space Available
4       GLINTD - When Set, all Interrupts Are
        Disabled
3       _TO - Set after Power Up or clrwdt
        Instruction. When Reset a Watchdog Timeout
        has occurred
2       _PD - Set after Power Up or clrwdt
        Instruction. Reset by a "sleep" instruction
1       _POR - Reset After Power Up in PIC17C5x. Not
        Available in All PIC17Cxx devices
0       _BOR - Reset After Brown Out Reset. Not
        Available in All PIC17Cxx devices
```

Program counter

The PIC17Cxx's processor can access 64k 16-bit words of program memory, either internally or externally to the chip. Each instruction word is given a single address; so to address the 64k words (or 128k bytes), 16 bits are required. From the application developer's perspective, these 16 bits can be accessed via the "PCL" and "PCLATH" registers in exactly the same way as the low-end and mid-range PICmicro® MCUs. The PIC17Cxx's program counter block diagram is shown in Fig. 5.8.

The block diagram in Fig. 5.8 differs from the mid-range PICmicro® MCU's program counter block diagram in one important respect; when the "goto" and "call" instructions are executed, the upper 5 bits of the specified instruction overwrite the lower 5 bits of the PCLATH register. After execution of a "goto" or "call" instruction PCLATH has been changed to the current address.

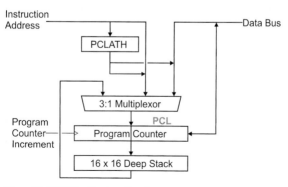

Figure 5.8   PIC17Cxx Program Counter

Interrupt operation

The PIC17Cxx can have four different interrupt vector addresses, depending on their source and priority. The Interrupts and their vectors are listed below:

PIC17Cxx Interrupt Vector Address and Priorities for Different Sources		
Priority	Vector Address	Source
High	0x0008	RAO/INT Pin Interrupt
	0x00010	TMRO Overflow Interrupt
	0x00018	TOCKI Pin Interrupt
Low	0x00020	Peripheral Device Interrupt

PIC17Oxx CPU STA Register Definition			
Any	0x007	INTSTA	Interrupt Status and Control Register
			Bit Function
			7 PEIE - Set when Peripheral Interrupt is Pending
			6 TOCKIF - Set when RA1/TOCKI Pin has Interrupt Source. Cleared by Hardware when Interrupt Vector 0x0018 is executed
			5 TOIF - Set when TMR0 Overflows. Cleared by Hardware when Interrupt Vector 0x0010 is executed
			4 INTF - Set when RA0/INT Pin Interrupt Request Active. Cleared by Hardware when Interrupt Vector 0x0008 is executed
			3 PEIE - Set to Enable Peripheral Interrupt Requests

PIC17Cxx CPUSTA Register Definition (*Continued*)		
	2	T0CKIE - Set to Enable RA1/T0CKI Interrupt Request
	1	T0IE - Set to Enable TMR0 Overflow Interrupt Request
	0	INTE - Set to Enable RA0/INT Pin Interrupt Request

Interrupt handler skeleton

```
org    ??              ; Vector According to Source
 movpf  ALUSTA, _alusta ; Save Context Registers
 movpf  WREG, _w
 movpf  BSR, _bsr
 movpf  PCLATH, _pclath
 clrf   PCLATH          ; Make sure Execution in
                        ;   Page 0
 :                      ; Execute Interrupt Handler
 movfp  _pclath, PCLATH ; Restore the Context
                        ;   Registers
 movfp  _bsr, BSR
 movfp  _w, WREG
 movfp  _alusta, ALUSTA
 retfie
```

PIC18Cxx

The PIC18Cxx architecture encompasses parts with the part numbers:

   PIC18Cxx2

The unique features of the PIC18Cxx, as compared to the other PICmicro® MCU's, include:

1. A built-in 8×8 multiplier.
2. Up to 3,840 file registers in up to 16 banks.

3. Up to 1,048,576 words of program memory address space.

4. The ability to read and write program memory.

5. Prioritized Interrupt Requests.

The PIC18Cxx processor has a block diagram as shown in Fig. 5.9. The important differences between the PIC18Cxx architecture and the low-end and mid-range PICmicro® MCU architectures are as follows:

1. The accumulator, "WREG", can be addressed in the register space.

2. The "Access Bank", which is used to allow access to the first 128 file registers and the Hardware I/O registers without involving the BSR.

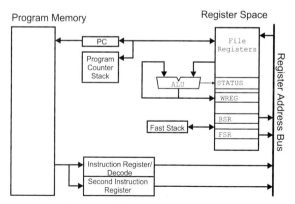

Figure 5.9 PIC18Cxx Architecture Block Diagram

3. The program counter works slightly differently from the other architectures.

4. The "registers" are accessed differently and accesses can bypass the "WREG".

The reset address of the PIC18Cxx is 0x00000.

Register access

The PIC18Cxx can access up to 4,096 8-bit registers that are available in a contiguous memory space. Twelve address bits are used to access each address within the "Register Map" space shown in Fig. 5.10.

To access a register directly, the PIC18Cxx's "BSR" ("Bank Select Register") register must be set to the bank the register is located in. The BSR register contains the upper 4 bits of the register's address, with the

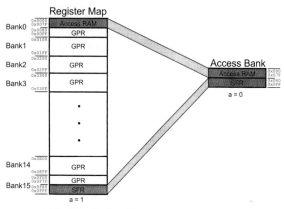

Figure 5.10   PIC18Cxx Register Format

lower 8 bits explicitly specified within the instruction. The direct address is calculated using the formula:

```
Address = (BSR << 8) + Direct Address
```

The index register operation of the PIC18Cxx is very well organized and will make it much easier for compiler writers to create PIC18Cxx compilers than for other PICmicro® MCUs. Along with the three 12-bit-long FSR registers, when data is accessed it can result in the FSR being incremented before or after the data access, decremented after or access to the address of the FSR contents added to the contents of the "w" register. A specific access option is selected by accessing different "INDF" register addresses. The table below lists the different INDF registers and their options concerning their respective FSR registers:

PIC18Cxx FSR Change Access Registers	
INDF Register	Operation
INDF#	Access the Register Pointed to by FSR#
POSTINC#	Access the Register Pointed to by FSR# and then Increment FSR#
POSTDEC#	Access the Register Pointed to by FSR# and then Decrement FSR#
PREINC#	Increment FSR# and then Access the Register Pointed to by FSR#
PLUSW#	Access the Register Pointed to by the Contents of the WREG added to FSR#

To simulate a "push" of the contents of the "WREG" using FSR0 as a Stack Pointer, use the operation:

```
POSTDEC0 = WREG;
```

A "pop WREG" could be implemented as:

```
WREG = PREINC0;
```

Specific elements relative to the start of the stack could be accessed using the code:

```
WREG = 3;
WREG = PLUSW0;
```

STATUS register

Two registers contain the status information for the PIC18Cxx and control the operation of the PICmicro® MCU.

PIC18Cxx STATUS Register Definition

```
Bit   Function
7-5   Unused
  4   N - Set when the two's complement result after
      addition/subtraction is negative
  3   OV - Set when there is a two's complement
      overflow after addition/subtraction
  2   Z - Set when the eight bit result is equal to
      zero
  1   DC - Set when the low Nybble of
      addition/subtraction result carries to the
      high nybble
  0   C - Set when the addition/subtraction result
      carries to the next byte. Also used with the
      Rotate Instructions
```

PIC18Cxx RCON Register Definition

```
Bit   Function
  7   IPEN - When Set Interrupt Priority Levels are
      enabled
  6   LWRT - When Set, Enable writes to internal
      program memory
  5   Unused
  4   _RI - When Reset, the "Reset" Instruction was
      executed. This bit must be set in Software
```

PIC18Cxx RCON Register Definition (Continued)

```
3   _TO - Time Out Bit. Reset after a Watchdog
    Timer Reset
2   _PD - Power-down Active Bit. Reset after sleep
    instruction
1   _POR - Reset after a "Power On" Reset has
    occurred. This bit must be Set in Software
0   _BOR - Reset after a "Brown Out" Reset has
    occurred. This bit must be Set in Software
```

Program counter

The PIC18Cxx program counter and stack is similar to the hardware used in the other devices except for three important differences. The first difference is the need for accessing more than 16 address bits for the maximum one million possible instructions of program memory. The second difference is the availability of the "fast stack", which allows interrupt context register saves and restores to take place without requiring any special code. The last difference is the ability to read and write from the stack. These differences add a lot of capabilities to the 18Cxx that allow applications that are not possible in the other PICmicro® MCU architectures to be implemented.

In the PIC18Cxx, when handling addresses outside the current program counter, not only does a "PCLATH" register (or "PA" bits as in the low-end devices) update as required, but it is also a high-order register for addresses above the first 64 instruction words. This register is known as "PCLATU". "PCLATU" works identically to the "PCLATH" register and its contents are loaded into the PIC18Cxx PICmicro® MCUs program counter when "PCL" is updated.

Each instruction in the PIC18Cxx starts on an "even" address. This means that the first instruction starts at ad-

Program Counter Stack

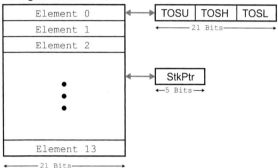

Figure 5.11 PIC18Cxx Program Counter Stack

dress zero, the second at address two, the third at address four and so on. Setting the program counter to an odd address will result in the MPLAB simulator halting and the PIC18Cxx working unpredictably. Changing the convention used in the previous PICmicro® MCUs to one, where each byte is addressed, means that some rules about addressing will have to be relearned for the PIC18Cxx.

The stack itself, at 31 entries, is deeper than the other PICmicro® MCU stacks and the hardware monitoring the stack is available as the "STKPTR" register. A block diagram of the stack is shown in Fig. 5.11.

The STKPTR register is defined as:

PIC18Cxx STKPTR Register Bit Definitions
Bit   Description
7   STKFUL - Stack Full Flag which is set when the Stack is Full or Overflowed

PIC18Cxx STKPTR Register Bit Definitions (*Continued*)

```
 6    STKUNF - Stack Underflow Flag which is set when
       more Stack Elements have been Popped than
       Pushed.
 5    Unused
4-0   SP4:SP0 - Stack Pointer
```

The "STKUNF" and "STKFUL" bits will be set if their respective conditions are met. If the "STVREN" bit of the configuration fuses is set, then when the STKUNF and STKFUL conditions are true the PICmicro® MCU will be reset.

The "fast stack" is used to simplify subroutine calls in applications that don't have interrupts enabled as well as working with interrupt handlers. To use the fast stack in the "call" and "return" instructions a "1" parameter is put at the end of the instructions. To prevent the fast stack from being used, a "0" parameter is put at the end of the "call" and "return" instructions. The "fast stack" is a 3-byte memory location where the "w", "STATUS" and "BSR" registers are stored automatically when an interrupt request is acknowledged and execution jumps to the interrupt vector. If interrupts are not used in an application, then these registers can be saved or restored with a "call" and "return", for instance:

```
Call sub, 1  ; Call "sub" after saving "w", "STATUS"
             ;  and "BSR"
   :
sub          ; Execute "Sub", Ignore "w", "STATUS"
   :         ;  and "BSR"

return 1     ; Restore "w", "STATUS" and "BSR" before
             ; Return to Caller
```

The reason the "fast" option is not recommended in applications in which interrupts are enabled is due to the interrupt overwriting the saved data when it executes. For this reason, the "fast" option cannot be used with nested subroutines or interrupts.

Interrupt operation

Interrupt Operation works similarly to the mid-range PICmicro® MCU's except for the addition of priority levels to the interrupt sources. If the "P" bit for the interrupt source is specified along with the "IPEN" bit of the RCON register, then the interrupt handler at address 0x00008 will execute. If the "IPEN" bit of the RCON register is set and the "P" bit for the interrupt source is reset, the interrupt handler at address 0x00018 will execute.

If "IPEN" is reset, then all interrupts will execute at address 0x00008.

Interrupt handler skeleton

If the "Fast Stack" is not used with PIC18Cxx interrupts, the code shown below can be used for the handler entry and exit code.

```
Int
  movwf    _w                        ; Save Context Registers
  movff    STATUS, _status
  movff    BSR, _bsr

    :                                 ; Interrupt Handler Code

  movff    _bsr, BSR                 ; Restore Context
                                          Registers
  movf     _w, w
  movff    _status, STATUS
  retfie
```

6

PICmicro® MCU Register Mappings

While the register addresses are very similar between PICmicro® MCUs of the same architecture family, remember that the bits in the different registers may change function with different PICmicro® MCU part numbers. To be absolutely sure of the bits and their function inside a register, consult the Microchip part datasheet.

Low-End PICmicro® MCUs

The low-end PICmicro® MCU devices have five register bank address bits for up to 32 unique file register addresses in each bank. Up to four register banks can be available in a low-end PICmicro® MCU with the first 16 addresses of each bank being common throughout the banks and the second 16 addresses being unique to the bank. This is shown in Fig. 6.1.

Using this scheme, low-end PICmicro® MCUs have anything from 25 to 73 unique file registers available for an application.

There are a few things to note with the low-end register addressing:

1. The "OPTION" and "TRIS" registers can only be written to by the "option" and "tris" instructions, respectively.

2. If the device has a built-in oscillator, the "OSCCAL" register is located in address five, which is normally the "PORTA" address.

Bank 0	Bank 1	Bank 2	Bank 3	
Addr – Reg	Addr – Reg	Addr – Reg	Addr – Reg	
00 – INDF 01 – TMR0 02 – PCL 03 – STATUS 04 – FSR 05 – PORTA* 06 – PORTB 07 – PORTC	20 – INDF 21 – TMR0 22 – PCL 23 – STATUS 24 – FSR 25 – PORTA* 26 – PORTB 27 – PORTC	40 – INDF 41 – TMR0 42 – PCL 43 – STATUS 44 – FSR 45 – PORTA* 46 – PORTB 47 – PORTC	60 – INDF 61 – TMR0 62 – PCL 63 – STATUS 64 – FSR 65 – PORTA* 66 – PORTB 67 – PORTC	Shared Registers
08-0F Shared File Regs	28-2F Shared File Regs	28-2F Shared File Regs	68-8F Shared File Regs	
10-1F Bank 0 File Regs	30-3F Bank 1 File Regs	50-4F Bank 2 File Regs	70-7F Bank 3 File Regs	Bank Unique Registers

* - "OSCCAL" may take place of "PORTA" in PICmicro® MCUs
    with Internal Oscillators

OPTION - Accessed via "option" Instruction

TRIS# - Accessed via "TRIS PORT#" Instruction

Figure 6.1 Low-end PICmicro® MCU Register Map

3. The "STATUS" and "OPTION" registers are always the same for Low-End Devices.

```
Address   Register   Bits   Bit Function
0x003     STATUS     7      GPWUF - in PIC12C5xx and
                            PIC16C505: when Set, Reset
                            from Sleep on Pin Change.
                            When Set, power up or
                            _MCLR reset. In other
                            Devices the Bit 7 is
                            Unused.
                     6-5    PA1-PA0 - Select the Page
                            to execute out of:
                      00 - Page 0 (0x0000 to 0x01FF)
                      01 - Page 1 (0x0200 to 0x03FF)
                      10 - Page 2 (0x0400 to 0x05FF)
                      11 - Page 3 (0x0600 to 0x07FF)
                     4      _TO - Set after Power Up,
                            clrwdt and sleep
                            instructions
                     3      _PD - Set after Power Up,
                            clrwdt instruction. Reset
                            after sleep instruction
                     2      Z - Set if the eight bit
                            result is equal to zero
                     1      DC - Set for low order
                            Nybble carry after
                            addition or subtraction
                            instruction
                     0      C - Set for Carry after
                            addition or subtraction
                            instruction
N/A       OPTION     7      _GPWU - In PIC12C5xx or
                            PIC16C505: Reset to Enable
                            Wake Up on Pin Change. In
                            other devices, Bit 7 is
                            Unused
                     6      _GPPU - In PIC12C5xx or
                            PIC16C505: Enable Pin
                            Pull-Ups. In other
                            devices, Bit 6 is Unused
                     5      T0CS - TMR0 Clock Source
                            Select. When Set, T0CKI
                            pin is Source. When Reset,
                            Instruction Clock
```

```
4    T0SE - TMR0 Edge Select.
     When Reset, increment TMR0
     on Rising Edge. When Set,
     increment TMR0 on Falling
     Edge
3    PSA - Prescaler Assignment
     Bit. When Set, the
     Prescaler is assigned to
     the Watchdog Timer else
     TMR0
2-0  PS2:PS0 - Prescaler Rate
     Select Bits
     Bits   TMR0 Rate   WDT Rate
     111     256:1       128:1
     110     128:1        64:1
     101      64:1        32:1
     100      32:1        16:1
     011      16:1         8:1
     010       8:1         4:1
     001       4:1         2:1
     000       2:1         1:1
```

4. The Low-End PICmicro® MCU FSR register can never equal zero.

Mid-Range PICmicro® MCUs

The standard mid-range PICmicro® MCU device addresses are as follows:

```
Offset  Bank 0   Bank 1   Comments
0x000   INDF     INDF
0x001   TMR0     OPTION
0x002   PCL      PCL
0x003   STATUS   STATUS
0x004   FSR      FSR
0x005   PORTA    TRISA
0x006   PORTB    TRISB
0x007   PORTC    TRISC    Available in 28/40 Pin Parts
0x008   PORTD    TRISD    Available in 40 Pin Parts
0x009   PORTE    TRISE    Available in 40 Pin Parts
0x00A   PCLATH   PCLATH
0x00B   INTCON   INTCON
```

From these basic addresses, peripheral I/O registers (discussed below) are added to the register banks with file registers starting at either offset 0x00C or 0x020. For most modern mid-range PICmicro® MCUs, the file registers start at address 0x020 of the bank.

The specific part number datasheets will have to be checked to find where the file registers that are shared across the banks are located.

The STATUS Register, in the mid-range PICmicro® MCU is defined as:

```
Bit    Function
 7     IRP - FSR Select Between the High and Low
         Register Banks
6-5    RP1:RP0 - Direct Addressing Select Banks (0
         through 3)
 4     _TO - Time Out Bit  Reset after a Watchdog
         Timer Reset
 3     _PD - Power-down Active Bit. Reset after sleep
         instruction
 2     Z - Set when the eight bit result is equal to
         zero
 1     DC - Set when the low Nybble of addition/
         subtraction result carries to the high Nybble
 0     C - Set when the addition/subtraction result
         carries to the next byte. Also used with the
         Rotate Instructions
```

The OPTION Register (which has the label "OP-TION_REG" in the Microchip include files) is defined as:

```
Bit    Function
 7     _RBPU - when reset, the PORTB Pin Pull Up is
         Enabled
 6     INTEDG - When Set, Interrupt Request on Rising
         Edge of RB0/INT Pin
 5     T0CS - When Set, TMR0 is incremented from the
         T0CKI Pin, else by the internal instruction
         clock
 4     T0SE - When Set, TMR0 is Incremented on the
         High to Low ("Falling Edge") of T0CKI
```

```
3     PSA - Prescaler Assignment Bit. When Set, the
      Prescaler is assigned to the Watchdog Timer
      else to TMR0
2-0   PS2:PS0 - Prescaler Rate Select.
      Bits    TMR0 Rate    WDT Rate
      111      256:1        128:1
      110      128:1         64:1
      101       64:1         32:1
      000       32:1         16:1
      011       16:1          8:1
      010        8:1          4:1
      001        4:1          2:1
      000        2:1          1:1
```

Many devices have the "PCON" register that enhances the returned information contained in the "_TO" and "_PD" bits of the STATUS Register

```
Bit   Function
 7    MPEEN - Set if there is a Memory Parity Error.
      This capability is built into a small number
      of PICmicro® MCUs
6-3   Unused
 2    _PER - Reset when there was a Program Memory
      Parity Error. This capability is built into a
      small number of PICmicro® MCUs
 1    _POR - Reset when execution is from a Power On
      Reset takes place
 0    _BOR - Reset when execution is from a Brown
      Out Reset
```

The PCLATH Register's contents are written to the Program Counter each time a "goto" or "call" instruction is executed or if the contents of PCL are changed.

```
Bit   Function
7-5   Unused
 4    Select High and Low Pages
 3    Select Odd or Even Pages
2-0   Select the 256 Instruction Address Block
      within Current Page. This data is used when
      PCL is written to
```

Some mid-range devices are now available with built-in RC oscillators. To make the operation of the oscillators more accurate, the "OSCCAL" register is written to with a factory specified "Calibration Value".

```
Bit    Function
7-4    CAL3:CAL0 - Sixteen Bit Calibration Value
  3    CALFST - Increase the speed of the RC
         Oscillator
  2    CALSLW - Decrease the speed of the RC
         Oscillator
1-0    Unused
```

Interrupts are controlled from the "INTCON" register, which controls the basic mid-range PICmicro® MCU interrupts as well as access to enhanced interrupt features.

```
Bit    Function
  7    GIE - Global Interrupt Enable. For any
         Interrupt Requests to be acknowledged, this
         bit must be set
  6    Device Specific Interrupt Enable. See Below
  5    T0IE - TMR0 Interrupt Overflow Request Enable
  4    INTE - RB0/INT Pin Interrupt Request Enable
  3    RBIE - PORTB Change Interrupt Request Enable
  2    T0IF - TMR0 Interrupt Overflow Request
  1    INTF - RB0/INT Pin Interrupt Request
  0    RBIF - PORTB Change Interrupt Request
```

Bit 6 of INTCON may be a peripheral device interrupt enable/request bit or it can be "PEIE", which when set will enable Peripheral Interrupts set in "PIR" and "PIE" registers. The "PIR" register(s) contains the "F" bits (Interrupt Request Active), while "PIE" contains the "E" bits (Interrupt Request Enable). As I work through the different peripherals, the "E" and "F" bits will be listed, but their actual location is part number specific and the datasheet will have to be consulted.

Data EEPROM is accessed via the EEADR and

EEDATA registers with EECON1 and EECON2 providing the access control. EECON2 is a "pseudo-register" and the act of writing to it is used to "verify" that the operation request is valid. EECON1 is defined as:

```
Bit    Function
7-5    Unused
  4    EEIF - EEPROM Write Complete Interrupt Request
  3    WRERR - Bit Set when EEPROM Write was invalid
  2    WREN - Set to Enabling Writing to EEPROM
  1    WR - Write control Bit
  0    RD - Set to Allow an EEPROM Data Read
```

The Data EEPROM Write Interrupt Request Bit ("EEIE") is either in a PIE register or INTCON.

The Parallel Slave Port ("PSP", available only in 40-Pin mid-range PICmicro® MCUs) is enabled by setting the PSPMODE bit. Interrupt requests are enabled by the PSPIE flag and requested by the PSPIF flag of the PIE and PIR registers, respectively. The Parallel Slave Port is controlled from "TRISE". Note that when the Parallel Slave Port is enabled, PORTD and PORTE cannot be used for I/O.

```
Bit    Function
  7    IBF - Bit Set when a Word has been Written
         into the PICmicro® MCU and has not been read
  6    OBF - Bit Set when a Byte has been written to
         the PORTD Output Register and has not been
         read
  5    IBOV - Bit Set when a Word has been Written
         into the PICmicro® MCU before the previous one
         has been read
  4    PSPMODE - Bit set to enable Parallel Slave
         Port
  3    Unused
  2    TRISE2 - TRIS Bit for E2
  1    TRISE1 - TRIS Bit for E1
  0    TRISE0 - TRIS Bit for E0
```

Along with TMR0, some mid-range PICmicro® MCU's have TMR1 and TMR2, which are used for basic timing operations as well as "CCP" ("Compare, Capture, and PWM") I/O. TMR1 is a 16-bit-wide register (accessed via "TMR1L" and "TMR1H") that will request an interrupt on overflow ("TMR1IF") if the "TMR1IE" bit is set. The T1CON register that is defined below controls the operation of TMR1:

```
Bit     Function
7-6     Unused
5-4     T1CKPS1:T1CKPS2: TMR1 Input Prescaler Select
  3     T1OSCEN - Set to Enable External TMR1
        Oscillator
  2     _T1SYNC - If External Clock used for TMR1,
        then Synchronize to it when this bit is Reset
  1     TMR1CS - When Set, TMR1 is driven by External
        Clock/TMR1 Oscillator
  0     TMR1ON - Set to Enable TMR1
```

TMR2 is an 8-bit register that is continually compared against a value in the PR2 register. To have TMR2 operate like TMR0 as an 8-bit Timer with a range of 0x000 to 0x0FF, then the "PR2" (the register TMR2 is compared against) is set to 0x000. The TMR2 output can be used to drive a PWM signal out. Interrupts ("TMR2IF") can be requested after the TMR2 overflow has passed through a Postscaler and "TMR2IE" is set. The T2CON register controls the operation of TMR2:

```
Bit     Function
  7     Unused
6-3     TOUTPS3:TOUTPS0 - TMR2 Output Postscaler
        Select
        Bits          Postscaler
        1111          16:1
        1110          15:1
        1101          14:1
        1100          13:1
```

```
        1011        12:1
        1010        11:1
        1001        10:1
        1000        9:1
        0111        8:1
        0110        7:1
        0101        6:1
        0100        5:1
        0011        4:1
        0010        3:1
        0001        2:1
        0000        1:1
2       TMR2ON - Set to Enable TMR2
1-0     T2CKPS1:T2CKPS0 - TMR2 Input Prescaler Select
        Bits        Prescaler
        1x          16:1
        01          4:1
        00          1:1
```

TMR1 and TMR2 are used with one of the two "CCP" Modules for advanced I/O. TMR1 is used for Capture and Compare and TMR2 is used for PWM Output. The CCPR2x Registers are used for Storing Compare/Capture Values and the CCPx register specifies the Pin used for CCP. The CCPxCON register is used for controlling the CCP operation:

```
Bit     Function
7-6     Unused
5-4     DCxB1:DCxB0 - PWM Duty Cycle Bit 1 and Bit 0.
        These bits are only accessed by the PWM for
        its low output values
3-0     CCPxM3:CCPxM0 - CCPx Mode Select
        Bits    Function
        11xx    PWM Mode
        1011    Compare Mode, Trigger Special Event
        1010    Compare Mode, Trigger on Compare Match
        1001    Compare Mode, Initialize CCP Pin High,
                On Compare Match force CCP Low
        1000    Compare Mode, Initialize CCP Pin Low,
                On Compare Match force CCP High
        0111    Capture on Every 16th Rising Edge
        0110    Capture on Every 4th Rising Edge
```

```
0101    Capture on Every Rising Edge
0100    Capture on Every Falling Edge
001x    Unused
0001    Unused
0000    Capture/Compare/PWM Off
```

CCP Interrupts are requested via the "CCPxIF" flag and enabled by the "CCPXIE" flag where "x" is "1" or "2" depending on the active CCP Module.

There are three different "SSP" Modules built into the PICmicro® MCU. Each one provides somewhat different Options and understanding how they work will be critical to your applications and if I2C is going to be used with them. The basic SSP modules ("SSP" and "BSSP") provide a full SPI Interface and I2C "Slave Mode" Interface. The SSPBUF Register provides simple buffering while the SSPADD buffers the received address for comparing against I/O operations. To control the Operation of the SSP, the "SSPCON" register is used:

```
Bit    Function
 7     WCOL - Set if SSPBUF was written to while
       transmitting data or not in correct mode for
       transmit
 6     SSPOV - Set when SSP Receive overflow occurs
 5     SSPEN - Enables Pins for SSP Mode
 4     CKP - In SPI, Set for Idle Clock High. In I2C
       Mode, set to Enable Clock
3-0    SSPM3:SSPM0 - SSP Mode Select
       1111 - I2C Slave Mode, 10 Bit Address
       1110 - I2C Slave Mode, 7 Bit Address
       110x - Reserved
       1011 - I2C firmware controlled Master
       1010 - Reserved
       1001 - Reserved
       1000 - Reserved
       0111 - I2C Slave Mode, 10 Bit Address
       0110 - I2C Slave Mode, 7 Bit Address
       0101 - SSP Slave, _SS Disabled
       0100 - SSP Slave, _SS Enabled
       0011 - SPI Master, Clock = TMR2
```

```
0010 - SPI Master, Fosc/64
0001 - SPI Master, Fosc/16
0000 - SPI Master, Fosc/4
```

The SSPSTAT Register is also used to Control the SSP:

```
Bit    Function
 7     SMP - Data Sampled at end of data output time
         if Set, else middle
 6     CKE - Data transmitted on rising edge of SCK
         when Set
 5     D/_A - Used by I2C. When Set indicates last
         byte transferred was data. When Reset
         indicates last byte transferred was address
 4     P - Set when Stop Bit Detected
 3     S - Set when Start Bit Indicated
 2     R/_W - Set when command received was a Read
 1     UA - Set when application must update SSPADD
         Register
 0     BF - Set when Buffer is full in RX and when
         TX is in process
```

The "Master" SSP ("MSSP") accesses similar registers for the same functions, with a Second SSPCON Register. The important difference between the MSSP and the other SSP modules is the enabled I2C Master hardware in the MSSP. The MSSP's "SSPCON1" register is defined as:

```
Bit    Function
 7     WCOL - Set if SSPBUF was written to while
         transmitting data or not in correct mode for
         transmit
 6     SSPOV - Set when SSP Receive overflow occurs
 5     SSPEN - Enables Pins for SSP Mode
 4     CKP - In SPI, Set for Idle Clock High. In I2C
         Mode, set to Enable Clock
3-0    SSPM3:SSPM0 - SSP Mode Select
       1xx1 - Reserved
       1x1x - Reserved
       1000 - I2C Master Mode, Clock = Fosc/(4 *
              (SSPADD + 1))
       0111 - I2C Slave Mode, 10 Bit Address
```

```
0110 - I2C Slave Mode, 7 Bit Address
0101 - SSP Slave, _SS Disabled
0100 - SSP Slave, _SS Enabled
0011 - SPI Master, Clock = TMR2
0010 - SPI Master, Fosc/64
0001 - SPI Master, Fosc/16
0000 - SPI Master, Fosc/4
```

SSPCON2 is used for I2C Master mode and is defined as:

Bit	Function
7	GCEN - Set to Enable Interrupt when General Call Address is Received
6	ACKSTAT - Set when Acknowledge Received from I2C Slave Device
5	ACKDT - Reset to send Acknowledge at the end of a Byte Receive
4	ACKEN - Acknowledge I2C Sequence when Set
3	RCEN - Set to Enable I2C Receive Mode
2	PEN - Reset to Initiate Stop Condition on I2C Clock and Data
1	RSEN - Set to Initiate Repeated Start Condition on I2C Clock and Data
0	SEN - Set to Initiate Start Condition on I2C Clock and Data

The SSPSTAT Register for MSSP is

Bit	Function
7	SMP - Data Sampled at end of data output time if Set, else middle
6	CKE - Data transmitted on rising edge of SCK when Set
5	D/_A - Used by I2C. When Set indicates last byte transferred was data. When Reset indicates last byte transferred was address
4	P - Set when Stop Bit Detected
3	S - Set when Start Bit Indicated
2	R/_W - Set when command received was a Read
1	UA - Set when application must update SSPADD Register
0	BF - Set when Buffer is full in RX and when TX is in process

Interrupts are Requested from the SSP via the "SSPIF" bit and enabled by the "SSPIE" bit.

"Non-Return to Zero" ("NRZ") Asynchronous Serial Communications are accomplished by the built-in "USART". This circuit can also be used for Synchronous Serial Communications. The clock Speed is determined by the SPBRG. The TXREG and RCREG registers are used to transfer data. The "RCSTA" is the primary USART control Register

```
Bit    Function
 7     SPEN - Set to Enable the USART
 6     RX9 - Set to Enable 9-Bit Serial Reception
 5     SREN - Set to enable single receive for
         Synchronous Mode
 4     CREN - Set to Enable Continuous Receive Mode
 3     ADDEN - Enables Address Detection in
         Asynchronous Mode
 2     FERR - Framing Error Bit
 1     OERR - Set after Overrun Error
 0     RX9D - Ninth bit of Data Received
```

and the "TXSTA" is used to control the Serial Output.

```
Bit    Function
 7     CSRC - Set for Synchronous Clock Generated
         Internally
 6     TX9 - Set to Enable Nine Bit Data Transmission
 5     TXEN - Set to Enable Transmit
 4     SYNC - Set to Select Synchronous Mode
 3     Unused
 2     BRGH - Set to Select the High Baud Rate
 1     TRMT - Set when Transmit Shift Register is
         Empty
 0     TX9D - Ninth bit of Transmit Data
```

The RCIF interrupt request bit, when set, means there is a character received in the USART. RCIF is enabled by RCIE. TXIF is set when the TX Holding Register is empty and is enabled by TXIE.

Comparator Equipped PICmicro® MCUs have a built-in Reference Voltage Source that is controlled by the VRCON Register:

```
Bit    Function
7      VREN - Set to Turn on Voltage Reference
       Circuit
6      VROE - Set to Output Voltage Reference
       Externally
5      VRR - Set for Low Voltage Reference Range,
       Reset for High Voltage Reference Range
4      Unused
3-0    VR3:VR0 - Select the Reference Voltage Output
```

The Voltage Reference Output is defined by the formula:

```
Vref = (1/4*Vdd*(1-VRR)) + Vdd*(VR3:VR0/(24+(8*(1
     - VRR))))
```

For Vdd equal to 5.0 Volts, the following table lists different Vref values:

VR3:VR0	VRR = 1	VRR = 0
1111	3.13 Volts	3.59 Volts
1110	2.92 Volts	3.44 Volts
1101	2.71 Volts	3.28 Volts
1100	2.50 Volts	3.13 Volts
1011	2.29 Volts	2.97 Volts
1010	2.08 Volts	2.81 Volts
1001	1.88 Volts	2.66 Volts
1000	1.67 Volts	2.50 Volts
0111	1.46 Volts	2.34 Volts
0110	1.25 Volts	3.19 Volts
0101	1.04 Volts	2.03 Volts
0100	0.83 Volts	1.88 Volts
0011	0.63 Volts	1.72 Volts
0010	0.42 Volts	1.56 Volts
0001	0.21 Volts	1.41 Volts
0000	0.00 Volts	1.25 Volts

The Voltage Reference is normally used with the Voltage Comparator, which is controlled by the "CMCON" Register.

```
Bit    Function
7      C2OUT - Set when C2Vin+ is Greater than C2Vin-
6      C1OUT - Set when C1Vin+ is Greater than C1Vin-
5-4    Unused
3      CIS - Comparator Input Switch, See CM2:CM0
2-0    CM2:CM0 - Comparator Mode Select
```

Bits	CIS	C1Vin+	C1Vin-	C2Vin+	C2Vin-	Comments
111	x	Gnd	Gnd	Gnd	Gnd	Comparators Off
110	x	AN2	AN0	AN2	AN1	AN3 = C1OUT RA4 = C2OUT
101	x	Gnd	Gnd	AN2	AN1	
100	x	AN3	AN0	AN2	AN1	
011	x	AN2	AN0	AN2	AN1	
010	1	Vref	AN3	Vref	AN2	
010	0	Vref	AN0	Vref	AN1	
001	1	AN2	AN3	AN2	AN1	
001	0	AN2	AN0	AN2	AN1	
000	x	AN3	AN0	AN2	AN1	Comparators Off

Interrupts Requested by Change on Comparator Outputs are specified CMIF and enabled by CMIE.

There are also some Analog to Digital Converter ("ADC") Options that can be used with the PICmicro® MCU. The operation of the ADC is controlled by AD-CON0 Register:

```
Bit    Function
7-6    ADCS1:ADCS0 - ADC Conversion Clock Select.
          11 - Internal RC Oscillator
          10 - Divide PICmicro® MCU clock by 32
          01 - Divide PICmicro® MCU clock by 8
          00 - Divide PICmicro® MCU clock by 2
5-3    CHS2:CHS0 - ADC Conversion Channel Select Bits
          111 - AN7
          110 - AN6
          101 - AN5
          100 - AN4
          011 - AN3
          010 - AN2
          001 - AN1
          000 - AN0
```

```
2      GO/_DONE - Set to Start A/D Conversion, Reset
       by Hardware when Conversion Before
1      Unused
0      ADON - Set to Turn on the ADC Function
```

Selecting the PORTA, Analog/Digital Functions, there are a number of different formats of ADCON1 that you should be aware of. For basic 18-pin PICmicro® MCU's ADCs, ADCON1 is defined as:

```
Bit    Function
7-2    Unused
1-0    PCFG1:PCFG0 - A/D Select
       Bits     AN3        AN2        AN1        AN0
       11       D          D          D          D
       10       D          D          A          A
       01       Vref+      A          A          A
       00       A          A          A          A
```

For more advanced 18-pin PICmicro® MCUs, ADCON1 is defined as:

```
Bit    Function
7-3    Unused
2-0    PCFG2:PCFG0 - A/D Select
       Bits     AN3        AN2        AN1        AN0
       111      D          D          D          D
       110      D          D          D          A
       101      D          D          Vref+      A
       100      D          D          A          A
       011      D          A          Vref+      A
       010      D          A          A          A
       001      A          A          Vref+      A
       000      A          A          A          A
```

28- and 40-pin PICmicro® MCUs have the ADCON1 Register:

```
Bit    Function
7-3    Unused
2-0    PCFG2:PCFG0 - A/D Select
       Bits  AN7  AN6  AN5  AN4  AN3    AN2  AN1  AN0
       11x   D    D    D    D    D      D    D    D
       101   D    D    D    D    Vref+  D    A    A
```

100	D	D	D	D	A	D	A	A
011	D	D	A	A	Vref+	A	A	A
010	D	D	D	A	A	A	A	A
001	A	A	A	A	Vref+	A	A	A
000	A	A	A	A	A	A	A	A

The result of the ADC Operation is stored in ADRES and ADIF is set upon completion of the ADC operation to request an interrupt if ADIE is set.

Ten-bit ADCs are also available in the PICmicro® MCU, with a different ADCON1 Register:

```
Bit   Function
7-6   Unused
  5   ADFM - When Set, the Result is "Right
      Justified" else "Left Justified"
  4   Unused
3-0   PCFG3:PCFG0 - A/D Select
```

Bits	AN7	AN6	AN5	AN4	AN3	AN2	AN1	AN0	VR+	VR-
1111	D	D	D	D	VR+	VR-	D	A	AN3	AN2
1110	D	D	D	D	D	D	D	A	Vdd	Vss
1101	D	D	D	D	VR+	VR-	A	A	AN3	AN2
1100	D	D	D	A	VR+	VR-	A	A	AN3	AN2
1011	D	D	A	A	VR+	VR-	A	A	AN3	AN2
1010	D	D	A	A	VR+	A	A	A	AN3	Vss
1001	D	D	A	A	A	A	A	A	Vdd	Vss
1000	A	A	A	A	VR+	VR-	A	A	AN3	AN2
011x	D	D	D	D	D	D	D	D	N/A	N/A
0101	D	D	D	D	VR+	D	A	A	AN3	Vss
0100	D	D	D	D	A	D	A	A	Vdd	Vss
0011	D	D	D	D	VR+	A	A	A	AN3	Vss
0010	D	D	D	A	A	A	A	A	Vdd	Vss
0001	A	A	A	A	VR+	A	A	A	AN3	Vss
0000	A	A	A	A	A	A	A	A	Vdd	Vss

In the case of 10-bit ADCs, the result is stored in ADRESL and ADRESH.

This mid-range register list does not include the PIC16C92x's LED Control Registers. This, as well as any other I/O Hardware registers that were not available when this was written, can be found in the Microchip Datasheets.

PIC17Cxx

The PIC17Cxx banking scheme has register addresses 0x010 to 0x017 and 0x020 to 0x0FF being banked and accessed by the BSR register separately. All other register addresses are common regardless of the banks selected within the BSR. The register space of 0x010 to 0x017 consists of the I/O Hardware registers listed according to bank.

Address Range 0x000 to 0x01F is considered to be the "Primary" Register set ("p") in the PIC17Cxx "move" instructions. For PIC17C4x devices there are four "Primary Banks" (address 0x010 to 0x017); in the PIC17C5x, there are eight "Primary Banks".

BSR	Addr	Register	Function/Bit Definition
Any	0x000	INDF0	Register Pointed to by FSR0
Any	0x001	FSR0	Index Register 0
Any	0x002	PCL	Low Byte of the Program Counter
Any	0x003	PCLATH	Latched High Byte of the Program Counter
Any	0x004	ALUSTA	Processor Status and Control Register

Bit   Function
7-6   FSR1 Mode Select
       1x - FSR1 Does not Change after Access
       01 - Post Increment FSR1
       00 - Post Decrement FSR1
5-4   FSR0 Mode Select
       1x - FSR0 Does not Change after Access
       01 - Post Increment FSR0
       00 - Post Decrement FSR0
3   OV - Set when there is a two's complement overflow
       after addition/subtraction
2   Z - Set when the eight bit result is equal to Zero
1   DC - Set for low order Nybble carry after addition
       or subtraction instruction
0   C - Set for Carry after addition or subtraction
       instruction

| Any | 0x005 | TOSTA | TMR0 Status and Control Register |

Bit   Function
7   INTEDG - Select the RA0/INT Pin
       Interrupt request Edge. When Reset, Rising Edge
       Increments TMR0 else, Falling Edge Increments TMR0
6   T0SE - TMR0 Clock Input Edge Select - When Set, the
       Rising edge of the incoming clock increments TMR0
       else Falling Edge increments TMR0

5   TOCS - TMR0 Clock Source Select - When Set the Instruction Clock is used, else the TOCKI pin is used

4-1   PS3-PS0 - TMR0 Prescaler Selection.

Bits	Prescaler
1xxx	256:1
0111	128:1
0110	64:1
0101	32:1
0100	16:1
0011	8:1
0010	4:1
0001	2:1
0000	1:1

0   Unused

Any   0x006 CPUSTA   Processor Operating Status Register

Bit Function

7-6   Unused

5   STKAV - When Set, there is Program Counter Stack Space Available

4   GLINTD - When Set, all Interrupts Are Disabled

3   _TO - Set after Power Up or clrwdt Instruction. When Reset a Watchdog Timeout has occurred

2   _PD - Set after Power Up or clrwdt Instruction. Reset by a "sleep" instruction

1   _POR - Reset After Power Up in PIC17C5x. Not Available in All PIC17Cxx devices

0   _BOR - Reset After Brown Out Reset. Not Available in All PIC17Cxx devices

BSR	Addr	Register	Function/Bit Definition
Any	0x007	INTSTA	Interrupt Status and Control Register
			Bit Function
			7 PEIE – Set when Peripheral Interrupt is Pending
			6 T0CKIF – Set when RA1/T0CKI Pin has Interrupt Source. Cleared by Hardware when Interrupt Vector 0x0018 is executed
			5 T0IF – Set when TMR0 Overflows. Cleared by Hardware when Interrupt Vector 0x0010 is executed
			4 INTF – Set when RA0/INT Pin Interrupt Request Active. Cleared by Hardware when Interrupt Vector 0x0008 is executed
			3 PEIE – Set to Enable Peripheral Interrupt Requests
			2 T0CKIE – Set to Enable RA1/T0CKI Interrupt Request
			1 T0IE – Set to Enable TMR0 Overflow Interrupt Request
			0 INTE – Set to Enable RA0/INT Pin Interrupt Request
Any	0x008	INDF1	Register Pointed to by FSR1
Any	0x009	FSR1	Index Register 1
Any	0x00A	WREG	Processor Accumulator
Any	0x00B	TMR0L	Low Byte of TMR0
Any	0x00C	TMR0H	High Byte of TMR0
Any	0x00D	TBLPTRL	Low Byte of the Table Pointer
Any	0x00E	TBLPTRH	High Byte of the Table Pointer
Any	0x00F	BSR	Bank Select Register
			Bit Function
			7-4 Select General Purpose RAM Bank at Addresses 0x020 to 0x0FF
			3-0 Select the I/O Hardware Register Bank at Addresses 0x010 to 0x017

0	0x010 PORTA	PORTA I/O Bits
		Bit Function
		7 _RBPU - When Reset, Pull Up on PORTB is Enabled
		6 Unused
		5 RA5/TX - Input or USART TX Pin. Schmidt Trigger Input
		4 RA4/RX - Input or USART RX Pin. Schmidt Trigger Input
		3 RA3 - Schmidt Trigger Input/Open Drain Output
		2 RA2 - Schmidt Trigger Input/Open Drain Output
		1 RA1/T0CKI - Bit Input or TMR0 Input. Schmidt Trigger Input
		0 RA0/INT - Bit Input or External Interrupt
0	0x011 DDRB	PORTB Data Direction Port. When Bit Reset, Pin is in "Output" Mode
0	0x012 PORTB	PORTB I/O Bits
		Bit Function
		7-6 RB7:RB6 - I/O Pin with Interrupt in Input Change. Schmidt Trigger Input
		5 RB5 - I/C Pin with TMR3 Clock Input. Interrupt on Input Change. Schmidt Trigger Input
		4 RB4 - I/C Pin with TMR1/TMR2 Clock Input. Interrupt on Input Change. Schmidt Trigger Input
		3 RB3 - I/O Pin with CCP2 PWM Output. Schmidt Trigger Input
		2 RB2 - I/O Pin with CCP1 PWM Output. Schmidt Trigger Input
		1 RB1 - I/O Pin with CCP2 Capture Input. Schmidt Trigger Input
		0 RB0 - I/O Pin with CCP1 Capture Input. Schmidt Trigger Input

BSR	Addr	Register	Function/Bit Definition
0	0x013	RCSTA	USART Receive Status and Control Register
			Bit Function
			7 SPEN – Set to Enable the USART
			6 RX9 – Set to Enable 9-Bit Serial Reception
			5 SREN – Set to enable single receive for Synchronous Mode
			4 CREN – Set to Enable Continuous Receive Mode
			3 Unused
			2 FERR – Framing Error Bit
			1 OERR – Set after Overrun Error
			0 RX9D – Ninth bit of Data Received
0	0x014	RCREG	USART Receiver Holding Register
0	0x015	TXSTA	USART Transmit Status and Control Register
			Bit Function
			7 CSRC – Set for Synchronous Clock Generated Internally
			6 TX9 – Set to Enable Nine Bit Data Transmission
			5 TXEN – Set to Enable Transmit
			4 SYNC – Set to Select Synchronous Mode
			3 Unused
			2 BRGH – Set to Select the High Baud Rate
			1 TRMT – Set when Transmit Shift Register is Empty
			0 TX9D – Ninth bit of Transmit Data
0	0x016	TXREG	USART Transmit Holding Register
0	0x017	SPBRG	USART Clock Divisor Register
1	0x010	DDRC	PORTC Data Direction Port. When bit is reset, PORTC bit is in "Output" mode
1	0x011	PORTC	PORTC I/O Pins or External Memory Data/Address Pins
1	0x012	DDRD	PORTD Data Direction Port. When bit is reset, PORTD bit is in "Output" mode

1	0x013	PORTD	PORTD I/O Pins or External Memory Data/Address Pins
1	0x014	DDRE	PORTE Data Direction Port. When bit is reset, PORTE bit is in "Output" mode
1	0x015	PORTE	PORTE I/O Pins or External Memory Data/Address Pins Control

Pins

Bit   Function
2   RE2/_WR - I/O Pin or System Bus Write
1   RE1/_OE - I/O Pin or System Bus Read
0   RE0/ALE - I/O Pin or System Bus Address Latch Enable

| 1 | 0x016 | PIR1 | Interrupt Status Register 1. This may be the only Interrupt Status Register in Some Devices (in which case it is labeled "PIR") |

Bit   Function
7   RBIF - Set if PORTB Interrupt on Change Active
6   TMR3IF - Set if TMR3 has Overflowed or Capture Timer has rolled over
5   TMR2IF - Set if TMR2 has Overflowed
4   TMR1IF - Set if TMR1 has Overflowed
3   CA2IF - Set if Capture2 Event Occurred
2   CA1IF - Set if Captured Event Occurred
1   TXIF - USART Transmit Interrupt Request
0   RCIF - USART Receive Interrupt Request

| 1 | 0x017 | PIE1 | Interrupt Control Register 1. This may be the only Interrupt Control Register is some Devices (in which case it is labeled "PIE") |

Bit   Function
7   RBIE - Set to Enable PORTB Interrupt on Change
6   TMR3IE - Set to Enable TMR3 Interrupt
5   TMR2IE - Set to Enable TMR2 Interrupt

BSR	Addr	Register	Function/Bit Definition
			4 TMR1E - Set to Enable TMR1 Interrupt
			3 CA2IE - Set to Enable Capture2 Event Interrupt
			2 CA1IE - Set to Enable Capture1 Even Interrupt
			1 TXIE - Set to Enable USART Transmit Interrupt Request
			0 RCIE - Set to Enable USART Receive Interrupt Request
2	0x010	TMR1	TMR1 Data Register
2	0x011	TMR2	TMR2 Data Register
2	0x012	TMR3L	Low Byte of the TMR3 Data Register
2	0x013	TMR3H	High Byte of the TMR3 Data Register
2	0x014	PR1	TMR1 Period Register
2	0x015	PR2	TMR2 Period Register
2	0x016	PR3L	Low Byte of the TMR3 Period Register
2	0x017	PR3H	High Byte of the TMR3 Period Register
3	0x010	PW1DCL	PWM1 Least Significant two Compare Bits
			Bit Function
			7 DC1 - Bit 1 of the PWM Compare
			6 DC0 - Bit 0 of the PWM Compare
			5-0 Unused
3	0x011	PW2DCL	PWM2 Least Significant two Compare Bits
			Bit Function
			7 DC1 - Bit 1 of the PWM Compare
			6 DC0 - Bit 0 of the PWM Compare
			5 TM2PW2 - Set to Select PWM2 Clock Source as TMR2 and PR2 else PWM2 Clock Source is TMR1 and PR1
			4-0 Unused
3	0x012	PW1DCH	High Eight Bits of PWM1 Compare
3	0x013	PW2DCH	High Eight Bits of PWM2 Compare
3	0x014	CA2L	Low Byte of Capture 2 Data

| 3 | 0x015 | CA2H | High Byte of Capture 2 Data |
| 3 | 0x016 | T1CON | TMR1 and TMR2 Control Register |

Bit   Function

7-6   CA2ED1:CA2ED0 - Capture 2 Mode Select
         11 - Capture on 16th Rising Edge
         10 - Capture on 4th Rising Edge
         01 - Capture on Every Rising Edge
         00 - Capture on Every Falling Edge

5-4   CA1ED1:CA1ED0 - Capture 1 Mode Select
         11 - Capture on 16th Rising Edge
         10 - Capture on 4th Rising Edge
         01 - Capture on Every Rising Edge
         00 - Capture on Every Falling Edge

3   TMR2:TMR1 Mode Select. When Set, TMR2:TMR1 are a 16
         bit Timer else two separate eight bit Timers

2   TMR3CS - When Set TMR3 Increments from Falling Edge
         of RB3/TCLK3 Pin else Increments from Instruction
         Clock

1   TMR2CS - When Set TMR2 Increments from Falling Edge
         of the RB4/TCLK12 Pin else Increments from
         Instruction Clock

0   TMR1CS - When Set TMR1 Increments from Falling Edge
         of the RB4/TCKL12 Pin else Increments from
         Instruction Clock

| 3 | 0x017 | TCON2 | TMR1 and TMR2 Control Register 2 |

Bit   Function

7   CA2OVF - Set if Overflow Occurred in Capture2
         Register

6   CA1OVF - Set if Overflow Occurred in Capture1
         Register

BSR	Addr	Register	Function/Bit Definition
			5 PWM2ON - Set if PWM2 is Enabled
			4 PWM1ON - Set if PWM1 is Enabled
			3 CA1/_PR3 - Set to Enable Capture1 else Enables the Period Register
			2 TMR3ON - Set to Enable TMR3
			1 TMR2ON - Set to Enable TMR2. Must be Set if TMR2/TMR1 are combined
			0 TMR1ON - Set to Enable TMR1. When TMR2/TMR1 are combined, controls operation of 16 bit Timer
4	0x010	PIR2	Interrupt Status Register 2
			Bit Function
			7 SSPIF - Set if SSP Interrupt has been Requested
			6 BCLIF - Set if there is a Bus Collision Interrupt Request
			5 ADIF - Set if there is an ADC Interrupt Request
			4 Unused
			3 CA4IF - Set if Capture4 Event has Requested an Interrupt
			2 CA3IF - Set if Capture4 Event has Requested an Interrupt
			1 TX2IF - Set if USART2 Transmit Interrupt Requested
			0 RC2IF - Set if USART2 Receive Interrupt Requested
4	0x011	PIE2	Interrupt Control Register 2
			Bit Function
			7 SSPIE - Set to Enable SSP Interrupt
			6 BCLIE - Set to Enable Bus Collision Interrupt
			5 ADIE - Set to Enable ADC Interrupts
			4 Unused

		3	CA4IE - Set to Enable Capture4 Interrupt
		2	CA3IE - Set to Enable Capture3 Interrupt
		1	TX2IE - Set to Enable USART2 Transmit Interrupt
		1	RC2IE - Set to Enable USART2 Receive Interrupt
4	0x012	RCSTA2	USART2 Receive Status and Control Register
			Bit Function
		7	SPEN - Set to Enable the USART
		6	RX9 - Set to Enable 9-Bit Serial Reception
		5	SREN - Set to enable single receive for Synchronous Mode
		4	CREN - Set to Enable Continuous Receive Mode
		3	Unused
		2	FERR - Framing Error Bit
		1	OERR - Set after Overrun Error
		0	RX9D - Ninth bit of Data Received
4	0x014	RCREG2	USART2 Receiver Holding Register
4	0x015	TXSTA2	USART2 Transmit Status and Control Register
			Bit Function
		7	CSRC - Set for Synchronous Clock Generated Internally
		6	TX9 - Set to Enable Nine Bit Data Transmission
		5	TXEN - Set to Enable Transmit
		4	SYNC - Set to Select Synchronous Mode
		3	Unused
		2	BRGH - Set to Select the High Baud Rate
		1	TRMT - Set when Transmit Shift Register is Empty
		0	TX9D - Ninth bit of Transmit Data
4	0x016	TXREG2	USART2 Transmit Holding Register
4	0x017	SPBRG2	USART2 Clock Divisor Register
5	0x010	DDRF	PORTF Data Direction Port. When bit is reset, PORTF bit is in "Output" mode

BSR	Addr	Register	Function/Bit Definition
5	0x011	PORTF	PORTF I/O Pins or Analog Inputs 4 through 11
5	0x012	DDRG	PORTG Data Direction Port. When bit is reset, PORTG bit is in "Output" mode
5	0x013	PORTG	PORTG I/O Pins

Bit	Function
7	RG7/TX2 - Schmidt Trigger I/O or USART2 TX Pin
6	RG6/RX2 - Schmidt Trigger I/O or USART2 RX Pin
5	RG5/PWM3 - Schmidt Trigger I/O or PWM3 Output
4	RG4/CAP3 - Schmidt Trigger I/O or Capture3 Pin
3	RG3/AN0 - Schmidt Trigger I/O or Analog Input
2	RG2/AN1 - Schmidt Trigger I/O or Analog Input
1	RG1/AN2 - Schmidt Trigger I/O or Analog Input
0	RG0/AN3 - Schmidt Trigger I/O or Analog Input

BSR	Addr	Register	Function/Bit Definition
5	0x014	ADCON0	ADC Control Register 1

Bit	Function
7-4	CHS2:CHS0 - Analog Channel Select
	11xx Reserved
	1011 AN11
	1010 AN10
	1001 AN9

		1000	AN8
		0111	AN7
		0110	AN6
		0101	AN5
		0100	AN4
		0011	AN3
		0010	AN2
		0001	AN1
		0000	AN0

Bit	
3	Unused
2	GO/_DONE - Set to Start A/D Conversion, Reset by Hardware when Finished
1	Unused
0	ADON - Set to Turn on the ADC

5 0x015 ADCON1 ADC Control Register 2

Bit	Function
7-6	ADCS1:ADCS0 - ADC Clock Select
	11 - Internal RC Clock
	10 - Fosc / 64
	01 - Fosc / 32
	00 - Fosc / 8
5	ADFM - Set for Right Justified Result Format, else Left Justified Result Format
4	Unused
3-1	PCFG3:PFG1 - Specify A/D Pins

Bits	AN11	AN10	AN9	AN8	AN7	AN6	AN5	AN4	AN3	AN2	AN1	AN0
111	D	D	D	D	D	D	D	D	D	D	D	D
110	D	D	A	A	D	D	D	D	D	D	A	A

BSR	Addr	Register	Function/Bit Definition							
	101		A	A	A	A	D	D	D	D
	100		A	A	A	A	D	D	D	D
	011		A	A	A	A	D	D	D	A
	010		A	A	A	A	D	D	A	A
	001		A	A	A	A	D	A	A	A
	000		A	A	A	A	A	A	A	A

0 PCFG0 - When Set VR+ and VR- Pins are used for Vref+ and Vref- else Vdd and Vss

BSR	Addr	Register	Function/Bit Definition
5	0x016	ADRESL	Low Byte of ADC Result
5	0x017	ADRESH	High Byte of ADC Result
6	0x010	SSPADD	MSSP Address Compare Register
6	0x011	SSPCON1	MSSP Control Register1

Bit   Function

7   WCOL - Set if SSPBUF was written to while transmitting data or not in correct mode for transmit

6   SSPOV - Set when SSP Receive overflow occurs

5   SSPEN - Enables Pins for SSP Mode

4   CKP - In SPI, Set for Idle Clock High. In I2C Mode, set to Enable Clock

3-0   SSPM3:SSPM0 - SSP Mode Select
       1111 - I2C Slave Mode, 10 Bit Address
       1110 - I2C Slave Mode, 7 Bit Address
       110x - Reserved
       1011 - I2C firmware controlled Master
       1010 - Reserved
       1001 - Reserved

		1000 - I2C Master, Fosc/(4*(SSPAD+1))
		0111 - I2C Slave Mode, 10 Bit Address
		0110 - I2C Slave Mode, 7 Bit Address
		0101 - SSP Slave, _SS Disabled
		0100 - SSP Slave, _SS Enabled
		0011 - SPI Master, Clock = TMR2
		0010 - SPI Master, Fosc/64
		0001 - SPI Master, Fosc/16
		0000 - SPI Master, Fosc/4
6	0x012 SSPCON2	MSSP Control Register2
		Bit Function
		7 GCEN - Set to Enable Interrupt when General Call Address is Received
		6 ACKSTAT - Set when Acknowledge Received from I2C Slave Device
		5 ACKDT - Reset to send Acknowledge at the end of a Byte Receive
		4 ACKEN - Acknowledge I2C Sequence when Set
		3 RCEN - Set to Enable I2C Receive Mode
		2 PEN - Reset to Initiate Stop Condition on I2C Clock and Data
		1 RSEN - Set to Initiate Repeated Start Condition on I2C Clock and Data
		0 SEN - Set to Initiate Start Condition on I2C Clock and Data
6	0x013 SSPSTAT	MSSP Status Register
		Bit Function
		7 SMP - Data Sampled at end of data output time if Set, else middle

BSR	Addr	Register	Bit	Function/Bit Definition
			6	CKE – Data transmitted on rising edge of SCK when Set
			5	D/A – When Set indicates last byte transferred was data. When Reset indicates last byte transferred was address
			4	P – Set when Stop Bit Detected
			3	S – Set when Start Bit Indicated
			2	R/_W – Set when command received was a Read
			1	UA – Set when application must update SSPADD Register
			0	BF – Set when Buffer is full in RX and when TX is in process
6	0x014	SSPBUF		MSSP Data Buffer
7	0x010	PW3DCL		PWM3 Least Significant two Compare Bits
			Bit	Function
			7	DC1 – Bit 1 of the PWM Compare
			6	DC0 – Bit 0 of the PWM Compare
			5	TM2PW3 – Set to Select PWM3 Clock Source as TMR2 and PR2 else PWM3 Clock Source is TMR1 and PR1
			4-0	Unused
7	0x011	PW3DCH		High Eight Bits of PWM3 Compare
7	0x012	CA3L		Low Byte of Capture 3 Data
7	0x013	CA3H		High Byte of Capture 3 Data
7	0x014	CA4L		Low Byte of Capture 4 Data
7	0x015	CA4H		High Byte of Capture 4 Data

7	0x016	TCON3	CCP3/CCP4 Control Register

	Bit	Function
	7	Unused
	6	CA4OVF - Set if Overflow on Capture4
	5	CA3OVF - Set if Overflow on Capture3
	4-3	CA4ED1:CA4ED) - Capture4 Select
		11 - Capture on 16th Rising Edge
		10 - Capture on 4th Rising Edge
		01 - Capture on Every Rising Edge
		00 - Capture on Every Falling Edge
	2-1	CA3ED1:CA3ED0 - Capture3 Select
		11 - Capture on 16th Rising Edge
		10 - Capture on 4th Rising Edge
		01 - Capture on Every Rising Edge
		00 - Capture on Every Falling Edge
	0	PWM3ON - Set to Enable PWM3

| Any | 0x018 | PRODL | Low Byte of Multiplication Product |
| Any | 0x019 | PRODH | High Byte of Multiplication Product |

PIC18Cxx

The hardware registers built into the PIC18Cxx are defined in the following table. Note that these registers are either accessed via the "Access Bank" or using the BSR set to 0x0F.

Address	Register	Function/Bit Definition
0x0#80	PORTA	PORTA Read/Write Register. Pin options are defined below:

PORTA Read/Write Register. Pin options are defined below:

Bit Function
7 Unused
6 OSC2
5 Slave Select/Optional AN4
4 Open Drain Output/Schmidt Trigger Input
3-0 Optional AN3-AN0

0x0#81 PORTB

PORTB Read/Write Register. I/O Pins can be pulled by software. Pin options are defined below:

Bit Function
7-6 ICSP Programming Pins/Interrupt on Pin Change
5 Interrupt on Pin Change
4 Interrupt on Pin Change
3 CCP2 I/O and PWM Output
2 Interrupt Source 3
1 Interrupt Source 2
0 Interrupt Source 1

0x0#82 PORTC

PORTC Read/Write Registers. I/O Pins have Schmidt Trigger Inputs. Pin options are defined below:

Bit Function
7 UART Receive Pin
6 UART Transmit Pin
5 Synchronous Serial Port Data
4 SPI Data or I2C Data
3 SPI Clock or I2C Clock

Address	Register	Function/Bit Definition
		2 CCP1 I/O and PWM Output/TMR1 Clock Output
		0 TMR1 Clock Input
0x0#83	PORTD	PORTD Only Available on 40 Pin18Cxx Devices. Schmidt Trigger Inputs. Used for Data Slave Port.
0x0#84	PORTE	PORTE Only Available on 40 Pin PIC18Cxx. Schmidt Trigger Inputs for I/O Mode. Used for Data Slave Port as Defined below:
		Bit Function
		7-3 Unused
		2 Negative Active Chip Select
		1 Negative Active Write Enable to PIC18Cxx
		0 Negative Active Output Enable ("_RD") from PIC18Cxx
0x0#89	LATA	Data Output Latch/Bypassing PORTA
0x0#8A	LATB	Data Output Latch/Bypassing PORTB
0x0#8B	LATC	Data Output Latch/Bypassing PORTC
0x0#8C	LATD	Data Output Latch/Bypassing PORTD. Only available on 40 Pin PIC18Cxx
0x0#8D	LATE	Data Output Latch/Bypassing PORTE. Only available on 40 Pin PIC18Cxx
0x0#92	TRISA	I/O Pin Tristate Control Register. Set bit to "0" for output mode
0x0#93	TRISB	I/O Pin Tristate Control Register. Set bit to "0" for output mode
0x0#94	TRISC	I/O Pin Tristate Control Register. Set bit to "0" for output mode
0x0#95	TRISD	I/O Pin Tristate Control Register. Only available on 40 Pin PIC18Cxx. Set bit to "0" for output mode

0x0#96	TRISE	I/O Pin Tristate Control Register. Only available on 40 Pin PIC18Cxx. Set bit to "0" for output mode. Special function bits Specified below:
		Bit Function
		7 IBF - Set when PSP is enabled and a byte has been written to the PICmicro® MCU
		6 OBF - Set when PSP is enabled and a byte output has not been read from the PICmicro® MCU
		5 IBOV - Set when PSP is enabled and the byte written to the PICmicro® MCU has been overwritten by a subsequent Byte
		4 PSPMODE - Set to Enable PICmicro® MCU's PSP I/O Port
		3 Unused
		2 TRISE2 - TRIS Bit for RE2
		1 TRISE1 - TRIS Bit for RE1
		0 TRISE0 - TRIS Bit for RE0
0x0#9D	PIE1	Peripheral Interrupt Enable Register
		Bit Function
		7 PSPIE - Set to Enable PSP Interrupt Request on Read/Write
		6 ADIE - Set to Enable Interrupt Request on Completion of A/D Operation
		5 RCIE - Set to Enable Interrupt Request on USART Data Receive
		4 TXIE - Set to Enable Interrupt Request on USART Transmit Holding Register Empty
		3 SSPIE - Master Synchronous Serial Port Interrupt Enable Bit

Address	Register	Function/Bit Definition
		2 CCP1IE – Set to Enable CCP1 Interrupt Request Enable
		1 TMR2IE – Timer2 to PR2 Match Interrupt Request Enable
		0 TMR1IE – TMR1 Overflow Interrupt Request Enable
0x0#9E	PIR1	Peripheral Interrupt Request Register
		Bit Function
		7 PSPIF – Set on PSP Read/Write
		6 ADIF – Set when A/D Complete
		5 RCIF – Set on USART Data Receive
		4 TXIF – Set on USART Transmit Holding Register Empty
		3 SSPIF – Set on Synchronous Serial Port Data Transmission/Reception Complete
		2 CCP1IF – Set on TMR1 Capture or Compare Match
		1 TMR2IF – Set on Timer2 to PR2 Match
		0 TMR1IF – Set on TMR1 Overflow
0x0#9F	IPR1	Peripheral Interrupt Priority Register
		Bit Function
		7 PSPIP – Set to Give PSP Interrupt Request on Read/Write Priority
		7 ADIP – Set to Give Interrupt Request on Completion of A/D Operation Priority
		5 RCIP – Set to Give Interrupt Request on USART Data Receive Priority
		4 TXIP – Set to Enable Interrupt Request on USART Transmit Holding Register Empty Priority
		3 SSPIP – Master Synchronous Serial Port Interrupt Priority when Set
		2 CCP1IP – Set to Give CCP1 Interrupt Request Priority

Address	Register	Bit	Function
		1	TMR2IP - Timer2 to PR2 Match Interrupt Request Priority when Set
		0	TMR1IF - TMR1 Overflow Interrupt Request Priority when Set
0x0#9A	PIE2		Peripheral Interrupt Enable Register
		Bit	Function
		7-4	Unused
		3	BCLIE - Bus Collision Interrupt Request Enabled when Set
		2	LVDIE - Low Voltage Detect Interrupt Request Enabled when Set
		1	TMR3IE - TMR3 Overflow Interrupt Request Enabled when Set
		0	CCP2IE - CCP2 Interrupt Request Enabled when Set
0x0#9B	PIR2		Peripheral Interrupt Request Register
		Bit	Function
		7-4	Unused
		3	BCLIF - Set for Bus Collision Interrupt Request
		2	LVDIF - Set for Low Voltage Detect Interrupt Request
		1	TMR3IF - Set for TMR3 Overflow Interrupt Request
		0	CCP2IF - Set for CCP2 Interrupt Request
0x0#9C	IPR2		Peripheral Interrupt Priority Register
		Bit	Function
		7-4	Unused
		3	BCLIF - Set for Bus Collision Interrupt given Priority
		2	LVDIF - Set for Low Voltage Detect Interrupt given Priority

Address	Register	Function/Bit Definition	
		1	TMR3IF – Set for TMR3 Overflow Interrupt Request given Priority
		0	CCP2IF – Set for CCP2 Interrupt Request given Priority
0x0#AB	RCSTA	USART Receive Status and Control Register Bit Function	
		7	SPEN – Set to Enable the USART
		6	RX9 – Set to Enable 9-Bit Serial Reception
		5	SREN – Set to enable single receive for Synchronous Mode
		4	CREN – Set to Enable Continuous Receive Mode
		3	ADDEN – Enables Address Detection in Asynchronous Mode
		2	FERR – Framing Error Bit
		1	OERR – Set after Overrun Error
		0	RX9D – Ninth bit of Data Received
0x0#AC	TXSTA	USART Transmit Status and Control Register Bit Function	
		7	CSRC – Set for Synchronous Clock Generated Internally
		6	TX9 – Set to Enable Nine Bit Data Transmission
		5	TXEN – Set to Enable Transmit
		4	SYNC – Set to Select Synchronous Mode
		3	Unused
		2	BRGH – Set to Select the High Baud Rate
		1	TRMT – Set when Transmit Shift Register is Empty
		0	TX9D – Ninth bit of Transmit Data
0x0#AD	TXREG	USART Transmit Buffer Register	

Address	Register	Description
0x0#AE	RCREG	USART Receive Holding Register
0x0#AF	SPBRG	USART Clock Divisor Register
0x0#B1	T3CON	TMR3 Control Register

Bit Function

7 RD16 - Enable Read/Write of TMR3 as a 16 Bit Operation

6,3 T3CCP2:T3CCP2 - TMR3 and TMR1 to CCPx Enable Bits
  1x - TMR3 is CCP Clock Source
  01 - TMR3 is CCP2 Clock Source/TMR1 is CCP1 Clock Source
  00 - TMR1 is CCP Clock Source

5-4 T3CKPS1:T3CKPS0 - TMR3 Input Clock Prescaler Control
  11 - 1:8 Prescaler
  10 - 1:4 Prescaler
  01 - 1:2 Prescaler
  00 - 1:1 Prescaler

2 T3SYNC - When Reset, TMR3 External Clock is Synchronized

1 TMR3CS - Set to Select External Clock for TMR3. Reset to Select Instruction Clock

0 TMR3ON - Set to Enable TMR3

0x0#B2	TMR3L	Low Byte of TMR3
0x0#B3	TMR3H	High Byte of TMR3
0x0#BA	CCP2CON	CCP2 Control Register

Bit Function

7-6 Unused

5-4 DC2BX1:DC2BX0 - Two Least Significant Bits for the 10 Bit PWM

Address	Register	Function/Bit Definition
		3-0 CCP2M3:CCP2M0 - CCP2 Mode Select Bits
		11xx - PWM Mode
		1011 - Trigger Special Event Compare Mode
		1010 - Generate Interrupt on Compare Match
		1001 - Initialize CCP2 High and Force Low on Compare Match
		1000 - Initialize CCP1 High and Force High on Compare Match
		0111 - Capture on Every 16th Rising Edge
		0110 - Capture on Every 4th Rising Edge
		0101 - Capture on Every Rising Edge
		0100 - Capture on Every Falling Edge
		0011 - Reserved
		0010 - Toggle output on Compare Match
		0001 - Reserved
		0000 - Capture/Compare/PWM off
0x0#BB	CCPR2L	Least Significant Capture/Compare/PWM2 Register
0x0#BC	CCPR2H	Most Significant Capture/Compare/PWM2 Register
0x0#BD	CCP1CON	CCP1 Control Register
		Bit Function
		7-6 Unused
		5-4 DC1BX1f:DC1BX0 - Two Least Significant Bits for the 10 Bit PWM
		3-0 CCP1M3:CCP1M0 - CCP1 Mode Select Bits
		11xx - PWM Mode
		1011 - Trigger Special Event Compare Mode
		1010 - Generate Interrupt on Compare Match
		1001 - Initialize CCP2 High and Force Low on Compare Match

1000 - Initialize CCP1 High and Force High on Compare Match
0111 - Capture on Every 16th Rising Edge
0110 - Capture on Every 4th Rising Edge
0101 - Capture on Every Rising Edge
0100 - Capture on Every Falling Edge
0011 - Reserved
0010 - Toggle output on Compare Match
0001 - Reserved
0000 - Capture/Compare/PWM off

Addr	Register	Description
0x0#BE	CCPR1L	Least Significant Capture/Compare/PWM1 Register
0x0#BF	CCPR1H	Most Significant Capture/Compare/PWM1 Register
0x0#C1	ADCON1	A/D Control Register1

Bit	Function
7	ADFM - Set to Return Result in Right Justified Format, Reset to Return Result in Left Justified Format
6	ADCS2 - Upper Bit of A/D Conversion Clock Select. See "ADCON0" for Bit Definition
5-4	Unused
3-0	PCFG3:PCFG0 - A/D Pin Configuration Select Bits

Bits	AN7	AN6	AN5	AN4	AN3	AN2	AN1	AN0	VR+	VR-
1111	D	D	D	D	VR+	VR-	D	A	AN3	AN2
1110	D	D	D	D	D	D	D	A	Vdd	Vss
1101	D	D	D	D	VR+	VR-	A	A	AN3	AN2
1100	D	D	D	A	VR+	VR-	A	A	AN3	AN2
1011	D	D	A	A	VR+	VR-	A	A	AN3	AN2
1010	D	D	A	A	VR+	A	A	A	AN3	Vss

Address	Register	Function/Bit Definition									
		1001	D	D	A	A	A	A	A	Vdd	Vss
		1000	A	A	A	VR+	VR–	A	A	AN3	AN2
		011x	D	D	D	D	D	D	A	N/A	N/A
		0101	D	D	D	D	D	A	A	Vdd	Vss
		0100	D	D	A	VR+	D	A	A	Vdd	Vss
		0011	D	D	D	VR+	A	A	A	AN3	Vss
		0010	D	D	D	A	A	A	A	Vdd	Vss
		0001	A	A	A	VR+	A	A	A	AN3	Vss
		0000	A	A	A	A	A	A	A	Vdd	Vss

0x0#C2   ADCON0   A/D Control Register2

```
Bit    Function
7-6    ADCS1:ADCS0 - ADC Conversion Clock
       Select, with ADCS2 from "ADCON1"
       111 - Internal RC Oscillator
       110 - Divide PICmicro® MCU clock by 64
       101 - Divide PICmicro® MCU clock by 16
       100 - Divide PICmicro® MCU clock by 4
       011 - Internal RC Oscillator
       010 - Divide PICmicro® MCU clock by 32
       001 - Divide PICmicro® MCU clock by 8
       000 - Divide PICmicro® MCU clock by 2
5-3    CHS2:CHS0 - ADC Conversion Channel
       Select Bits
       111 - AN7
       110 - AN6
       101 - AN5
       100 - AN4
       011 - AN3
```

```
                                    010 - AN2
                                    001 - AN1
                                    000 - AN0
                               2    GO/_DONE - Set to Start A/D Conversion,
                                    Reset by Hardware when Conversion Before
                                    Unused
                               1
                               0    ADON - Set to Turn on the ADC Function
0x0#C3   ADRESL    Low Byte of the ADC Result
0x0#C4   ADRESH    High Byte of the ADC Result
0x0#C5   SSPCON2   MSSP Control Register2
                   Bit  Function
                               7    GCEN - Set to Enable Interrupt when
                                    General Call Address is Received
                               6    ACKSTAT - Set when Acknowledge Received
                                    from I2C Slave Device
                               5    ACKDT - Reset to send Acknowledge at the
                                    end of a Byte Receive
                               4    ACKEN - Acknowledge I2C Sequence when Set
                               3    RCEN - Set to Enable I2C Receive Mode
                               2    PEN - Reset to Initiate Stop Condition
                                    on I2C Clock and Data
                               1    RSEN - Set to Initiate Repeated Start
                                    Condition on I2C Clock and Data
                               0    SEN - Set to Initiate Start Condition on I2C Clock
                                    and Data
0x0#C6   SSPCON1   MSSP Control Register1
                   Bit  Function
                               7    WCOL - Set if SSPBUF was written to while
                                    transmitting data or not in correct mode for
                                    transmit
```

Address	Register	Function/Bit Definition
		6 SSPOV - Set when SSP Receive overflow occurs
		5 SSPEN - Enables Pins for SSP Mode
		4 CKP - In SPI, Set for Idle Clock High. In I2C Mode, set to Enable Clock
		3-0 SSPM3:SSPM0 - SSP Mode Select
		1111 - I2C Slave Mode, 10 Bit Address
		1110 - I2C Slave Mode, 7 Bit Address
		110x - Reserved
		1011 - I2C firmware controlled Master
		1010 - Reserved
		1001 - Reserved
		1000 - I2C Master, $Fosc/(4*(SSPAD+1))$
		0111 - I2C Slave Mode, 10 Bit Address
		0110 - I2C Slave Mode, 7 Bit Address
		0101 - SSP Slave, _SS Disabled
		0100 - SSP Slave, _SS Enabled
		0011 - SPI Master, Clock = TMR2
		0010 - SPI Master, Fosc/64
		0001 - SPI Master, Fosc/16
		0000 - SPI Master, Fosc/4
0x0#C7	SSPSTAT	MSSP Status Register
		Bit Function
		7 SMP - Data Sampled at end of data output time if Set, else middle
		6 CKE - Data transmitted on rising edge of SCK when Set
		5 D/_A - When Set indicates last byte transferred was data. When Reset indicates last byte transferred was address

4   P – Set when Stop Bit Detected
3   S – Set when Start Bit Indicated
2   R/_W – Set when command received was a Read
1   UA – Set when application must update SSPADD Register
0   BF – Set when Buffer is full in RX and when TX is in process

0x0#C8	SSPADD	MSSP Address Compare Register
0x0#C9	SSPBUF	MSSP Data Buffer
0x0#CA	T2CON	TMR2 Control Register

Bit   Function
7   Unused
6:3   TOUTPS3:TOUTPS0 – TMR2 Output Postscaler
      1111 – 16x
      1110 – 15x
      1101 – 14x
      1100 – 13x
      1011 – 12x
      1010 – 11x
      1001 – 10x
      1000 – 9x
      0111 – 8x
      0110 – 7x
      0101 – 6x
      0100 – 5x
      0011 – 4x
      0010 – 3x
      0001 – 2x
      0000 – 1x
2   TMR2ON – Set to Enable TMR2

Address	Register	Function/Bit Definition
		1-0 T2CKPS1:T2CKPS0 - TMR2 Prescaler Select Bits
		1x - Prescaler is 16
		01 - Prescaler is 4
		00 - Prescaler is 1
0x0#CB	PR2	TMR2 Period Compare Register
0x0#CC	TMR2	TMR2 Register
0x0#CD	T1CON	TMR1 Control Register
		Bit Function
		7 RD16 - When Set, Enables 16 Bit TMR1 Operations
		6 Unused
		5:4 T1CKPS1:T1CKPS0 - TMR1 Input Clock Prescaler Select
		11 - 1:8 Prescaler
		10 - 1:4 Prescaler
		01 - 1:2 Prescaler
		00 - 1:1 Prescaler
		3 T1OSCEN - Set to Enable TMR1 Oscillator
		2 _T1SYNC_ - Set to Synchronize External Clock Input
		1 TMR1CS - TMR1 Clock Source Select. Set to Select
		External Clock
		0 TMR1ON - Set to Enable TMR1
0x0#CE	TMR1L	Low Byte of TMR1
0x0#CF	TMR1H	High Byte of TMR1
0x0#D0	RCON	Power Up Status Register
		Bit Function
		7 IPEN - Set to Enable Priority Levels on Interrupts
		6 LWRT - Set to Enable "TBLWT" Instruction to Internal
		Memory
		5 Unused

		Bit	Function	
0x0#D1	WDTCON			

0x0#D1 WDTCON Watchdog Timer Control Register

Bit	Function
7-1	Unused
0	SWDTEN - Set to Enable the Watchdog Timer if "__WDT_ON" is specified in "__CONFIG"

4	__RI - Reset when the "Reset" Instruction in Software instructions
3	__TO - Set after Power Up, clrwdt or sleep instructions
2	__PD - Set by Power Up or clrwdt Instruction. Reset by sleep instruction
1	__POR - Reset if a Power On Reset has Occurred
0	__BOR - Reset if a Brown Out Reset has Occurred

0x0#D2 LVDCON Low Voltage Detect Control Register

Bit	Function
7-6	Unused
5	IRVST - Set to indicate Low Voltage Detect Logic will Generate Interrupt
4	LVDEN - Set to Enable Low Voltage Detect
3-0	LVDL3:LVDL0 - Specify the Low Voltage Detect Limits

```
1111 - External Voltage Used (LVDIN)
1110 - 4.5V Min - 4.77V Max
1101 - 4.2V Min - 4.45V Max
1100 - 4.0V Min - 4.24V Max
1011 - 3.8V Min - 4.03V Max
1010 - 3.6V Min - 3.82V Max
1001 - 3.5V Min - 3.71V Max
1000 - 3.3V Min - 3.50V Max
0111 - 3.0V Min - 3.18V Max
0110 - 2.8V Min - 2.97V Max
```

Address	Register	Function/Bit Definition
		0101 - 2.7V Min - 2.86V Max
		0100 - 2.5V Min - 2.65V Max
		0011 - 2.4V Min - 2.54V Max
		0010 - 2.2V Min - 2.33V Max
		0001 - 2.0V Min - 2.12V Max
		0000 - 1.8V Min - 1.91V Max
0x0#D3	OSCCON	Select PICmicro® MCU Clock Source
		Bit Function
		7-1 Unused
		0 SCS - Reset to use Primary Oscillator. Set to use TMR1's Oscillator
0x0#D5	T0CON	TMR0 Control Register
		Bit Function
		7 TMR0ON - Set to Enable TMR0
		6 T08Bit - Set to Enable TMR0 as an 8 Bit Timer. Reset to Enable TMR0 as a 16 Bit Timer
		5 T0CS - Set to make TMR0 Clock Source T0CKI pin. Reset to use Instruction Clock
		4 T0SE - Set to Make TMR0 Increment on Falling Edge of Clock
		3 PSA - Reset to Assign TMR0 Prescaler
		2-0 T0PS2:T0PS0 - TMR0 Prescaler Value
		111 - 1:256 Prescaler
		110 - 1:128 Prescaler
		101 - 1:64 Prescaler
		100 - 1:32 Prescaler
		011 - 1:16 Prescaler
		010 - 1:8 Prescaler

Address	Register	Description
0x0#D6	TMR0L	Low Byte of TMR0
0x0#D7	TMR0H	High Byte of TMR0
0x0#D8	STATUS	PICmicro® MCU Processor Status Register

```
                                                  001 - 1:4 Prescaler
                                                  000 - 1:2 Prescaler
```

Bit   Function
7-6   Unused
4   N - Set when the Result has bit seven set
3   OV - Set when the Result overflows a two's complement
            number (bit seven changes polarity inadvertently)
2   Z - Set when the Least Significant eight bits of the
            Result are all Zero
1   DC - Set when the Lower Nybble of the
            addition.subtraction overflows
0   C - Set im Addition when the result is greater than
            0xFF. Reset in Subtraction when the result is
            negative

Address	Register	Description
0x0#D9	FSR2L	Low Byte of FSR Register 2
0x0#DA	FSR2H	High Byte of FSR Register 2
0x0#DB	PLUSW2	INDF2 Consisting of FSR2 + WREG for Address
0x0#DC	PREINC2	INDF2 With FSR2 Incremented Before Access
0x0#DD	POSTDEC2	INDF2 With FSR2 Decremented After Access
0x0#DE	POSTINC2	INDF2 With FSR2 Incremented After Access
0x0#DF	INDF2	Register Pointed to by FSR2
0x0#E0	BSR	Bank Select Register - Select Register Bank

Bit   Function
7-4   Unused
3-0   BSR3:BSR0, Bank Select Register Bits

Address	Register	Description
0x0#E1	FSR1L	Low Byte of FSR Register 1

Address	Register	Function/Bit Definition
0x0#E2	FSRIH	High Byte of FSR Register 1
0x0#E3	PLUSW1	INDF1 Consisting of FSR1 + WREG for Address
0x0#E4	PREINC1	INDF1 With FSR1 Incremented Before Access
0x0#E5	POSTDEC1	INDF1 With FSR1 Decremented After Access
0x0#E6	POSTINC1	INDF1 With FSR1 Incremented After Access
0x0#E7	INDF1	Register Pointed to by FSR1
0x0#E8	WREG	PICmicro® MCU Accumulator
0x0#E9	FSR0L	Low Byte of FSR Register 0
0x0#EA	FSR0H	High Byte of FSR Register 0
0x0#EB	PLUSW0	INDF0 Consisting of FSR0 + WREG for Address
0x0#EC	PREINC0	INDF0 With FSR0 Incremented Before Access
0x0#ED	POSTDEC0	INDF0 With FSR0 Decremented After Access
0x0#EE	POSTINC0	INDF0 With FSR0 Incremented After Access
0x0#EF	INDF0	Register Pointed to by FSR0
0x0#F0	INTCON3	Interrupt Control Register 3

Bit Function

7	INT2IP – INT2 External Interrupt Priority. Set for "High"
6	INT1IP – INT1 External Interrupt Priority. Set for "High"
5	Unused
4	INT2IE – Set to Enable External Int2
3	INT1IE – Set to Enable External Int1
2	Unused
1	INT2IF – Set when External Int2 Requested
0	INT1IF – Set when External Int1 Requested

| 0x0#F1 | INTCON2 | Interrupt Control Register 2 |

Bit Function

		7	_RBPU - Reset to Enable PORTB Pull Ups
		6	INTEDG0 - Set for External Int0 on Rising Edge
		5	INTEDG1 - Set for External Int1 on Rising Edge
		4	INTEDG2 - Set for External Int2 on Rising Edge
		3	Unused
		2	TMR0IP - High TMR0 Interrupt Request Priority when Set
		1	Unused
		0	RBIP - High PORTB Change Interrupt Request Priority when Set
0x0F2	INTCON		Interrupt Control Register
		Bit	Function
		7	GIE/GIEH - When Set, Enables all Interrupt Request Sources (unmasked sources when "IPEN" is reset)
		6	PEIE/GEIL - Enables all Low Priority Interrupt Request Sources when Set
		5	TMR0IE - When Set Enable TMR0 Interrupt Requests
		4	INT0IE - When Set Enable INT0 Interrupt Requests
		3	RBIE - When Set Enable PORTB Change
		2	TMR0IF - When Set TMR0 Interrupt Request Active
		1	INT0IF - When Set INT0 External Interrupt Request Active
		0	RBIF - When Set PORTB Change on Interrupt Request Active
0x0F3	PRODL		Low Byte of "Multiply" Instruction Product
0x0F4	PRODH		High Byte of "Multiply" Instruction Product
0x0F5	TABLAT		Table Read and Write Buffer
0x0F6	TBLPTRL		Low Byte of Program Memory Table Pointer
0x0F7	TBLPTRH		Middle Byte of Program Memory Table Pointer

Address	Register	Function/Bit Definition
0x0#F8	TBLPTRU	High Byte of Program Memory Table Pointer
0x0#F9	PCL	Low Byte of PICmicro® MCU Program Counter
0x0#FA	PCLATH	Latched Middle Byte of PICmicro® MCU Program Counter
0x0#FB	PCKATHU	Latched High Byte of PICmicro® MCU Program Counter
0x0#FC	STKPTR	Stack Pointer/Index
		Bit Function
		7 STKFUL - Bit Set when Stack is Full or Overflowed
		6 STKUNF - Bit Set when Stack Underflows
		5 Unused
		4-0 SP4:SP0 - Stack Pointer Location Bits
0x0#FD	TOSL	Low Byte Access to Top of Program Counter Stack
0x0#FE	TOSL	Middle Byte Access to Top of Program Counter Stack
0x0#FF	TOSU	High Byte Access to Top of Program Counter Stack

7

Built-In Hardware Features

Configuration Registers

The Configuration Register Fuses are responsible for specifying:

- Oscillators Mode Used
- Program Memory Protection
- reset parameters
- Watchdog Timer
- 16F87x debug mode

The configuration register fuses are unique to each PICmicro® MCU part number. The addresses for the different registers are shown in the table below:

```
Device Family       Configuration Register Address(es)
Low-End             0x0FFF
Mid-Range           0x02007
PIC17Cxx            0x0FE00-0x0FE07 Low Byte
                    0x0FE0F-0x0FE08 High Byte
PIC18Cxx            0x0300000-0x0300007
```

In each PICmicro® MCU's MPLAB device ".inc" files, there is a list of parameters for the different options. These parameters are used with the "__CONFIG" statement of an assembler file. I have a few recommendations about this that I will repeat throughout the book. For the PIC18Cxx, there are multiple "__CONFIG#" statements (where "#" is "0" through "7") and each statement is given a set of bits that can specify different functions.

The "__CONFIG" options are ANDed together to form a word that is programmed into the configuration addresses.

Oscillators

The basic oscillator options are as follows:

1. Internal Clocking

2. R/C Networks

3. Crystals

4. Ceramic Resonators

5. External Oscillators

The "Internal Clocking" option is available in many new PICmicro® MCUs and consists of a capacitor and variable resistor for the oscillator. The "OSCCAL" register shown in Fig. 7.1 is a register that is loaded with a "calibration value", which is provided by Microchip. This type of oscillator will have an accuracy of 1.5% or better while running at 4 MHz.

The second type of oscillator is the external "RC" oscillator in which a resistor/capacitor network provides the clocking for the PICmicro® MCU as is shown in Fig. 7.2.

The resistor capacitor charging/discharging voltage is buffered through a Schmidt Trigger noninverting buffer, which is used to enable or disable an N-Channel MOSFET transistor pull-down switch. The values for the resistor and capacitor can be found in the Microchip documentation.

Crystals and ceramic resonators use a similar connection scheme for operation. The crystal or ceramic resonator is wired into the circuit as shown in Fig. 7.3. The two capacitors are used to add impedance to the

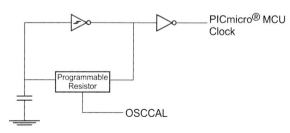

Figure 7.1 PICmicro® MCU Built-In Oscillator

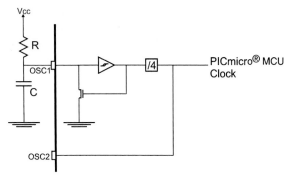

Figure 7.2 PICmicro® MCU RC Oscillator

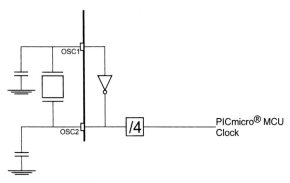

Figure 7.3 PICmicro® MCU Crystal Oscillator

crystal/resonator circuit and their values for reliable operation are specified by Microchip and their ranges are presented elsewhere in this book. As well, for best results, a "parallel circuit" crystal should be used.

There are three speed ranges defined for each device, with the speed specification defining the current output in the PICmicro® MCUs crystal/resonator oscillator circuit.

The speed ranges are

PICmicro® MCU Oscillator Frequency Ranges	
Range	Frequency
LP	0 – 200 kHz
XT	200 kHz – 4 MHz
HS	4 MHz – 20 MHz (or the Device Maximum)

These speed ranges are selected in the "configuration register".

Using the crystal or ceramic resonator, the OSC2 pin can be used to drive one CMOS input as is shown in Fig. 7.4.

The last type of oscillator is the external oscillator and is driven directly into the OSC1 pin as shown in Fig. 7.5.

The PIC18Cxx has seven different oscillator modes that are available to the application designer. Along with the standard modes described above, there is a PLL clock four time multiplier circuit available, which allows the PICmicro® MCU to run with one instruction cycle per clock cycle. There is also the ability to run from the TMR1 Clock, which can be a slow-speed, power saving clock option for the application.

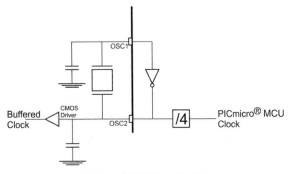

Figure 7.4 Buffered PICmicro® MCU Crystal Oscillator

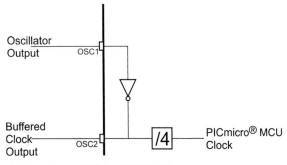

Figure 7.5 External PICmicro® MCU Oscillator

1. RC oscillator

2. LP oscillator

3. XT oscillator

4. HS oscillator

5. 4x HS oscillator

6. External oscillator

7. TMR1 clock

The "external oscillator" option will take in an external clock signal and output a one-quarter speed clock on OSC2 unless the OSC2 pin is to be used as "RA6" (like the RC oscillator mode and known as "ECIO"). The external oscillator will work for all data speeds from DC to 40MHz that the 18Cxx can run at.

When the TMR1 oscillator is enabled (by setting the "SCS" bit), execution moves over immediately to the TMR1 clock and the standard oscillator is shut down. This transition is very fast, with only eight TMR1 clock cycles lost before execution resumes with TMR1 as the clock source.

When transitioning from TMR1 to the standard oscillator, the oscillator is restarted with a 1,024 cycle delay for the clock to stabilize before resuming execution. The oscillator circuit in the PIC18Cxx appears in block diagram form as shown in Fig. 7.6.

Sleep

The PICmicro® MCU's "sleep" function and instruction provides the capability of "shutting down" the PICmicro® MCU by turning off the oscillator and making the PICmicro® MCU wait for reset ("_MCLR" or "WDT"),

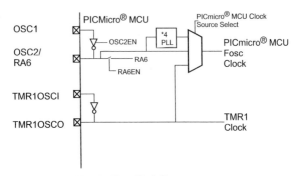

Figure 7.6 PIC18Cxx Oscillator Block Diagram

an external interrupt, or an externally clocked timer interrupt. Most internal timer interrupt requests are not able to become active because the PICmicro® MCU instruction clock driving the internal hardware clocks is shut off.

Entering "sleep" is accomplished by simply executing the "sleep" instruction. "Sleep" will be terminated by the following events:

Sleep Termination Events and Execution Resume Addresses	
Event	Execution Resume
_MCLR Reset	Reset Vector
WDT Reset	Reset Vector
External Interrupt	Next/Instructions or Interrupt Vector
TMR1 Interrupt	Next Instructions or Interrupt Vector

The interrupt requests can only wake the device if the appropriate "IE" bits are set. After the "sleep" instruction, the next instruction is always executed, even if the "GIE" bit is set. For this reason it is recommended that a "nop" be always placed after the sleep instruction to ensure no invalid instruction is executed before the interrupt handler:

Two Instruction Sequence used to Initiate "Sleep"

```
sleep
nop
```

The clock restart from "sleep" will be similar to that of a power-on reset, with the clock executing for 1,024 cycles before the "nop" instruction is executed ("Inst(PC + 1)" in the diagram below). This is shown in Fig. 7.7.

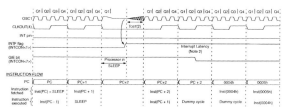

Note   1:   XT, HS or LP oscillator mode assumed.
           2:   Tost = 1024Tosc (drawing not to scale) This delay will not be there for RC osc mode.
           3:   GIE = '1' assumed. In this case after wake- up, the processor jumps to the interrupt routine. If GIE = '0', execution will continue in-line.
           4:   CLKOUT is not available in these osc modes, but shown here for timing reference.

Figure 7.7   Sleep Wave Form

Option Register

In the low-end devices, the option register is defined as:

```
┌─────────────────────────────────────────────────────────┐
│ Low-End PICmicro® MCU "OPTION" Register Definition       │
├─────────────────────────────────────────────────────────┤
│ Bit   Label/Function                                     │
│  7    _GPWU - Enable pull-up wakeup on pin change   ]    │
│  6    _GPPU - Enable I/O PORTB Weak Pull-ups        ]    │
│               Device Specific                            │
│  5    TOCS - TMRO clock source select                    │
│               1 - Tock1 pin                              │
│               0 - Instruction clock                      │
│  4    TOSE - TMRO Increment Source Edge Select           │
│               1 - High to Low on Tock1 Pin               │
│               0 - Low to High on Tock1 Pin               │
│  3    PSA - Prescaler Assignment Bit                     │
│               1 - Prescaler Assigned to Watchdog Timer   │
│               0 - Prescaler Assigned to TMRO             │
│ 2-0   PS2-PS0 - Prescaler Rate Select                    │
│               000 - 1:1                                  │
│               001 - 1:2                                  │
│               010 - 1:4                                  │
│               011 - 1:8                                  │
│               100 - 1:16                                 │
│               101 - 1:32                                 │
│               110 - 1:64                                 │
│               111 - 1:128                                │
└─────────────────────────────────────────────────────────┘
```

Updating the OPTION register in the low-end is accomplished by the "option" instruction, which moves the contents of "w" into the OPTION_REG (which is the MPLAB label for the option register).

The mid-range devices option register is quite similar, but does not have any device specific bits:

```
Mid-Range PICmicro® MCU "OPTION" Register Definition

Bit   Label/Function
 7    _RBPU - Enable PORTB Weak Pull-ups
          1 - Pull-ups Disabled
          0 - Pull-ups Enabled
 6    INTEDG - Interrupt Request On:
          1 - low to high on RB0/INT
          0 - high to low on RB0/INT
 5    TOCS - TMR0 clock source select
          1 - Tock1 Pin
          0 - Instruction Clock
 4    TOSE - TMR0 Update Edge Select
          1 - Increment on High to Low
          0 - Increment on Low to High
 3    PSA - Prescaler Assignment Bit
          1 - Prescaler Assigned to Watchdog Timer
          0 - Prescaler Assigned to TMR0
2-0   PS2-PS0 - prescaler rate select
          000 - 1:1
          001 - 1:2
          010 - 1:4
          011 - 1:8
          100 - 1:16
          101 - 1:32
          110 - 1:64
          111 - 1:128
```

The 17Cxx PICmicro® MCU's do not have an option register as many of the functions continued by option are either not present (such as the prescaler and PORTB weak pull-ups) or are provided in other registers. The 18Cxx provides a mid-range "compatible" option register, but it is not at the same address as the mid-range devices and cannot be written to using an "option" instruction.

Input/Output Ports and TRIS Registers

The block diagram of a "typical" PICmicro® MCU I/O pin is shown in Fig. 7.8. Each register "port" is made up of a

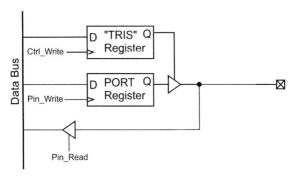

Figure 7.8 Standard PICmicro® MCU I/O Pin Block Diagram

number of these circuits, one for each I/O bit. The convention used for accessing I/O pins is

```
R%#
```

where "%" is the port letter (port A, port B, etc.) and "#" is the bit number of the port.

The TRIS ("TRI–state buffer enable") register is used to control the output capabilities of the I/O pin. When the register is loaded with a "1" (which is the power-up default), the pin is input only (or in "input mode"), with the tristate buffer disabled and not driving the pin. When a "0" is loaded into a pin's TRIS bit, the tristate buffer is enabled ("output mode") and the value that is in the "data out" register is driven onto the pin.

The use of the "tris" instruction is not recommended in the mid-range PICmicro® MCU as the instruction can only access PORTA, PORTB, and PORTC. PORTD and PORTE cannot be controlled by the "tris" instruction.

The recommended way of accessing the mid-range PICmicro® MCU's TRIS registers is to change the "RP0" bit of the STATUS register and read or write the register directly as is shown below:

```
bsf      STATUS, RP0
movlw    NewTRISA
movwf    TRISA ^ 0x080
bcf      STATUS, RP0
```

Note, Pin 4 of PORTA ("RA4") in the mid-range PICmicro® MCUs is an "open drain" only output and its design is shown in Fig. 7.9. This pin cannot source a positive voltage out unless it is pulled up.

The weak pull-up on the PORTB pins is enabled by the "_RPBU" bit of the OPTION register and Is enabled when this bit is reset and the bit itself is set for out-

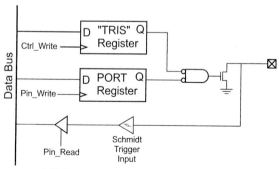

Figure 7.9 PORTA Bit 4 I/O Pin Block Diagram

put. The "weak" pull-up is approximately 50 k and can simplify button inputs, eliminating the need for an external pull-up resistor. The port B pin block diagram is shown in Fig. 7.10. When the built-in oscillator is selected by the "_IntRC_OSC" parameter of the "__CONFIG" statement in your source file the pins used for the oscillator are available for IO. When the PICmicro® MCU is programmed, a value for the "calibration register" ("OSCCAL") has to be inserted. By convention, a

```
movlw OSCCAL-value
```

instruction is put in at the reset address and then at address zero (when the program counter overflows), this value is saved into the OSCCAL register using a

```
movwf OSCCAL
```

instruction.

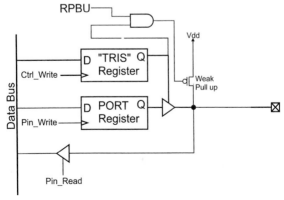

Figure 7.10 PICmicro® MCU PORTB I/O Pin Block Diagram

```
__CONFIG _MCLRE_OFF & _IntRC_OSC ;  Add Application
                                       Specific
                                 ;  "CP" and "WDT"
                                       parameters
  org 0
  movf   OSCCAL
  movlw  0x0FF ^ (1 << T0CS)
  option
; All I/O pins are NOW Available and Internal 4 MHz
   Clock is Running
; - Start Application
```

Watchdog Timer

The Watchdog Timer is an 18 msec (approximately) RC delay, which will reset the PICmicro® MCU if it times out. Normally in an application, it is reset before timing out by executing a "clrwdt" instruction. The block diagram of the WDT is shown in Fig. 7.11.

TMR0

TMR0 is an 8-bit incrementing counter that can be "preset" (loaded) by application code with a specific

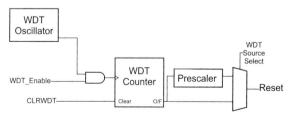

Figure 7.11 PICmicro® MCU Watchdog Timer Block Diagram

value. The counter can either be clocked by an external source or by the instruction clock. Each TMR0 input is matched to two instruction clocks for "synchronization." This feature limits the maximum speed of the timer to one half the instruction clock speed. The TMR0 block diagram is shown in Fig. 7.12. The "TOCS" and "TOCE" bits are used to select the clock source and the clock edge, which increments TMR0 (rising or falling edge). These bits are located in the "OPTION" register.

TMR0 can be driven by external devices through the "T0CKI" pin. The "T0CKI" pin is dedicated to this function in the low-end devices (although in the 12C5xx and 16C505 PICmicro® MCUs the pin can be used for digital I/O). In the other PICmicro® MCU architectures, the pin can also be used to provide digital I/O. When a clock is driven into the TMR0 input, the input is buffered by an internal "Schmidt Trigger" to help minimize noise-related problems with the input.

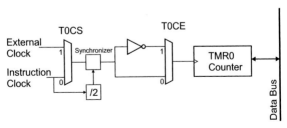

Figure 7.12 TMR0 Block Diagram

TMR0 in the mid-range PIC17Cxx and PIC18Cxx can be used to request interrupts when it "overflows" to 0x000 from 0x0FF.

Input to TMR0 can be made with and without the "prescaler", which provides a "divide by" feature to the TMR0 input. For the low-end and mid-range PICmicro® MCU's, TMR0 is located at register address 0x001. The contents of TMR0 can be read from and written to directly.

Delays (time from which TMR0 is initialized until it overflows) are calculated by using the formula:

```
TMR0 Initial = 256 - (Delay Time * Clock Frequency / 8)
```

Prescaler

The "prescaler" is a power-of-two counter that can be selected for use with either the Watchdog Timer or TMR0. Its purpose is to divide the incoming clock signals by a software selectable power-of-two value to allow the 8-bit TMR0 to time longer events or increase the watchdog delay from 18 msecs to 2.3 seconds (Fig. 7.13).

The prescaler's operation is controlled by the four "PSA" bits in the OPTION register. "PSA" selects whether the watchdog timer uses the prescaler (when PSA is "set") or TMR uses the prescaler (when PSA is "reset"). Note that the prescaler has to be assigned to either the watchdog timer or TMR0. Both functions are able to execute with no prescaler or with the prescaler's delay count set to one.

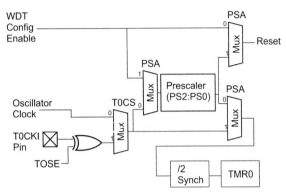

Figure 7.13 PICmicro® MCU Prescaler Circuit

Prescaler Values to Delays	
PS2 - PS0	prescaler delay
000	1 cycle
001	2 cycles
010	4 cycles
011	8 cycles
100	16 cycles
101	32 cycles
110	64 cycles
111	128 cycles

TMR1

TMR1 is 16 bits and can have four different inputs as is shown in Fig. 7.14.

To access TMR1 data, the "TMR1L" and "TMR1H" registers are read and written. If the TMR1 value registers are written, the TMR1 prescaler is reset. A TMR1 inter-

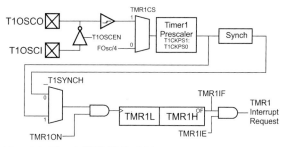

Figure 7.14 Timer1 ("TMR1") Block Diagram

rupt request ("TMR1IF") is made when TMR1 overflows and the TMR1IE bit is set.

TMR1IF and TMR1IE are normally located in the "PIR" and "PIE" registers. To request an interrupt, along with TMR1IE and "GIE" being set, the INTCON "PIE" bit must also be set.

To control the operation of TMR1, the T1CON register is accessed with its bits defined as:

T1CON Bit Definition	
Bit	Description
7-6	Unused
5-4	T1CPS1:T1CPS0 - Select TMR1 Prescaler Value
	11 - 1:8 prescaler
	10 - 1:4 prescaler
	01 - 1:2 prescaler
	00 - 1:1 prescaler
3	T1OSLEN - Set to Enable TMR1's built in Oscillator
2	T1SYNCH - when TMR1CS reset the TMR1 clock is synchronized to the Instruction Clock
1	TMR1CS - When Set, External Clock is Used
0	TMR1ON - When Set, TMR1 is Enabled

The external oscillator is designed for fairly low-speed real-time clock applications. Normally a 32.768 kHz watch crystal is used along with two 33 pF capacitors. 100 kHz or 200 kHz crystals can be used with TMR1, but the capacitance required for the circuit changes to 15 pF. The TMR1 oscillator circuit is shown in Fig. 7.15.

In the PIC18Cxx devices, TMR1 can be specified as the processor clock to allow low-speed, low-power application execution without putting the PICmicro® MCU to "sleep".

The TMR1 prescaler allows 24-bit instruction cycle delay values to be used with TMR1. These delays can either be a constant value or an "overflow", similar to TMR0. To calculate a delay, use the formula:

```
Delay = (65,536 - TMR1Init)
                          × prescaler / T1frequency
```

where the "T1frequency" can be the instruction clock, TMR1 oscillator or an external clock driving TMR1.

```
TMR1Init = 65,536 -
                   (Delay × T1Frequency / prescaler)
```

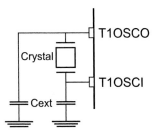

Figure 7.15 Timer1 ("TMR1") Oscillator Circuit

TMR2

TMR2 (Fig. 7.16) is used as a recurring event timer. When it is used with the CCP module, it is used to provide a Pulse Width Modulated timebase frequency. In normal operations, it can be used to create a 16-bit instruction cycle delay.

TMR2 is continually compared against the value in "PR2". When the contents of TMR2 and PR2 match, TMR2 is reset, the event is passed to the CCP as "TMR2 Reset". If the TMR2 is to be used to produce a delay within the application, a postscaler is incremented when TMR2 overflows and eventually passes an interrupt request to the processor.

TMR2 is controlled by the T2CON register, which is defined as:

T2CON Bit Definition	
Bit	Description
7	Unused
6–5	TOUTPS3:TOUTPS0 – TMR2 Postscaler Select
	1111 – 16:1 Postscaler
	1110 – 15:1 Postscaler
	:
	0000 – 1:1 Postscaler

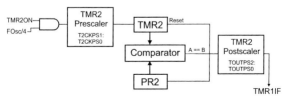

Figure 7.16 Timer2 ("TMR2") Block Diagram

T2CON Bit Definition (*Continued*)

```
2       TMR2ON - When Set, TMR2 Prescaler is Enabled
1-0     T2CKPS1:T2CKPS0 - TMR2 Prescaler Select
        1x - 16:1 prescaler
        01 -  4:1 prescaler
        00 -  1:1 prescaler
```

The "TMR2" register can be read or written at any time with the caution that writes may cause the prescaler to be zeroed. Updates to TMR2 do not reset the TMR2 prescaler. The timer itself is not synchronized with the instruction clock because it can only be used with the instruction clock.

PR2 contains the reset, or count up to, value. The delay before reset is defined as:

```
Delay = prescaler × (PR2 + 1) / ( Fosc / 4)
```

If "PR2" is equal to zero, the delay is:

```
Delay = (prescaler × 256) / ( Fosc / 4)
```

Interrupts use the "TMR2IE" and "TMR2IF" bits that are similar to the corresponding bits in TMR1. These bits are located in the "PIR" and "PIE" registers. Because of the exact interrupt frequency, TMR2 is well suited for applications that provide "bit banging" functions like asynchronous serial communications or Pulse Width Modulated signal outputs.

Compare/Capture/PWM (CCP) Module

Included with TMR1 and TMR2 is a control register and a set of logic functions (known as the "CCP"), which enhances the operation of the timers and can simplify applications. This hardware may be provided singly or in pairs, which allows multiple functions to execute at the same time. If there are two CCP modules built into the PICmicro® MCU, then one is known as "CCP1" and the other as "CCP2". In the case where there are two CCP modules built-in, then all the registers are identified with the "CCP1" or "CCP2" prefix.

The CCP hardware is controlled by the "CCP1CON" (or "CCP2CON") register, which is defined as:

```
CCPxCON Bit Definition

Bit    Function
7-6    Unused
5-4    DC1B1 :DC1B0 - CEPST significant 2 bits of
       the PWM compare value.
3-0    CCP1M3 :CCP1M0 - CCP module operating mode.
       11xx - PWM Mode
       1011 - Compare Mode - Trigger Special Event
       1010 - Compare Mode - Generate Software
                Interrupt
       1001 - Compare Mode - on Match CCP pin low
       1000 - Compare Mode - on Match CCP pin high
       0111 - Capture on every 16th rising edge
       0110 - Capture on every 4th rising edge
       0101 - Capture on every rising edge
       0100 - Capture on every falling edge
       00xx - CCP off
```

"Capture mode" loads the CCPR registers ("CCPR14", "CCPR1C", "CCPR2H", and "CCPR2L") according to the mode the CCP register is set in. This function is shown in Fig. 7.17 and shows that the current TMR1 value is saved when the specified compare condition is met.

Before enabling the capture mode, TMR1 must be enabled (usually running with the PICmicro® MCU clock). The "edge detect" circuit in Fig. 7.17 is a four-to-one multiplexor, which chooses between the prescaled ris-

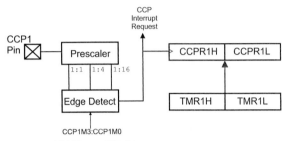

Figure 7.17 CCP "Capture" Module

ing edge input or a falling edge input and passes the selected edge to latch the current TMR1 value and optionally request an interrupt.

In capture mode, TMR1 is running continuously and is loaded when the condition on the CCPx pin matches the condition specified by the CCPxMS:CCPxM0 bits. When a capture occurs, then an interrupt request is made. This interrupt request should be acknowledged and the contents of CCPRxH and CCPRxL saved to avoid having them written over and the value in them lost.

"Compare" mode changes the state of the CCPx pin of the PICmicro® MCU when the contents of TMR1 match the value in the CCPRxH and CCPRxL registers as shown in Fig. 7.18. This mode is used to trigger or control external hardware after a specific delay.

"PWM" CCP mode outputs a PWM signal using the TMR2 reset at a specific value capability. The block diagram of PWM mode is shown in Fig. 7.19. The mode is a combination of the normal execution of TMR2 and capture mode; the standard TMR2 provides the PWM period while the compare control provides the "on" time specification.

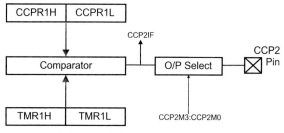

Figure 7.18 CCP "Compare" Module

When the PWM circuit executes, TMR2 counts until
its most significant 8 bits are equal to the contents of
PR2. When TMR2 equals PR2, TMR2 is reset to zero and
the CCPx pin is set "high". TMR2 is run in a 10 bit mode

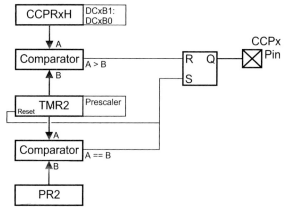

Figure 7.19 CCP "PWM" Module

(the 4:1 prescaler is enabled before PWM operation). This 10-bit value is then compared to a program value in CCPRxM (along with the two DCxBx bits in CCPxCON) and when they match, the CCPx output pin is reset low.

Code to Setup up a 65% Duty Cycle PWM

```
   movlw  199
   movwf  PR2                      ; Set up TMR2
                                       Operation
   movlw  (1 << TMR2 on) +1
   movwf  T2CON                    ; Start it Running
                                   ;  with a 50 msec
                                   ;  Period
   movlw  130                      ; 65% of the Period
   movwf  CCPRxH
   movlw  (1 << DC xB1) +0x00F
   movwf  CCPxCON                  ; Start PWM
;  PWM is operating
```

The table below gives the fractional DCxBX bit values:

CCP DCxBX Bit Definition

Fraction	DCxB1:DCxB0
0.00	00
0.25	01
0.50	10
0.75	11

USART Module

There are three modules to the USART, the clock generator, the serial data transmission unit and the serial data reception unit. The two serial I/O units require the clock generator for shifting data out at the write interval. The clock generator's block diagram is Fig. 7.20.

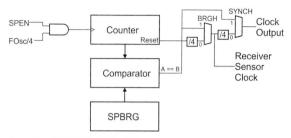

Figure 7.20 USART Clock Block Diagram

In the clock generator circuit, the "SPBRG" register is used as a comparison value for the counter. When the counter is equal to the "SPBRG" register's value, a clock "tick" output is made and the counter is reset. The counter operation is gated and controlled by the "SPEN" ("serial port enable") bit along with the "synch" (which selects whether the port is in synchronous or asynchronous mode) and "BRGH" which selects the data rate.

For asynchronous operation, the data speed is specified by the formula:

```
Data Rate = Fosc / (16 × (4 ** (1 - BRGH))
                                    × (SPBRG + 1))
```

This formula can be rearranged so that the SPBRG value can be derived from the desired data rate:

```
SPBRG = Fosc / (Data Rate × 16 × (4**(1 - BRGH)) - 1
```

The transmission unit of the USART can send 8 or 9 bits in a clocked (synchronous) or unclocked (asynchronous) manner. Data transmission is initiated by sending a byte to the "TXREG" register. The block diagram of the hardware is shown in Fig. 7.21.

The transmit hold register can be loaded with a new value to be sent immediately following the passing of the byte in the "Transmit shift register". This single buffering of the data allows data to be sent continuously without the software polling the TXREG to find out when is the correct time to send out another byte. USART transmit interrupt requests are made when the TX holding register is empty. This feature is available for both synchronous and asynchronous transmission modes.

The USART receive unit is the most complex of the USART's three parts. This complexity comes from the need for it to determine whether or not the incoming asynchronous data is valid or not using the "Pin Buffer and Control" unit built into the USART receive pin. The block diagram for the USART's receiver is shown in Fig. 7.22.

If the port is in synchronous mode, data is shifted in

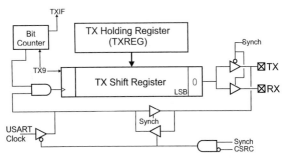

Figure 7.21 USART Transmit Hardware Block Diagram

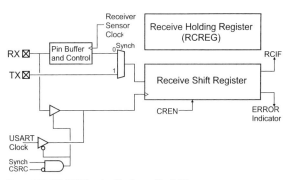

Figure 7.22 USART Receive Hardware Block Diagram

either according to the USART's clock or using an external device's clock.

Like the TX unit, the RX unit has a "holding register", so if data is not immediately processed, and an incoming byte is received, the data will not be lost. But, if the data is not picked up by the time the next byte has been received, then an "overrun" error will occur. Another type of error is the "framing error", which is set if the "stop" bit of the incoming NRZ packet is not zero. These errors are recorded in the "RCSTA" (receiver status) register and have to be reset by software.

For asynchronous data, the "Receiver Sensor Clock" is used to provide a polling clock for the incoming data. This sixteen time data rate clock's input into the "Pin Buffer and Control" unit provides a polling clock for the hardware. When the input data line is low for three Receive Sensor Clock periods, data is then read in from the "middle" of the next bit as is shown in Fig. 7.23. When data is being received, the line is polled three

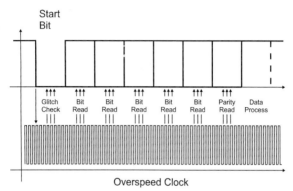

Figure 7.23 Reading an Asynch Data Packet

times and the majority of states read is determined to be the correct data value. This repeats for the 8 or 9 bits of data with the stop bit being the final check.

In some PICmicro® MCUs, the USART receive unit can also be used to receive two asynchronous bytes in the format "Data:Address", where "Address" is a byte destined for a specific device on a bus. When the "AD-DEN" bit of the "RCSTA" register is set, no interrupts will be requested until both the address and data bytes have been received. To distinguish between the bytes, the ninth address bit is set (while the ninth bit of data packets are reset). When this interrupt request is received, the interrupt handler checks the device address for its value before responding to the data byte.

To control the USART, two registers are used explicitly. The "TXSTA" ("transmitter status") register is located at address 0x098 in the mid-range PICmicro® MCUs and has the bit definitions:

```
USART TXSTA Bit Definition

Bit    Definition
 7     CSRC - Clock Source Select used in
          Synchronous Mode. When Set, the USART Clock
          Generator is Used
 6     TX9 - Set to Enable nine bit Serial I/O
 5     TXEN - Set to Enable Data Transmission
 4     SYNC - Set to Enable Synchronous Transmission
 3     Unused
 2     BRGH - Used in Asynchronous Mode to Enable
          Fast Data Transmission. It is Recommended to
          keep this bit Reset
 1     TRMT - Set if the Transmission Shift Register
          is Empty
 0     TXD - Nine bit of Transmitted Data
```

The "SPBRG" register is usually at address 0x099 for the mid-range PICmicro® MCUs.

The "RCSTA" (receiver status) register is at address 0x018 in the mid-range PICmicro® MCUs and is defined as:

```
USART RCSTA Bit Definition

Bit    Definition
 7     SPEN - Set to Enable the USART
 6     RX9 - Set to Enable nine bit USART Receive
 5     SREN - Set to Enable Single Byte Synchronous
          Data Receive. Reset when data has been
          received
 4     CREN - Set to Enable Continuous Receive
 3     ADDEN - Set to Receive Data:Address
          Information. May be unused in many PICmicro®
          MCU Part Numbers
 2     FERR - "Framing Error" bit
 1     OERR - "Overrun Error" bit
 0     RX9D - Received ninth bit
```

The TXREG is normally at address 0x019 and RCREG is normally at address 0x01A for the mid-range PICmicro® MCUs. The TXIF, TXIE, RCIE, and RCIF bits are in different interrupt enable request registers and bit numbers are specific to the part being used.

To set up asynchronous serial communication transmit, the following code is used:

```
Code to set up USART Asynchronous Serial
Transmission

   bsf    STATUS, RP0
   bcf    TXSTA, SYNCH     ; Not in Synchronous mode
   bcf    TXSTA, BRGH      ; BRGH =0

   movlw  DataRate         ; Set USART Data Rate
   movwf  SPBRG

   bcf    STATUS, RP0      ; Enable serial port
   bsf    RCSTA ^ 0x080, SPEN
   bsf    STATUS, RP0
   bcf    TXSTA, TX9       ; Only 8 bits to send
   bsf    TXSTA, TXEN      ; Enable Data Transmit
   bcf    STATUS, RP0
```

To send a byte in "w", use the code:

```
USART Asynchronous Serial Transmission Byte Send
Code

   btfss  TXSTA, TRMT
   goto   $ - 1            ; Wait for Holding Register
                          ;  to become Free/Empty
   movwf  TXREG           ; Load Holding Register

                          ; If Transmit Shift Register
                          ; is Empty, byte will be sent
```

To set up an asynchronous read, the following code is used:

```
bsf     STATUS, RPO
bcf     TXSTA, SYNCH          ; Want Asynch
                                 Communications
bcf     TXSTA, BRGH           ; Low Speed Clock
movlw   DataRate              ; Set Data Rate
movwf   SPBRG
bsf     RCSTA ^ 0x080, SPEN   ; Enable Serial Port
bcf     TCSTA ^ 0x080, RX9    ; Eight Bits to
                                 Receive
```

To receive data, use the code:

```
btfss  PIR1, RXIF    ; Wait for a Character to be
 goto $ - 1          ;  Received
movf   RCREG, w      ; Get the byte Received
bcf    RXIF          ; Reset the RX byte Interrupt
                     ;  Request Flag
```

SSP Module

The Synchronous Serial Protocol Module is used to send and receive data serially using a synchronous (with a clock) protocol like the data stream shown in Fig. 7.24.

SPI operation

SPI is an 8-bit synchronous serial protocol that uses three data bits to interface to external devices. Data is clocked out, with the most significant bit first, on rising or falling edges of the clock. The clock itself is generated

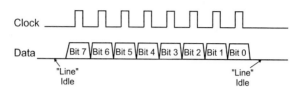

Figure 7.24 SPI Synchronous Serial Data Waveform

within the PICmicro® MCU ("master mode"), or it is provided by an external device and used by the PICmicro® MCU ("slave mode") to clock out the data. The SPI data stream looks like Fig. 7.24.

The clock can be "positive" as shown in Fig. 7.24 with a "0" "idle" or negative (high "line idle") with a "1" idle and the clock pulsing to "0" and back again. The Data receive latch is generally on the return to idle state transition.

The "BSSP" module is the "Basic SSP" module and provides data pulling on the return to idle clock edge. The original SSP module provides the ability to vary when data is output and read. Controlling the operation of the different SSP modules is the "SSPCON" register.

SSP/BSSP SSPCON Bit Definition

```
Bit  Function
  7  WCOL - Write collision, set when new byte
        written to SSPBUF while transfer is taking
        place
  6  SSPOV - Receive Overflow, indicates that the
        unread byte is SSPBUF Over written while in
        SPI slave mode
  5  SSPEN - Set to enable the SSP module
  4  CKP - Clock polarity select, set to have a
        high idle
```

SSP/BSSP SSPCON Bit Definition (*Continued*)

```
3-0   SSPM3:SSPM0 SPI mode select
        1xxx - I2C and reserved modes
        011x - I2C slave modes
        0101 - SPI slave mode, clock = SCK pin,
               _SS not used
        0100 - SPI slave mode, clock = SCK pin,
               _SS enabled
        0011 - SPI master mode, TMR2 clock used
        0010 - SPI master mode, INSCK/16
        0001 - SPI master mode, INSCK/4
        0000 - SPI master mode, INSCK
```

The block diagram for the SSP module is shown in Fig 7.25.

In master mode, when a byte is written to SSPBUF, an 8-bit, most-significant-bit first data transfer process is initiated. The status of the transfer can be checked by the SSPSTAT register "BF" flag; the SSPSTAT register is defined as:

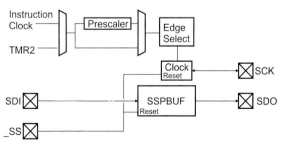

Figure 7.25 SSP SPI Module

SSP/BSSP SSPSTAT Bit Definition
Bit Function

7   SMP - Set to have data sampled after active to
      idle transition, reset to sample at active to
      idle transition, not available in BSSP

6   CKE - Set to TX data on idle to active
      transition, else TX data on active to idle
      transition, not available in BSSP

5   D/_A - Used by I2C
4   P - Used by I2C
3   S - Used by I2C
2   R/_W - Used by I2C
1   UA - Used by I2C
0   BF - Busy flag, reset while SPI operation   active

The SSP SPI transfers can be used for single byte synchronous serial transmits of receivers with serial devices. Figure 7.26 shows the circuit to transmit a byte to

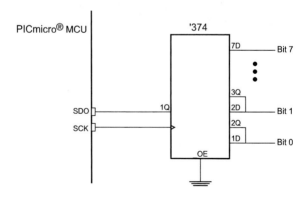

Figure 7.26 SSP SPI Module Used to Shift Data Out

a 74LS374 wired as a serial in/parallel out shift register. Figure 7.27 shows a 74LS374 being used with a 74LS244 as a synchronous parallel in/serial out register. Both of these operations are initiated by a write to SSPBUF.

```
bsf     IOpin                   ; Want to Latch Data
                                  into the '374
bcf     SCK
bsf     STATUS, RP0
bcf     IOpin
bcf     SCK
bcf     STATUS, RP0
bsf     SCK                     ; Latch the Data into
                                  the '374
bcf     SCK
bcf     IOpin                   ; Disable '244
                                  output, Enable '374
movlw   (I << SMP) + (I << CKE)
movwf   SSPSTAT                 ; Set up the SSP
                                  Shift In
movlw   (I << SSPEN) + (I << CKP) +0x000
```

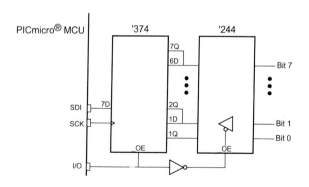

Figure 7.27 SSP SPI Module Used to Shift Data In

```
movwf    SSPCON
movf     TXData, f        ; Load the Byte to
                          ;   Send
movwf    SSPBUF           ; Start Data Transfer
btfss    SSPTAT, BF
 goto    $ - 1            ; Wait for Data
                          ;   Receive to
                          ;   Complete
                          ; Data Ready in
                          ;   SSPBUF when
                          ;   Execution
                          ;   Here
bcf      SSPCON, $SPEN    ; Turn off SSP
```

When using the SSP, the data rate can either be selected as a multiple of the executing clock or use the
TMR2 overflow output. The actual timing is dependent
on the hardware the PICmicro® MCU SSP master is
communicating with.

When in slave mode, along with an external clock
being provided, there is a transmit reset pin known
as "_SS". When this pin is asserted high, the SSP output
is stopped (the SDO TRIS bit is changed to input mode)
and the SSP is reset with a count of zero. When the bit
is reset, the clock will start up again, the original
most significant bit is reset, followed by the remaining 7
bits.

I2C operation

The enhanced MSSP will be designed into all new devices that have the SSP module. In this section, the single master I2C interface is focused on.

There are five registers that are accessed for MSSP I2C
operation, they are the SSP control registers ("SSPCON"
and "SSPCON2"), the SSP status register ("SSPSTAT"),

the SSP receive/transmit register ("SSPBUF") and the
SSP address register ("SSPADD"). These registers are
available in the SSP and BSSP, but are slightly different
for the MSSP.

The MSSP control registers are defined as:

MSSP SSPCON Bit Definition

Bit	Function
7	WCOL - Write collision, set when new byte written to SSPBUF while transfer is taking place
6	SSPOV - Receive Overflow, indicates that the unread byte is SSPBUF over written
5	SSPEN - Set to enable the SSP module
4	CKP - In I2C Modes, if bit is reset, the I2C "SCL" Clock Line is Low. Keep this bit set.
3-0	SSPM3:SSPM0 SPI mode select

```
      1111 - I2C 10 Bit Master Mode/Start and
        Stop Bit Interrupts
      1110 - I2C 7 Bit Master Mode/Start and
        Stop Bit Interrupts
      1101 - Reserved
      1100 - Reserved
      1011 - I2C Master Mode with Slave Idle
      1010 - Reserved
      1001 - Reserved
      1000 - I2C Master Mode with SSPADD Clock
        Definition
      0111 - I2C Slave Mode, 10 Bit Address
      0110 - I2C Slave Mode, 7 Bit Address
      0101 - SPI slave mode, clock = SCK pin,
        _SS not used
      0100 - SPI slave mode, clock = SCK pin,
        _SS enabled
      0011 - SPI master mode, TMR2 clock used
      0010 - SPI master mode, INSCK/16
      0001 - SPI master mode, INSCK/4
      0000 - SPI master mode, INSCK
```

```
MSSP SSPCON2 Bit Definition

Bit     Function
 7      GCEN - Enable Interrupt when "General Call
          Address" (0x0000) is Received
 6      ACKSTAT - Received Acknowledge Status. Set
          when Acknowledge was Received
 5      ACKDT - Acknowledge Value Driven out on Data
          Write
 4      ACKEN - Acknowledge Sequence Enable Bit which
          when Set will Initiate an Acknowledge
          sequence on SDA/SCL. Cleared by Hardware
 3      RCEN - I2C Receive Enable Bit
 2      PEN - Stop Condition Initiate Bit. When Set,
          Stop Condition on SDA/SCL. Cleared by
          Hardware
 1      RSEN - Set to Initiate the Repeated Start
          Condition on SDA/SCL. Cleared by Hardware
 0      SEN - When Set, a Start Condition is
          Initiated on the SDA/SCL. Cleared by
          hardware.
```

The status of the transfer can be checked by the SSPSTAT register "BF" flag; the SSPSTAT register is defined as:

```
MSSP SSPSTAT Bit Definition

Bit     Function
 7      SMP - Set to have data sampled after active
          to idle transition, reset to sample at
          active to idle transition, not available
          in BSSP
 6      CKE - Set to TX data on idle to active
          transition, else TX data on active to idle
          transition, not available in BSSP
 5      D/_A - Used by I2C
 4      P - Used by I2C
 3      S - Used by I2C
 2      R/_W - Used by I2C
 1      UA - Used by I2C
 0      BF - Busy flag, reset while SPI operation
          active
```

I2C connections between the PICmicro® MCU's I2C "SDA" (data) and "SCL" (clock) pins is very simple with just a Pull Up on each line as shown in Fig. 7.28. 1K resistors are recommended for 400 kHz data transfers and a 10K for 100 kHz data rates. Note that before any of the I2C modes are to be used, the "TRIS" bits of the respective "SDA" and "SCL" pins *must* be in input mode. Unlike many of the other built-in advanced I/O functions, MSSP does not control the TRIS bits. Not having the TRIS bits in input mode will not allow the I2C functions to operate.

In "Master Mode", the PICmicro® MCU is responsible for driving the clock ("SCL") line for the I2C network. This is done by selecting one of the SPI Master Modes and loading the SSPADD register with a value to provide a data rate that is defined by the formula:

```
I2C Data Rate = Fosc / (4 * (SSPADD + 1))
```

This can be rearranged to:

```
SSPADD = (Fosc / (4 * I2C Data Rate)) − 1
```

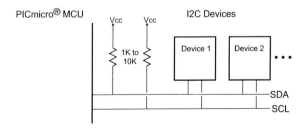

Figure 7.28 I2C Connection to PICmicro® MCU

To send data from the PICmicro® MCU to an I2C device using the MSSP, the following steps must be taken:

1. The SDA/SCL lines MUST be put into "Input Mode" (i.e., their respective "TRIS" bits must be set).

2. I2C Master Mode is enabled. This is accomplished by setting the "SSPEN" bit of SSPCON and writing 0b01000 to the SSPM3:SSPM0 bits of the SSPCON register.

3. A "Start Condition" is initiated by setting the "SEN" bit of SSPCON2. This bit is then polled until it is reset.

4. SSPBUF is loaded with the address of the device to access. Note that for many I2C devices, the least significant bit transmitted is the "Read/Write" bit. The "R/_W" bit of SSPSTAT is polled until it is reset (which indicates the transmit has been completed).

5. The ACK bit from the receiving device is checked by reading the "ACKDT" bit of the SSPCON2 register.

6. SSPBUF is loaded with the first 8 bits of data or a secondary address that is within the device being accessed. The "R/_W" bit of SSPSTAT is polled until it is reset.

7. The ACK bit from the receiving device is checked by reading the "ACKDT" bit of the SSPCON2 register.

8. A new "Start Condition" may have to be initiated between the first and subsequent data bytes. This is initiated by setting the "SEN" bit of SSPCON2. This bit is then polled until it is reset.

9. Operations six through eight are repeated until all data is sent or a "NACK" (negative Acknowledge) is received from the receiving device.

10. A "Stop Condition" is initiated by setting the "PEN" bit of SSPCON2. This bit is then polled until it is reset.

This sequence of operations is shown in Fig. 7.29. Note that in Fig. 7.29, the "SSPIF" interrupt request flag operation is shown. In the sequence above, I avoid interrupts, but the "SSPIF" bit can be used to either request an interrupt or to avoid the need to poll different bits to wait for the various operations to complete.

To receive data from a device requires a similar set of operations with the only difference being that after the address byte(s) have been sent, the MSSP is configured to receive data when the transfer is initiated:

1. The SDA/SCL lines MUST be put into "Input Mode" (i.e., their respective "TRIS" bits must be set).

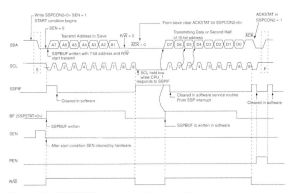

Figure 7.29 MSSP 12C data address/transmission

2. I2C Master Mode is enabled. This is accomplished by setting the "SSPEN" bit of SSPCON and writing 0b01000 to the SSPM3:SSPM0 bits of the SSPCON register.

3. A "Start Condition" is initiated by setting the "SEN" bit of SSPCON2. This bit is then polled until it is reset.

4. SSPBUF is loaded with the address of the device to access. Note that for many I2C devices, the least significant bit transmitted is the "Read/Write" bit. The "R/_W" bit of SSPSTAT is polled until it is reset (which indicates the transmit has been completed).

5. The ACK bit from the receiving device is checked by reading the "ACKDT" bit of the SSPCON2 register.

6. SSPBUF is optionally loaded with the secondary address within the device being read from. The "R/_W" bit of SSPSTAT is polled until it is reset.

7. If a secondary address was written to the device being read from, reading the "ACKDT" bit of the SSPCON2 register checks the ACK bit from the receiving device.

8. A new "Start Condition" may have to be initiated between the first and subsequent data bytes. This is initiated by setting the "SEN" bit of SSPCON2. This bit is then polled until it is reset.

9. If the secondary address byte was sent, then a second device address byte (with the "Read" indicated) may have to be sent to the device being read. The "R/_W" bit of SSPSTAT is polled until it is reset.

10. The "ACKDT" will be set ("NACK") or reset ("ACK") to indicate whether or not the data byte transfer is to be acknowledged in the device being read.

11. The "RCEN" bit in the SSPCON2 register is set to start a data byte receive. The "BF" bit of the SSPSTAT register is polled until the data byte has been received.

12. Operations ten through eleven are repeated until all data is received and a "NACK" (negative Acknowledge) is sent to the device being read.

13. A "Stop Condition" is initiated by setting the "PEN" bit of SSPCON2. This bit is then polled until it is reset.

Fig. 7.30 shows the data receive operation waveform.

Along with the single "Master" mode, the MSSP is also capable of driving data in "Multi-Master" mode. In this mode, if a data write "collision" is detected, it stops transmitting data and requests an interrupt to indicate there is a problem. An I2C "collision" is the case where

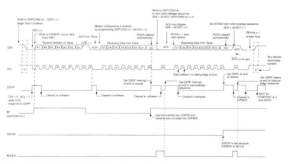

Figure 7.30 MSSP I2C data address/read

the current device is transmitting a "High" data value but there is a "Low" data value on the SDA line. This condition is shown in Fig. 7.31. The "WCOL" bit of the SSPCON register indicates that the collision has taken place.

When the collision occurs, the I2C software must wait some period of time before polling the SDA and SCL lines to ensure that they are high and then initiating a "Repeated Start Condition" operation. A "Repeated Start Condition" is the process of restarting the I2C data transfer right from the beginning (even if it was halfway through when the collision occurred).

Built-In ADC

All PICmicro® MCU devices that have a "seven" as the second to last character in the part number have a built-in analog to digital converter, which will indicate an analog voltage level from zero to Vdd, with 8- or 10-bit accuracy. The PORTA pins can be used as either digital I/O or analog inputs. The actual bit accuracy, utilization of pins and operating speed is a function of the

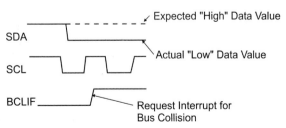

Figure 7.31 I2C MPPS "Collision" Response

PICmicro® MCU part number and the clock speed the PICmicro® MCU runs at.

When a pin is configured for analog input, it follows the models shown in Fig. 7.32.

"Rs" in the "Vsource" circuit is the in-line resistance of the power supply. In order to get reasonable times for charging the ADC's "holding capacitor", this value should be less than 10K.

The time required for the holding capacitor to load the analog voltage and to stabilize is

```
Tack = 5ms + [(temp - 25C) × 0.05 ms/C]
             + (3.19C × 10**7) × (8k + Rs)
```

which works out to anywhere from 7.6 usecs to 10.7 usecs at room temperature. For most applications, this calculation can be ignored and a "stabilization" time of 15 usecs can be used as a rule of thumb.

Once the voltage is stabilized at the capacitor, a test for each bit is made. 9.5 cycles are required to do an 8-bit conversion. The bit conversion cycle time (known as "TAD") can be anywhere from 1.6 to 6.4 usecs and can

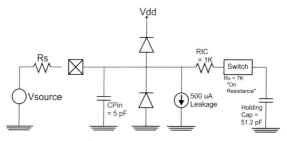

Figure 7.32 PICmicro® MCU Internal ADC Equivalent Input

either use the PICmicro® MCU's instruction clock or a built-in 250 kHz RC oscillator. To get a valid TAD time using the PICmicro® MCU's instruction clock, a two, eight, or thirty-two prescaler is built into the ADC.

A built-in 250 kHz oscillator is used to carry out the ADC conversion when the PICmicro® MCU is asleep or to avoid using the prescaler. For maximum ADC accuracy, Microchip recommends that the PICmicro® MCU be put to sleep during the ADC conversion for maximum accuracy (and minimum internal voltage or current upsets). If the PICmicro® MCU is put to sleep, then the minimum conversion time is much longer than what is possible using the built-in clock because the PICmicro® MCU has to restart when the ADC completion interrupt has been received.

The minimum conversion time is defined as the total time required for the holding capacitor to stabilize at the input voltage and for the ADC operation to complete.

To measure analog voltages, the analog input pins or the PICmicro® MCU, which are in "PORTA", have to be set to analog input on power up, the analog input pins are normally set to analog input and not digital I/O. To specify the modes, the "ADCON1" register is written to. In the following table, the two least significant bits (known as PCFG1:PCFG0) of the "ADCON1" register is shown with the types of I/O pin operation selected in a PIC16C71:

Sample ADCON1 Bit Definitions for the PIC16C71			
ADCON1 bits	AN3	AN2	AN1	AN0
11	D	D	D	D
10	D	D	A	A
01	Vref	A	A	A
00	A	A	A	A

The "ADCON 0" register is used to control the operation of the ADC. The bits of the register are typically defined as:

ADCON0 Bit Definitions

```
Bit     Function
7-6     ADCS1: ADCS0 bits used to select the TAD
        clock.
           11 - Internal 250 kHz Oscillator
           10 - FOSC/32
           01 - FOSC/8
           00 -FOSC/2
5-3     CHS2:CHS0 - Bits used to Select which Analog
        Input is to be Measured. These bits and
        their operation is Part Number Specific
 2      GO/_DONE - Set Bit to Start ADC Conversion,
        Reset by Hardware when ADC Conversion is
        Complete.
 1      ADIF - Set upon Completion of ADC Conversion
        and Requests an Interrupt.
 0      ADON - Set to Enable the ADC
```

The ADC consumes power even when it is not being used and for this reason, if the ADC is not being used "ADON" should be reset.

If the PICmicro® MCU's ADC is capable of returning a 10-bit result, the data is stored in the two "ADRESH" and "ADRES" registers. When 10-bit ADC results are available, the data can be stored in ADRESH/ADRESL in two different formats. The first is to store the data "right justified" with the most significant six bits of ADRESH loaded with "zero" and the least two significant bits loaded with the two most significant bits of the result. This format is useful if the result is going to be used as a 16-bit number, with all the bits used to calculate an average.

The second 10-bit ADC result format is "left justified" in which the eight most significant bits are stored in "ADRESH". This format is used when only an 8-bit value is required in the application and the two least significant bits can be "lopped" off or ignored.

To do an analog to digital conversion, the following steps are taken:

1. Write to ADCON1 indicating what are the digital I/O pins and which are the analog I/O pins. At this time, if a 10-bit conversion is going to be done, set the format flag in ADCON 1 appropriately.

2. Write to ADCON0, setting ADON, resetting ADIF and GO/_DONE and specifying the ADC TAD clock and the pin to be used.

3. Wait for the input signal to stabilize.

4. Set the GO/_DONE bit. If this is a high-accuracy measurement, ADIE should be enabled for interrupts and then the PICmicro® MCU put to "sleep".

5. Poll "GO/_DONE" until it is reset (conversion done).

6. Read the result form "ADRES" and optionally "ADRESH".

To read an analog voltage from the RA0 pin of a PIC167C1 running a 4-MHz PICmicro® MCU, the code would be

```
bsf     STATUS, RP0
movlw   0x002
movwf   ADCON1 ^ 0x080   ; AN1/AN0 are Analog Inputs
bcf     STATUS, RP0
movlw   0x041            ; Start up the ADC
movwf   ADCON0

movlw   5
addlw   0x0FF            ; Delay 20 usec for Holding
```

```
btfss    STATUS, Z       ;  Capacitor to Stabilize
  goto   $ - 2

bsf      ADCON0, GO      ; start the ADC conversion

btfsc    ADCON0, GO      ; Wait for the ADC
                           Conversion
  goto   $ - 1           ;  to End

movf     ADRES, w        ; Read the ADC result
```

Built-In Comparators

In the PIC16C2x, analog voltages can be processed by the use of comparators that indicate when a voltage is greater than another voltage. The inputs "compared" can be switched between different I/O pins as well as ground or a reference voltage that can be generated inside the PICmicro® MCU chip.

Enabling comparators is a very straightforward operation with the only prerequisite being that the pins used for the analog compare must be in "input" mode. Comparator response is virtually instantaneous, which allows "alarm" or other fast responses from changes in the comparator inputs (Fig. 7.33).

There are two comparators in the PIC16C62X controlled by the "CMCON" register, which is defined as:

CMCON Bit Definitions	
Bit	description
7	C2OUT – Comparator 2 Output (High if + > -)
6	C1OUT – Comparator 1 Output (High if + > -)
5–4	Unused
3	CIS – Comparator Input switch
2–0	CM2:CM0 – Comparator Mode

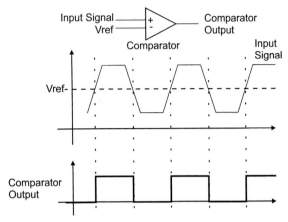

Figure 7.33 Comparator Response

The CIS and CM2:CM0 bits work together to select the operation of the comparators.

Comparator Module I/O Specification						
CM	CIS	Comp 1 + input	– input		Comp 2 + input	– input
000	X	RA0	RA3	(1)	RA2	RA1	(4)
001	0	RA2	RA0		RA2	RA1
001	1	RA2	RA3		RA2	RA1
010	0	Vref	RA3		Vref	RA1
010	1	Vref	RA3		Vref	RA2
011	X	RA2	RA0	(3)	RA2	RA1
100	X	RA3	RA0	(4)	RA2	RA1
101	X	DON'T	CARE		RA2	RA1
110	X	RA2	RA0	(5)	RA2	RA1	(6)
111	X	RA3	RA0	(7)	RA2	RA1	(8)

From these selections, there are some notes.

1. For CM2:CM0 equal to 000, RA3 through RA0 cannot be used for digital I/O.

2. For CM2:CM0 equal to 000, RA2 and RA1 cannot be used for digital I/O.

3. RA3 can be used for digital I/O.

4. RAO and RA3 can be used for digital I/O.

5. RA3 is a digital output, same as comparator 1 output.

6. RA4 is the open drain output of comparator 2.

7. RA0 and RA3 can be used for digital I/O.

8. RA1 and RA2 can be used for digital I/O.

Upon power up, the comparator CM bits are all reset, which means RA0 to RA3 are in analog input mode. If you want to disable analog input, the CM bits must be set (write 0x007 to CMCOM).

Interrupts can be enabled that will interrupt the processor when one of the comparator's output changes. This is enabled differently for each PICmicro® MCU with built-in comparators. Like the PORTB change on interrupt, after a comparator change interrupt request has been received, the CMCOM register must be read to reset the interrupt handler.

Along with comparing to external values, the PIC16C62x can also generate a reference voltage ("Vref" in the table above) using its own built-in 4-bit digital-to-analog converter. The digital-to-analog converter circuit is shown in Fig. 7.34.

The Vref control bits are found in the VRCON register and are defined as:

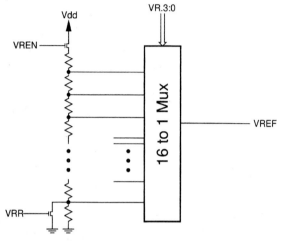

Figure 7.34 16C62x VRef circuit.

Comparator VRCON Bit Definitions	
Bit	description
7	VREN - Vref Enable when Set
6	VROE - Vref output enable when set
	RA2 - Vref
5	VRR - Vref Range Select
	1 = Low Range
	0 = High Range
4	Unused
3-0	VR3:VR0 - Voltage Selection Bits

The Vref output is dependent on the state of the "VRR" bit. The Vref voltage output can be expressed mathematically if VRR is set as:

```
Vref = Vdd*(VRCON & 0x00F)/24
```

Or, if it is reset as:

```
Vref = Vdd*(8 + (VRCON & 0x00F))/32
```

Note that when VRR is set, the maximum voltage of Vref is 15/24 of Vdd, or just less than two-thirds Vdd. When VRR is reset, Vref can be almost three-quarters of Vdd.

Parallel Slave Port

The PSP is very easy to wire up with separate chip select and read/write pins for enabling the data transfer. The block diagram of the PSP is shown in Fig. 7.35.

A read and write operation waveform is shown in Fig. 7.36.

The minimum access time is one clock (not "instruc-

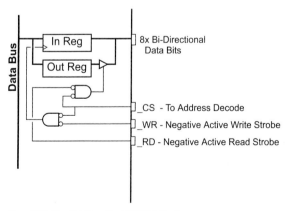

Figure 7.35 Parallel Slave Port ("PSP") Hardware

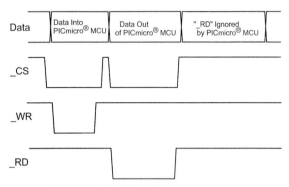

Figure 7.36 Parallel Slave Port Operation

tion clock") cycle. For a PICmicro® MCU running at 20 MHz, the minimum access time is 50 nsecs.

To enable the parallel slave port, the "PSP mode" bit of the TRISE register must be set. When this bit is set, port D becomes driven from the "_CS", "_RD", and "_WR" bits, which are RE2, RE1, and RE0, respectively. When the PSP mode bit is set, the values in PORTD, PORTE, TRISD, and TRISE are ignored.

When PSP mode is enabled and _CS and _RD are active, PORTD drives out the contents of "OUTREG". When "OUTREG" (which is at PORTD's address) is written to, the "OBF" ("Output Buffer Full") bit of TRISE is set. This feature, along with the input data flags in TRISE is not available in all devices. The PBF bit will become reset automatically, when the byte in the OUTREG is read by the device driving the external parallel bus.

When a byte is written into the parallel slave port (_CS and _WR are active), the value is saved in "INREG" until it is overwritten by a new value. If the optional status registers are available, the "IBF" bit is set when the

INREG is written to and cleared when the byte in INREG is read. If the byte is not read before the next byte is written into "INREG", the "IBOV" bit, which indicates the overwrite condition is set.

In older PICmicro® MCUs that have PSP port, the "IBF", "OBF", and "IBOV" bits are not available in TRISE.

Built-In EEPROM Data Memory Access

The "EECON1", "EECON2", "EEADR", and "EEDATA" are used to control access to the EEPROM. "EEADR" and "EEDATA" are used to provide the address and data interface into the up to 256 byte data EEPROM memory. "EECON" and "EECON2" are used to initiate the type of access as well as indicate that the operation has completed. "EECON2" is a "pseudo-register" that cannot be read from, but is written to with the data, 0x055/0x0AA to indicate the write is valid.

EECON1, contains the following bits for controlling the access:

Critical EECON1 Bits	
Bit	Function
EEPCD	Set to Access Program Memory. Reset to Access Data EEPROM only in 16F62x and 16F87x.
WRERR	Set if a write Error is Terminated early to indicate Data Write may not have been Successful.
WREN	When set, a write to EEPROM begins.
WR	Set to indicate an upcoming Write Operation. Cleared when the Write Operation is complete.
RD	Set to indicate Read Operation. Cleared by next Instruction Automatically.

Using these bits, a Read can be initiated as:

```
movf / movlw    address/ADDR, w
bcf             STATUS, RP0
movwf           EEADR
bsf             STATUS, RP0
bsf             EECON1, ^ 0x08, RD
bcf             STATUS, RP0
movf            EEDATA, w            ; w = EEPROM
                                         [address/ADDR]
```

Write operations are similar, but have two important differences. The first is that the operation can take up to ten milliseconds to complete, which means the "WR" bit of EECON1 has to be polled for completion, or in the EEPROM, interrupt request hardware enabled. The second difference as mentioned above, is that a "timed write" has to be implemented to carry out the operation.

```
movlw /movf    constant/DATA, w
bcf            STATUS, RP0
movwf          EEDATA
movlw /movf    address/ADDR, w
movwf          EEADR
bsf            STATUS,RP0
bsf            EECON1 ^ 0x080, WREN
bcf            INTCON,GIE
movlw          0x055                ] CRITICAL SECTION
movwf          EECON2 ^ 0x080       ]
movlw          0x0AA                ]
movwf          EECON2 ^ 0x080       ]
bsf            EECON1 ^ 0x080, WR ]
bsf            INTCON, GIE
btfsc          EECON1 ^ 0x080, WR ] Poll for
                                        Operation Ended
 goto          $ - 1                ]
bcf            EECON1 ^ 0x080, WREN
bcf            STATUS, RP0
bsf            INTCON, GIE
```

The EEPROM included PIC12CE5xx parts use the most significant bits of the "GPIO" ("general purpose I/O") register and its corresponding "TRIS" register. The PIC12CE5xx's EEPROM interface can be described as shown in the block diagram Fig. 7.37.

In Fig. 7.37, the GPIO bits six and seven do not have "TRIS" control bits. As well, bit six (the 12CEEPROM bit), "SDA") has an open-drain driver. This driver circuit is designed to let both the PICmicro® MCU and the EEPROM drive the data line at different intervals without having to disable the PICmicro® MCUs write of the EEPROM. Information is written to the EEPROM device using the waveform shown in Fig. 7.38.

The "start" and "stop" bits are used to indicate the beginning and end of an operation and can be used halfway through to halt an operation. The start and stop bits are actually invalid cases (data cannot change while one clock is active or "high").

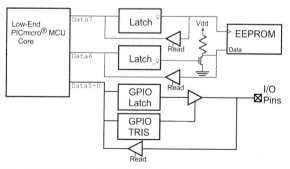

Figure 7.37 PIC12CE5xx EEPROM Interface

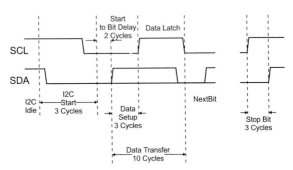

Figure 7.38 PIC12CE5xx EEPROM Interface Waveform

This operation means that the GPIO port must be accessed carefully; always make sure the SDA and SCL GPIO bits have a "one" in them or else the built-in EEPROM may be accessed incorrectly, causing problems with subsequent reads.

The instruction

```
clrf    GPIO
```

should never be used in applications that access low-end data EEPROM.

Data is written most significant bit first, which is probably backwards to most applications. Before any transfer, a "control byte" has to be written. The "control byte's" data is in the format:

```
0b01010000R
```

where "R" is the "Read/_Write" byte (indicating what is coming next). If the "read/write" bit is set, then a read of the EEPROM at the current address pointer will take place. If a write is to take place, the "read/write" bit is reset.

After a byte is sent, the SDA line is pulled low to indicate an "acknowledgment" ("ACK" or just "A" in the bit stream representations below). This bit is set low (as an acknowledgment) when the operation has completed successfully. If the acknowledgment bit is high ("NACK"), it does not necessarily mean there was a failure; if it is issued by the EEPROM then it indicates a previous write has not completed. The PICmicro® MCU will issue it to stop the EEPROM from preparing to send additional bytes out of its memory in a multi-byte read.

There are five operations that can be carried out with the EEPROM that is built into the PIC12CE5xx. They are

1. Current Address Set.

2. Current Address Set/Data Byte Write.

3. Data Byte Read at Current Address.

4. Sequential ("multi-byte") Read at Current Address.

5. Write completion poll.

The EEPROM in the PIC12CE5xx is only 16 bytes in size. Each byte is accessed using a 4-bit address. This address is set using a control byte with the "R" bit reset followed by the address. The bit stream looks like:

```
idle - Start - 1010000A - 0000addrA - DataByteA -
Stop - idle
```

In the second byte sent, the 0b00000addr pattern indicates that the four "addr" address bits become the address to set the EEPROM's internal address pointer to for subsequent operations. After the two bytes have been sent, the SCL and SDA lines are returned to "IDLE" for three cycles, using the instruction:

```
movlw     0x0C0

iorwf     GPIO, f        ; set SDA /SCL
```

before another operation can complete.

The address data write is similar to the address write, but does not force the two lines into IDLE mode and it passes along a data byte before stopping the transfer:

```
Idle - Start - 10100000A - 0000addrA - DataByteA
- Stop - idle
```

Data bytes can be read singly or sequentially depending on the state of "ACK" from the PICmicro® MCU to the EEPROM after reading a byte. To halt a read, when the last byte to be read has been received, the PICmicro® MCU issues a "NACK" (or "N" in the bitstream listing) to indicate that the operation has completed.

A single byte read looks like:

```
idle - Start - 10100001A - DataByteN - Stop - idle
```

while a 2-byte read looks like:

```
idle - Start - 10100001A - DataByteA - DataByteN
- Stop - idle
```

The last operation is sending dummy "write" control bytes to poll the EEPROM to see whether or not a byte write has completed (10 msecs are required). If the write has completed, then an "ACK" will be returned else a "NACK" will be returned.

EPROM Program Memory Access

To read from the EPROM (or external memory), the following code can be used:

```
movfp   SaveAddress + 1, TBLPTRH    ; Setup TBLPTR
                                       to the Data
movfp   SaveAddress, TBLPTRL       ; being Read

tablrd  0, 0, SaveData             ; Load TBLAT
                                       with Memory
tlrd    1, SaveData + 1            ; Contents
tlrd    0, SaveData

movfp   SaveData + 1, WREG         ; High
                                       Instruction
                                       Byte
  :
movfp   SaveData, WREG             ; Low
                                       Instruction
                                       Byte
  :
```

To write to the built-in EPROM of the PIC17Cxx, the "_MCLR" line will have to be driven to Vpp (13 to 14 volts). When the program memory is being written, all instruction execution in the PIC17Cxx stops. To resume operation after a program memory write, an interrupt, like returning from a TMR0 interrupt request, is executed. Sample code for writing to the PIC17Cxx's program memory is as follows:

```
org     0x00010
TMR0Int                                ; Timer
                                           Interrupt
                                           Request
```

```
                                       ;  Acknowledge

    retfie

      :

    movfp    SaveAddress, TBLPTRL       ; Point to the
                                          Memory being
    movfp    SaveAddress + 1, TBLPTRH   ; written to

    bcf      PORTA, 3                   ; Turn on
                                          Programming
                                          Voltage
    movlw    HIGH ((100000 / 5) + 256) ; Delay 100
                                          msecs for
    movwf    Dlay                       ; Programming
                                          Voltage to
                                          Stabilize

    movlw    LOW ((100000 / 5) + 256)
    addlw    0x0FF
    btfsc    ALUSTA, Z
     decfsz  Dlay, f
      goto   $ - 3

    movlw    HIGH (65536 - 10000)       ; Delay 10
                                          msecs for
                                          EPROM
                                          Write
    movwf    TMR0H
    movlw    LOW (65536 - 10000)
    movwf    TMR0L

    bsf      T0STA, T0CS                ; Start up the
                                          Timer

    movlw    1 << T0IE                  ; Enable
                                          Interrupts
    movwf    INTSTA
    bcf      CPUSTA, GLINTD

    tlwt     0, SaveData                ; Load Table
                                          Pointer with
                                          Data
    tlwt     1, SaveData + 1
    tablwt   1, 0, SaveData + 1         ; Write the
                                          Data In
```

```
nop
nop

clrf    INTSTA, f                   ; Turn Off
                                    ;   Interrupts
bsf     CPUSTA, GLINTD

movlw   2
call    SendMSG

bsf     PORTA, 3
```

Flash Program Memory Access

To read to program memory, the following code is used for the 16F87x. Note the two "nops" to allow the operation to complete before the instruction is available for reading:

```
bsf             STATUS, RP1
movlw /movwf    LOW address/ADDR, w
movwf           EEADR ^ 0x0100
movlw /movwf    HIGH address/ADDR, w
movwf           EEADRH ^ 0x0100
bsf             STATUS, RP0
bsf             EECON1 ^ 0x0180, EEPGD
bsf             EECON1 ^ 0x0180, RD
nop
nop
bcf             STATUS, RP0
movf            EEDATA, w
movwf           -----                   ; Store Lo
                                        ;   Byte of
                                        ;   Program
                                        ;   Memory
movwf           EEDATAH, w
movwf           -----                   ; Store Hi
                                        ;   Byte of
                                        ;   Program
                                        ;   Memory
bcf             STATUS, RP1
```

Writing to program memory is similar to writing to data, but also has the two nops in which the operation takes place. There are no polling or interrupts available for this operation, instead, the processor halts during this operation. Even though the processor has stopped for a program memory write, peripheral function (ADC's, serial I/O, etc.) are still active.

```
bsf            STATUS, RP1
movlw /movf    LOW address/ADDR, w
movwf          EEADR
movlw /movwf   HIGH address/ADDR, w
movwf          EEADRH
movlw /movwf   LOW Constant/DATA, w
movwf          EEDATA
movlw /movwf   HIGH Constant/DATA, w  ; Maximum 0x03F
movwf          EEDATAH
bsf            STATUS, RPO
bsf            EECON1 ^ 0x0180, EEPGO
bsf            EECON1 ^ 0x0180, WREN
bcf            INTCON, GIE            ]    Critically
movlw          0x055                 ]    timed
movwf          EECON2 ^ 0x0180       ]    code.
movlw          0x0AA                 ]
movwf          EECON2 ^ 0x0180, OR   ]
nop                                  ] operation
nop                                  ] executes
bcf            EECON1 ^ 0x0180, WREN
bsf            INTCON, GIE
```

External Parallel Memory

Parallel memory devices can be connected to the 17Cxx PICmicro® MCU devices to enhance the PICmicro® MCUs program memory space. The interface provided is up to 64k of 16 data bit "words" via a multiplexed address/data bus. The multiplexed bus may seem somewhat difficult to use, but it actually is not; memory devices can be added quite easily and quickly.

There are four memory modes available to the 17Cxx PICmicro® MCUs:

PIC17Cxx Memory Modes	
Mode	Program memory characteristics
Microcontroller	Internal to the PICmicro® MCU, able to read Configuration Fuses and Read and Write Program Memory
Protected Microcontroller	Internal to the PICmicro® MCU, able to read Configuration fuses Program Memory can be read but not Written
Extended Microcontroller	Program Memory Internal to PICmicro® MCU Accessible. External Memory in Address Space Above Read and Writeable as well. Unable to read Configuration Fuses.
Microprocessor	No internal Program Memory or Configuration Fuses Accessible. Whole 64k program memory space Accessible outside PICmicro® MCU

These modes can be seen in Fig. 7.39.

An unprogrammed PC17Cxx's configuration fuses sets the PICmicro® MCU into "microprocessor" mode that cannot access any internal program memory. This allows output devices to be placed into applications, with external program memory providing the application code. This feature allows a way of debugging an application before it is burned into the PICmicro® MCU.

External memory can be read from or written to, using the "TABLRD" and "TABLWT" instructions. In extended microcontrollers and microprocessor modes, the internal program memory can be read using the "TABLRD" instruction in the microcontroller modes. These "Table" instructions use the "Table Pointer" register ("TBLPTRH" for the high 8 bits and "TBLPTRL" for the low 8 bits) to address the operation. During table reads and writes, the "table latch" register ("TABLATH" for the high byte and "TABLATL" for the low byte) is

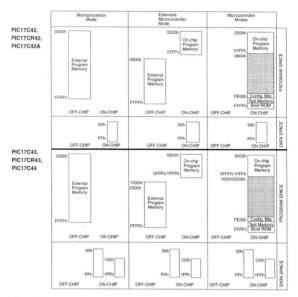

PIC17C42,
PIC17CR42,
PIC17C42A

PIC17C43,
PIC17CR43,
PIC17C44

Figure 7.39

used to buffer the 16 bits during the transfer because the 17Cxx PICmicro® MCUs processor can only access data 8 bits at a time.

The block diagram for accessing program memory in the 17Cxx family of PICmicro® MCUs is shown in Fig. 7.40.

To execute a read or write to program memory, the address in the table pointer has to be first set up. Writing to each of the two 8-bit registers does this. Next, if the operation is a read, the "TABLRD" instruction is executed with a dummy destination to update the table

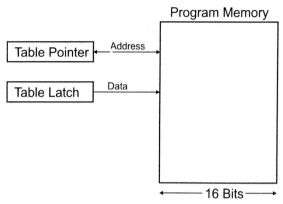

Program Memory

16 Bits

Figure 7.40 PIC17Cxx External Memory Access

latch register. Once this is done, two read instructions are carried out to read the 16 bits at the specified address. This instruction sequence is

```
PIC17Cxx Program Memory Table Read Code

  movlw    HIGH PM_address    ; Set up Table Pointer
  movwf    TBLPTRH
  movlw    LOW PM_address
  movwf    TABLPTRL
  tablrd   0, 0, WREG         ; Update Latch Register
  tlrd     1, WREG            ; Read High Byte
  movwf    HIGH Destination
  tablrd   0, 0, WREG         ; Read Low Byte
  movwf    LOW Destination
```

The external program memory read is identical to the internal EPROM program memory read.

PICmicro® MCU
Hardware Interfacing

Power

Connecting a PICmicro® MCU only requires a 0.01 to 0.1 uF "decoupling" cap across the "Vdd" and "Vss" pins. A typical Power connection is shown in Fig. 8.1. This capacitor should be of low "ESR" type (typically of "tantalum" type).

"Standard" PICmicro® MCUs are designed for anywhere from 4.0 to 6.0 volts of power. Some PICmicro® MCUs have been "qualified" to run from 2.0 to 6.0 volts and are identified for having this capability as being

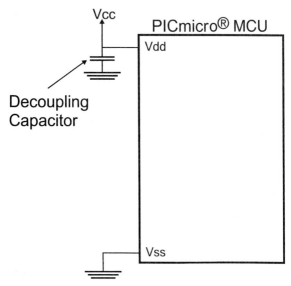

Figure 8.1 PICmicro® MCU Power Connections

"low-voltage" devices. These low-voltage parts are identical to the high-voltage supply parts except that they have been tested at the factory to run with input voltages down to 2.0 volts. Low voltage PICmicro® MCU parts are identified by the addition of the letter "L" before the "C" or "F" in the part number.

Note that the "brown out reset" built into many PICmicro® MCUs is designed to become active at 4.5 volts. This makes the brown out reset incompatible with most low-voltage applications, although there are some

PICmicro® MCUs that have a programmable brown out reset voltage level.

In Fig. 8.2, if Vdd goes below the brown out voltage of the Zener diode, then _MCLR will be pulled low and the PICmicro® MCU will become reset.

The PIC16HV540 has a built-in voltage regulator that allows the PICmicro® MCU to be driven without any external regulators for battery application or poorly regulated power input. The PICmicro® MCU itself is pin and program compatible with the PIC16F54, with PORTA and PORTB having different voltage outputs.

To connect a PIC16HV540 to a battery, the circuit can be as simple as is shown in Fig. 8.3, with "sleep" used for turning the device "off" and putting it in a low-power state.

The device's block diagram looks like Fig. 8.4.

The voltage regulator can work as either a 5- or 3-volt regulator by setting or resetting, respectively, the "RL"

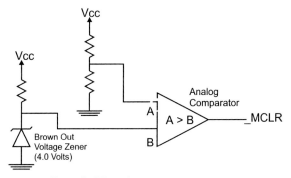

Figure 8.2 "Brown Out" Reset Circuit

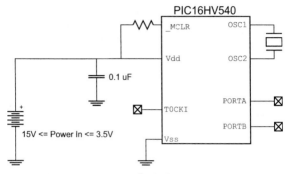

Figure 8.3 High-Voltage PICmicro® MCU Connections

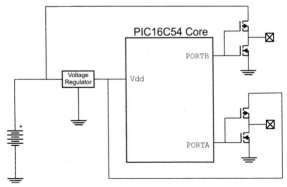

Figure 8.4 Actual High-Voltage PICmicro® MCU Circuit

bit of the "option 2" register, which is in the "OPTION/
TRIS" address space of the low-end PICmicro® MCU
processor. This register is an auxiliary configuration
fuses register, which can be modified within an application. The bits of the OPTION2 register are defined as:

PIC16HV540 "OPTION2" Register Definition

```
Bit    Description
7-6    Unused
 5     WPC - When set, device will Wake Up On RB0 -
       RB3 changing
 4     SWE - Software Watchdog Timer.  If the WDT is
       not Enabled in the Configuration Fuses,
       setting this bit will enable it in software
 3     RL - Regulated voltage select bit (Set for 5
       Volts, Reset for 3 Volts)
 2     SL - Sleep Voltage Level Setting (if Set, use
       "RL" Voltage, when Reset, use 3 Volts)
 1     BL - Brown Out Voltage Select.  When Set -
       3.1 volts for 5 Volt Operation and when
       Reset - 2.2 Volts for 3 Volt Operation
 0     BE - Brown Out Checking Enabled when Set.
```

OPTION2 is written using the TRIS instruction as:

```
TRIS   7
```

or

```
TRIS   OPTION2
```

Reset

If the simple reset shown in Fig. 8.5 is used for reset,
then the "PWRTE" option should be enabled to allow
the PICmicro® MCU's power input to stabilize before the
device starts executing.

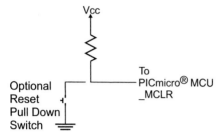

Figure 8.5 Simple External PICmicro® MCU Reset Circuit

Digital Logic Interfacing

Typical PICmicro® MCU output voltages are

```
Vol ("output low voltage") = 0.6 V (max)
Voh ("output high voltage") = VDD
        - 0.7 V (min)
```

The input "threshold" voltage, the point at which the input changes from an " I " to an "O" and vice versa, is also dependent on the input power "Vdd" voltage level. The threshold is different for different devices. For a number of different PICmicro® MCU part numbers, this value is specified as being in the range:

```
0.25 Vdd + 0.8V >= Vthreshold
        >= 0.48 Vdd
```

Parallel Bus Device Interfacing

Parallel busses can be created using PORTB for eight data bits and using other PORT pins for the "_RD" and "_WR" lines as shown in Fig. 8.6. Code to access the Parallel Bus Devices follows.

```
bsf     STATUS, RPO    ; Put PORTB into Input Mode
movlw   0x0FF
movwf   TRISB ^ 0X080
bcf     STATUS, RPO
bcf     PORTA, 0       ; Drop the "_RD" line
call    Dlay           ; Delay until Data Output
                            Valid
movf    PORT B, w      ; Read Data from the Port
bsf     PORT A, 0      ; "_RD" = 1 (disable "_RD"
                            Line)
```

Writing parallel bus devices is accomplished by the code:

```
bsf     STATUS, RPO
clrf    TRIS B ^ 0X080  ; PORTB Output
bcf     STATUS, RPO
```

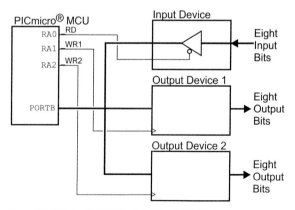

Figure 8.6 PICmicro® MCU Simulated Parallel IO Port

```
bcf    PORTA, 1       ;  Enable the "_WR1" Line
movwf  PORTB          ;   output the Data
call   Dlay           ;  Wait Data Receive Valid
bsf    PORTA          ;  "_WR1" = 1.
```

Button Interfacing

The typical button interface circuit is seen in Fig. 8.7.

The first button debouncing macro is inserted in the source code and waits for a Port Pin to reach a set state for a specific amount of time before continuing.

```
Debounce macro HiLo, Port, Bit
if HiLo == Lo
 btfss Port,Bit  ; Is the Button Pressed?
else
 btfsc Port,Bit
endif
  goto $ - 1     ;  Yes - Wait for it to be Released
 movlw InitDlay ; Wait for Release to be Debounced
 movwf Dlay     ; Have to Delay 20 msecs
 movlw 0
if HiLo == Lo
 btfss Port,Bit ; If Button Pressed, Wait Again for
                ; it
```

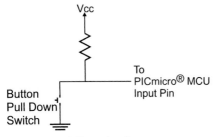

Figure 8.7 Simple Button Interface

```
  else
   btfsc Port,Bit
  endif
    goto $ - 6      ;  to be Released
  ifndef Debug      ; Skip Small Loop if "Debug"
                    ; Defined
   addlw 1          ; Increment the Delay Count
   btfsc STATUS, Z ;  Loop If Low Byte (w) Not Equal
                    ; to Zero
  else
   nop              ; Match the Number of Instructions
   nop
  endif
    decfsz Dlay
     goto $ - 5
  endm
```

The "InitDlay" constant is calculated using the formula:

```
TimeDelay = (((InitDlay - 1 ) * 256) * 7) /
                            (Frequency / 4)
```

or

```
InitDlay = ((TimeDelay * (Frequency / 4)) /
                            (256 * 7)) + 1
```

The second button debounce macro works similarly to
the Parallax Basic Stamp's PBASIC "Button" Function.

```
 Button macro Port, Pin, Down, Delay, Rate, Variable,
  Target, Address
  local ButtonEnd
   incf Variable, w           ; Increment the Counter
                                Variable
  if ((Down == 0) && (Target == 0)) || ((Down == 1)
     && (Target == 1))
   btfsc Port, Pin            ; If Low, then Valid Pin
  else
```

```
 btfss Port, Pin          ; If High, then Valid
                              Pin

 endif
 clrw                     ; Not Pressed, Clear the
                              Counter

 movwf Variable           ; Save the Counter Value
 movlw Delay & 0x07F
 subwf Variable, w        ; Button Debounced?
 btfsc STATUS, Z
  goto Address            ; If Equal, then "Yes"
if ((Delay & 0x080) != 0) ; Is Autorepeat used?
 btfsc STATUS, C
  decf Variable           ; No - Decrement if >
                              "Delay"

 else
 btfss STATUS, C
  goto ButtonEnd          ; Less than Expected -
                              End

 xorlw Rate               ; At the Autorepeat
                              Point yet?

 btfsc STATUS, Z
  goto ButtonEnd          ; No - Keep Incrementing
 movlw Delay              ; Yes, Reset back to the
                              Original

 movwf Variable           ;   Count and Repeat
 goto Address
 endif
ButtonEnd                 ; Macro Finished
 endm
```

The macro's parameters are defined as:

PicBasic "Button" Debounce Macro Code Parameters	
Parameter	Function
Port, Pin	The Button Pin (ie "PORTA, 0")
Down	The State When the Button is Pressed

PicBasic "Button" Debounce Macro Code Parameters (*Continued*)	
Delay	The number of iterations of the Macro code before the "Address" is jumped to (to 127). If Set to 0, then Jump if "Target" met without any debouncing. If Bit 7 of "Delay" is set, then no auto-repeats.
Rate	After the Initial jump to "address", the number of cycles (to 127) before autorepeating.
Target	The state ("1" or "0") to respond to.
Address	The Address to Jump to when the Button is pressed or Auto-repeats

Switch Matrix Keypad/Keyboard Interfacing

A switch matrix is simply a two-dimensional matrix of wires, with switches at each vertex. The switch is used to interconnect rows and columns (which are optionally pulled to ground) in the matrix, as can be seen in Fig. 8.8.

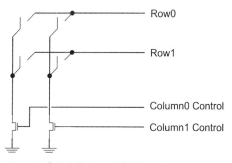

Figure 8.8 Switch Matrix with Pull Down Transistors

In this case, by connecting one of the columns to ground, if a switch is closed, the pull down on the row will connect the line to ground. When the row is polled by an I/O pin, a "0" or low voltage will be returned instead of a "1" (which is what will be returned if the switch in the row that is connected to the ground is open).

The PICmicro® MCU is well suited to implementing switch matrix keyboards with PORTB's internal pull-ups and the ability of the I/O ports to simulate the open-drain pull-downs of the columns as is shown in Fig. 8.9.

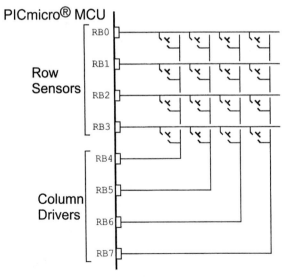

Figure 8.9 4×4 Switch Matrix Connected to PORTB

Normally, the pins connected to the columns are left in tristate (input) mode. When a column is being scanned, the column pin is output enabled driving a "0" and the four input bits are scanned to see if any are pulled low. In this case, the keyboard can be scanned for any closed switches (buttons pressed) using the code:

```
int KeyScan()            //  Scan the Keyboard and
{                            Return when a key is
                         //  pressed
int  i = 0;
int  key = -1;

   while (key == -1) {

     for (i = 0; (i < 4) & ((PORTB & 0x00F)
         == 0x0F0); i++);

     switch (PORTB & 0x00F) { //  Find Key that is
                                 Pressed
       case 0x00E:            //  Row 0
         key = i;
         break;
       case 0x00D:            //  Row1
       case 0x00C:
         key = 0x04 + i;
         break;
       case 0x00B:            //  Row2
       case 0x00A:
       case 0x009:
       case 0x008:
         key = 0x08 + i;
         break;
       else                   //  Row3
         key = 0x0C + i;
         break;
     }//end switch
   }// end while

   return key;

}  // End KeyScan
```

The "KeyScan" function will only return when a key has been pressed. This routine will not allow keys to be debounced or for other code to execute while it is executing.

These issues can be resolved by putting the key scan into an interrupt handler, which executes every 5 msecs:

```
Interrupt KeyScan( )      //  5 msec Interval Keyboard
                                  Scan
{

int   i = 0;
int   key = -1

  for (i = 0; (i <4) & ((PORTB & 0x00F) == 0x00F));
  i++);

 if (PORTB & 0x00F) != 0x00F)  {   //  Key Pressed
   switch (PORTB & 0x00F) {    //  Find Key that is
                                      Pressed
    case 0x00E:                 //  Row 0
      key = i;
      break;
    case 0x00D:                 //  Row1
    case 0x00C:
   -  key = 0x04 + i;
      break;
    case 0x00B:                 //  Row2
    case 0x00A:
    case 0x009:
    case 0x008:
      key = 0x08 + i;
      break;
    else                        //  Row3
      key = 0x0C+i;
      break;
   }//end switch
   if (key == KeySave) {
     keycount = keycount + 1;   //  Increment Count
     if (keycount == 4)
       keyvalid = key;          //  Debounced Key
   } else
     keycount = 0;              //  No match - Start
                                      Again
```

```
  KeySave = key;          //  Save Current key for
                                next 5 msec
  }                       //  Interval

} // End KeySave
```

This interrupt handler will set "keyvalid" variable to the row/column combination of the key button (which is known as a "scan code") when the same value comes up four times in a row. For time scan this is the debounce routine for the keypad. If the value doesn't change for four intervals (20 msecs in total), the key is determined to be debounced.

Combining Input and Output

When interfacing the PICmicro® MCU to a driver and receiver (such as a memory with a separate output and input), a resistor can be used to avoid bus contention at any of the pins as is shown in Fig. 8.10.

Buttons can also be put on PICmicro® MCU I/O lines as is shown in Fig. 8.11.

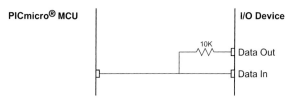

Figure 8.10 Combining "I/O" on One PICmicro® MCU Pin

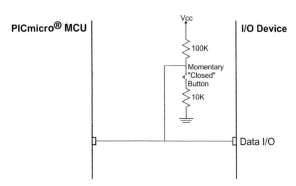

Figure 8.11 Combining Button Input with Digital I/O

Simulated "Open Collector"/"Open Drain" I/O

"Open Collector" ("Open Drain") I/O pins in the PICmicro® MCU are wired as in Fig. 8.12. These pins are available in different devices for different functions. This action can be simulated by using the code listed below that enables the I/O pin output as low if the Carry flag is reset. If the Carry flag is set, then the pin is put into input mode.

```
bcf     PORT#, pin    ;  Make Sure PORTB Pin Bit is
                             "0"
bsf     STATUS, RPO
btfss   STATUS, C     ;  If Carry Set, Disable Open
                             Collector
 goto   $ + 4         ;  Carry Reset, Enable Open
                             Collector
nop
bsf     TRIS ^ 0x080, pin
goto    $ + 3
```

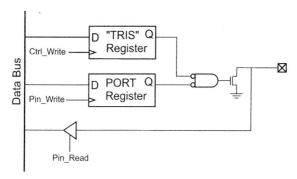

Figure 8.12 "Open Drain" I/O Pin Configuration

```
bcf     TRIS ^ 0x080, pin
goto    $ + 1
bcf     STATUS, RPO
```

LEDs

The typical circuit that used to control an LED from a PICmicro® MCU I/O pin is shown in Fig. 8.13. With this circuit, the LED will light when the microcontroller's output pin is set to "0" (ground potential). When the pin is set to input or outputs a "1", the LED will be turned off.

Multisegment LED displays

Seven Segment LED Displays (Fig. 8.14) can be added to a circuit without a lot of software effort. By turning on specific LEDs (each of which lights up a "segment" in

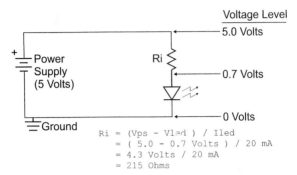

Voltage Level

5.0 Volts

0.7 Volts

0 Volts

$$Ri = (Vps - Vled) / Iled$$
$$= (5.0 - 0.7 \text{ Volts}) / 20 \text{ mA}$$
$$= 4.3 \text{ Volts} / 20 \text{ mA}$$
$$= 215 \text{ Ohms}$$

Figure 8.13 LED Circuit Operation

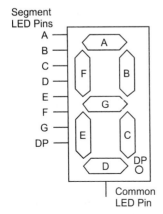

Segment
LED Pins

Figure 8.14 Organization of a 7-Segment LED Display

the display), the display can be used to output decimal numbers.

Each one of the LEDs in the display is given an identifier and a single pin of the LED is brought out of the package. The other LED pins are connected together and wired to a common pin. This common LED pin is used to identify the type of Seven Segment Display (as either "Common Cathode" or "Common Anode").

The typical method of wiring multiple Seven Segment LED Displays together is to wire them all in parallel and then control the current flow through the common Pin. Because the current is generally too high for a single microcontroller pin, a transistor is used to pass the current to the common power signal. This transistor selects which display is active as shown in Fig. 8.15. In this cir-

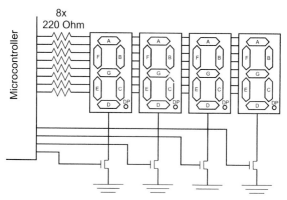

Figure 8.15 Wiring Four 7-Segment LED Displays

cuit, the PICmicro® MCU will shift between the displays showing each digit in a very short "time slice". This is usually done in a Timer Interrupt Handler. The basis for the interrupt handler's code is listed below:

```
Int
   - Save Context Registers
   - Reset Timer and Interrupt
   - LED_Display = 0                  ; Turn Off all the
                                        LEDs
   - LED_Output = Display[ Cur ]
   - Cur = (Cur + 1) mod #Displays ; Point to Next
                                        Sequence
                                        Display
   - LED_Display = 1 << Cur          ; Display LED for
                                        Current Display
   - Restore Context Registers
   - Return from Interrupt
```

This code will cycle through each of the digits (and displays), with current going through the transistors for each one. To avoid flicker, generally the code should run so that each digit is turned on/off at least 50 times per second. The more digits present, the faster you, the interrupt handler, will have to cycle the interrupt handler (i.e., eight Seven Segment Displays must cycle at least 2,000 digits per second, which is twice as fast as four displays).

LCD Interfaces

The most common connector used for the 44780-based LCDs is 14 pins in a row, with pin centers 0.100" apart. The pins are wired as:

Hitachi 44780 Based LCD Pinout	
Pin	Description
1	Ground
2	Vcc
3	Contrast Voltage
4	"R/S" - Instruction/Register Select
5	"R/W" - Read/Write LCD Registers
6	"E" - Clock
7-14	D0-D7 Data Pins

The contrast voltage to the display is typically controlled using a potentiometer wired as a voltage divider. This will provide an easily variable voltage between Ground and Vcc, which will be used to specify the contrast (or "darkness") of the characters on the LCD screen. This circuit is shown in Fig. 8.16.

The interface is a parallel bus, allowing simple and fast reading/writing of data to and from the LCD as shown in Fig. 8.17.

This waveform will write an ASCII byte out to the LCD's screen. The ASCII code to be displayed is 8-bits long and is sent to the LCD either 4- or 8-bits at a time. If 4-bit mode is used, two "nybbles" of data (sent high 4-bits and then low 4-bits with an "E" Clock pulse with each nybble) are sent to make up a full 8-bit transfer.

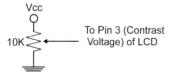

Figure 8.16 LCD Contrast Voltage Circuit

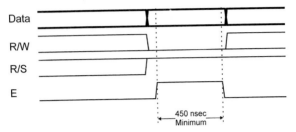

Figure 8.17 LCD Data Write Waveform

The "E" Clock is used to initiate the data transfer within the LCD.

Sending parallel data as either 4- or 8-bits are the two primary modes of operation. While there are secondary considerations and modes, deciding how to send the data to the LCD is the most critical decision to be made for an LCD interface application.

Eight bit mode is best used when speed is required in an application and at least ten I/O pins are available. Four bit mode requires a minimum of 6 bits. To wire a microcontroller to an LCD in 4-bit mode, just the top 4-bits (DB4-7) are written to.

The "R/S" bit is used to select whether data or an instruction is being transferred between the micro-controller and the LCD. If the bit is set, then the byte at the current LCD "Cursor" position can be read or written. When the bit is reset, either an instruction is being sent to the LCD or the execution status of the last instruction is read back (whether or not it has completed).

The different instructions available for use with the 44780 are shown in the following table:

Hitachi 44780 Based LCD Commands

R/S	R/W	D7	D6	D5	D4	D3	D2	D1	D0	Instruction/ Description
4	5	14	13	12	11	10	9	8	7	Pins
0	0	0	0	0	0	0	0	0	1	Clear Display
0	0	0	0	0	0	0	0	1	*	Return Cursor and LCD to Home Position
0	0	0	0	0	0	0	1	ID	S	Set Cursor Move Direction
0	0	0	0	0	0	1	D	C	B	Enable Display/ Cursor
0	0	0	0	0	1	SC	RL	*	*	Move Cursor/Shift Display
0	0	0	0	1	DL	N	F	*	*	Reset/Set Interface Length
0	0	0	1	A	A	A	A	A	A	Move Cursor to CGRAM
0	0	1	A	A	A	A	A	A	A	Move Cursor to Display
0	1	BF	*	*	*	*	*	*	*	Poll the "Busy Flag"
1	0	H	H	H	H	H	H	H	H	Write Hex Character to the Display at the Current Cursor Position
1	1	H	H	H	H	H	H	H	H	Read Hex Character at the Current Cursor Position on the Display

The bit descriptions for the different commands are

"*" - Not Used/Ignored. This bit can be either "1" or "0"

Set Cursor Move Direction:
  ID - Increment the Cursor after Each Byte Written
      to Display if Set
   S - Shift Display when Byte Written to Display

```
Enable Display/Cursor
    D - Turn Display On(1)/Off(0)
    C - Turn Cursor On(1)/Off(0)
    B - Cursor Blink On(1)/Off(0)

Move Cursor/Shift Display
   SC - Display Shift On(1)/Off(0)
   RL - Direction of Shift Right(1)/Left(0)

Set Interface Length
   DL - Set Data Interface Length 8(1)/4(0)
    N - Number of Display Lines 1(0)/2(1)
    F - Character Font 5x10(1)/5x7(0)

Poll the "Busy Flag"
   BF - This bit is set while the LCD is processing

Move Cursor to CGRAM/Display
    A - Address
Read/Write ASCII to the Display
    H - Data
```

Reading Data back is best used in applications that require data to be moved back and forth on the LCD (such as in applications which scroll data between lines). The "Busy Flag" can be polled to determine when the last instruction that has been sent has completed processing.

For most applications, there really is no reason to read from the LCD. "R/W" is tied to ground and the software simply waits the maximum amount of time for each instruction to complete. This is 4.1 msecs for clearing the display or moving the cursor/display to the "home position" and 160 usecs for all other commands. As well as making application software simpler, it also frees up a microcontroller pin for other uses.

One area of confusion is how to move to different locations on the display and, as a follow on, how to move to different lines on an LCD display. The following table shows how different LCD displays that use a single

44780 can be set up with the addresses for specific character locations. The LCDs listed are the most popular arrangements available and the "Layout" is given as number of columns by number of lines:

```
┌─────────────────────────────────────────────────────────────┐
│ Hitachi 44780 Based LCD Types and Character Locations         │
│                                                               │
│ LCD    Top    Ninth   Second  Third   Fourth  Comments        │
│        Left           Line    Line    Line                    │
│ 8x1    0      N/A     N/A     N/A     N/A     Note 1.         │
│ 16x1   0      0x040   N/A     N/A     N/A     Note 1.         │
│ 16x1   0      8       N/A     N/A     N/A     Note 3.         │
│ 8x2    0      N/A     0x040   N/A     N/A     Note 1.         │
│ 10x2   0      0x008   0x040   N/A     N/A     Note 2.         │
│ 16x2   0      0x008   0x040   N/A     N/A     Note 2.         │
│ 20x2   0      0x008   0x040   N/A     N/A     Note 2.         │
│ 24x2   0      0x008   0x040   N/A     N/A     Note 2.         │
│ 30x2   0      0x008   0x040   N/A     N/A     Note 2.         │
│ 32x2   0      0x008   0x040   N/A     N/A     Note 2.         │
│ 40x2   0      0x008   0x040   N/A     N/A     Note 2.         │
│ 16x4   0      0x008   0x040   0x020   0x040   Note 2.         │
│ 20x4   0      0x008   0x040   0x020   0x040   Note 2.         │
│ 40x4   0      N/A     N/A     N/A     N/A     Note 4.         │
│                                                               │
│ Note 1: Single 44780/No Support Chip.                        │
│ Note 2: 44780 with Support Chip.                             │
│ Note 3: 44780 with Support Chip. This is quite               │
│         rare.                                                 │
│ Note 4: Two 44780s with Support Chips. Addressing            │
│         is device specific.                                  │
└─────────────────────────────────────────────────────────────┘
```

Cursors for the 44780 can be turned on as a simple underscore at any time using the "Enable Display/Cursor" LCD instruction and setting the "C" bit. The "B" ("Block Mode") bit is not recommended as this causes a flashing full character square to be displayed and it really isn't that attractive.

The LCD can be thought of as a "Teletype" display because in normal operation, after a character has been sent to the LCD, the internal "Cursor" is moved one character to the right. The "Clear Display" and "Return Cursor and LCD to Home Position" instructions are used to reset the Cursor's position to the top right character on the display. An example of moving the cursor is shown in Fig. 8.18.

To move the Cursor, the "Move Cursor to Display" instruction is used. For this instruction, bit 7 of the instruction byte is set with the remaining 7 bits used as the address of the character on the LCD the cursor is to move to. These 7 bits provide 128 addresses, which matches the maximum number of LCD character addresses available. The table above should be used to determine the address of a character offset on a particular line of an LCD display. The LCD Character Set is shown in Fig. 8.19.

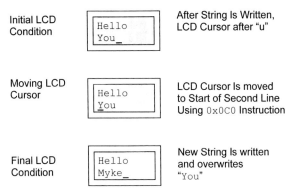

Initial LCD Condition

```
Hello
You_
```

After String Is Written, LCD Cursor after "u"

Moving LCD Cursor

```
Hello
You
```

LCD Cursor Is moved to Start of Second Line Using `0x0C0` Instruction

Final LCD Condition

```
Hello
Myke_
```

New String Is written and overwrites "You"

Figure 8.18 Moving an LCD Cursor

Char.code

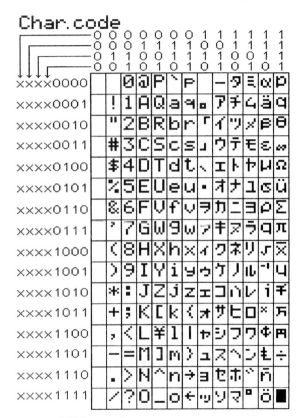

Figure 8.19 LCD Character Set

Eight programmable characters are available and use codes 0x000 to 0x007. They are programmed by pointing the LCD's "Cursor" to the Character Generator RAM ("CGRAM") Area at eight times the character address. The next 8 bytes written to the RAM are the line information of the programmable character, starting from the top. The "Character Box" is shown in Fig. 8.20.

The user defined character line information is saved in the LCD's "CGRAM" area. This 64 bytes of memory is accessed using the "Move Cursor into CGRAM" instruction in a similar manner to that of moving the cursor to a specific address in the memory with one important difference.

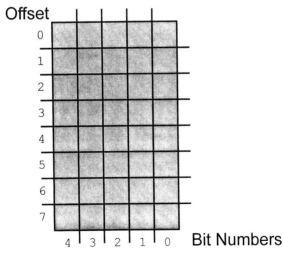

Figure 8.20 LCD Character "Box"

This difference is that each character starts at eight times its character value. This means that user definable character 0 has its data starting at address 0 of the CGRAM, character 1 starts at address 8, character 2 starts at address 0x010 (16) and so on. To get a specific line within the user definable character, its offset from the top (the top line has an offset of 0) is added to the starting address. In most applications, characters are written to all at one time with character 0 first. In this case, the instruction 0x040 is written to the LCD followed by all the user-defined characters.

Before commands or data can be sent to the LCD module, the module must be initialized. For 8-bit mode, this is done using the following series of operations:

```
1. Wait more than 15 msecs after power is applied.
2. Write 0x030 to LCD and wait 5 msecs for the
   instruction to complete.
3. Write 0x030 to LCD and wait 160 usecs for
   instruction to complete.
4. Write 0x030 AGAIN to LCD and wait 160 usecs or
   Poll the Busy Flag.
5. Set the Operating Characteristics of the LCD.
   - Write "Set Interface Length"
   - Write 0x010 to turn off the Display
   - Write 0x001 to Clear the Display
   - Write "Set Cursor Move Direction" Setting
     Cursor Behavior Bits
   - Write "Enable Display/Cursor" & enable Display
     and Optional Cursor
```

The first macro is "LCD8", which provides a basic interface to the LCD with "worst" case start-up delays. To invoke it, the statement

```
LCD8  DataPort,  EPort,  EPin,  RSPort,  RSPin,
  RWPort,  RWPin,  Frequency
```

is put in where "DataPort" is the 8-bit I/O port. "EPort" and "EPin" are the "E" clock Definition. "RSPort" and "RSPin" are the "RS" LCD Data Type Input. "RWPort" and "RWPin" are the pins used to poll the LCD for data reply (and are essentially unused). "Frequency" is the PICmicro® MCU operating speed and is used to calculate the delay values. The only variable required for the "LCD8" and "LCD8Poll" macros is the 8-bit variable "Dlay".

This macro should work with any low-end or mid-range PICmicro® MCU. Note that the "LCDPORTInit" subroutine cannot be used with low-end PICmicro® MCUs. To initialize the I/O ports the "TRIS" statements will have to be programmed in manually.

```
LCD8 Macro DataPort, EPort, EPin, RSPort, RSPin,
 RWPort, RWPin, Freq
 variable Dlay5Value, Dlay160Value, Dlay160Bit1 =
  -1, Dlay160Bit2 = -1
 variable BitCount = 0
 variable Value = 128, Bit = 7
Dlay5Value = ((5007 * (Freq / 1000) / 4000) / 7)
          + 256
Dlay160Value = (163 * (Freq / 1000) / 4000) / 3

 while (Bit >= 0)               ;  Find the Number of
                                ;    Bits and their
                                ;    positions in
                                ;    "Dlay160Value"
 if ((Dlay160Value & Value) != 0)
 if (Dlay160Bit1 == -1)         ;   Set the Upper Bit
Dlay160Bit1 = Bit
 else
 if (Dlay160Bit2 == -1)
Dlay160Bit2 = Bit
 endif
 endif
BitCount = BitCount + 1
 endif
```

```
Value = Value >> 1
Bit = Bit - 1
 endw
 if (BitCount > 2)              ;  Just Want max two
                               ;   Bits
 if ((Dlay160Bit1 - 1) == Dlay160Bit2)
Dlay160Bit1 = Dlay160Bit1 + 1  ;  Shift Top up by 1
Dlay160Bit2 = -1               ;  Delete Second
 else
Dlay160Bit2 = Dlay160Bit2 + 1  ;  Shift Bottom up by
                               ;   1
 endif
 endif

Dlay5                          ;  Delay 5 msecs
  movlw    (Dlay5Value & 0x0FF00) >> 8
  movwf    Dlay
  movlw    Dlay5Value & 0x0FF
  subwf    Dlay, w
  xorlw    0x0FF
  addwf    Dlay, w
  btfsc    STATUS, Z
  decfsz   Dlay, f
  goto     $ - 5
  return

LCDPORTInit                    ;  Initialize the I/O
                               ;   Ports
  bsf      STATUS, RP0         ;   ONLY used by mid-
                               ;   range
  movlw    0x000
  movwf    DataPort
  bcf      EPort, EPin
  bcf      RSPort, RSPin
  bcf      RWPort, RWPin
  bcf      STATUS, RP0
  bcf      EPort, EPin
  bcf      RSPort, RSPin
  bcf      RWPort, RWPin
  return

LCDIns                         ;  Send the
                               ;  Instruction to the
                               ;  LCD

  movwf    DataPort
  bcf      RSPort, RSPin
```

```
    if (Freq > 8000000)              ;  Make Sure Proper
                                     ;    Delay is In Place
    if (Freq < 16000000)
     nop
    else
     goto      $ + 1
    endif
    endif
     bsf       EPort, EPin
    if (Freq > 8000000)              ;  Make Sure Proper
                                     ;    Delay is In Place
    if (Freq < 16000000)
     nop
    else
     goto      $ + 1
    endif
    endif
     bcf       EPort, EPin
     bsf       Dlay, Dlay160Bit1     ;  Delay 160 usecs
    if (Dlay160Bit2 != -1)
     bsf       Dlay, Dlay160Bit2
    endif
     decfsz    Dlay, f
      goto     $ - 1
     andlw     0x0FC                 ;  Have to Delay 5
                                     ;    msecs?
     btfsc     STATUS, Z
      call     Dlay5
     return

    LCDChar                          ;  Send the Character
                                     ;    to the LCD
     movwf     DataPort
     bsf       RSPort, RSPin
    if (Freq > 8000000)              ;  Make Sure Proper
                                     ;    Delay is In Place
    if (Freq < 16000000)
     nop
    else
     goto      $ + 1
    endif
    endif
     bsf       EPort, EPin
    if (Freq > 8000000)              ;  Make Sure Proper
                                     ;    Delay is In Place
```

```
 if (Freq < 16000000)
  nop
 else
  goto      $ + 1
 endif
 endif
  bcf       EPort, EPin
  bsf       Dlay, Dlay160Bit1     ;  Delay 160 usecs
 if (Dlay160Bit-2 != -1)
  bsf       Dlay, Dlay160Bit2
 endif
  decfsz    Dlay, f
   goto     $ - 1
  return


LCDInit                           ;  Do the 8 Bit
                                  ;    Initialization
  call      Dlay5                 ;  Wait 15 msecs
  call      Dlay5
  call      Dlay5
  movlw     0x030
  call      LCDIns                ;  Send the Reset
                                  ;    Instruction
  call      Dlay5
  movlw     0x030
  call      LCDIns
  movlw     0x030
  call      LCDIns
  movlw     0x038                 ;  Set Interface
                                  ;    Length
  call      LCDIns
  movlw     0x010                 ;  Turn Off Display
  call      LCDIns
  movlw     0x001                 ;  Clear Display RAM
  call      LCDIns
  movlw     0x006                 ;  Set Cursor
                                  ;    Movement
  call      LCDIns
  movlw     0x00E                 ;  Turn on
                                  ;    Display/Cursor
  call      LCDIns
  return
 endm
```

The "LCD8Poll" macro is slightly more sophisticated than the "LCD8" macro. Instead of providing "hard-coded" delays in the application, the code "polls" the LCD to see if the Operation is complete before continuing. This is done by putting the "DataPort" into "Input Mode" and then strobing the "E" bit (with RS reset and RW set) and looking at bit 7 of the I/O port. The macro code is

```
LCD8Poll Macro DataPort, EPort, EPin, RSPort, RSPin,
 RWPort, RWPin, Freq
 variable Dlay5Value, Dlay160Value, Dlay160Bit1 =
  -1, Dlay160Bit2 = -1
 variable BitCount = 0
 variable Value = 128, Bit = 7
 errorlevel 0,-224
Dlay5Value = ((5007 * (Freq / 1000) / 4000) / 7)
          + 256
Dlay160Value = (163 * (Freq / 1000) / 4000) / 3

 while (Bit >= 0)                    ;  Find the Number of
                                     ;   Bits and their
                                     ;    Positions in
                                     ;    "Dlay160Value"
 if ((Dlay160Value & Value) != 0)
 if (Dlay160Bit1 == -1)             ;  Set the Upper Bit
Dlay160Bit1 = Bit
 else
 if (Dlay160Bit2 == -1)
Dlay160Bit2 = Bit
 endif
 endif
BitCount = BitCount + 1
 endif
Value = Value >> 1
Bit = Bit - 1
 endw
 if (BitCount > 2)                   ;  Just Want max two
                                     ;    Bits
 if ((Dlay160Bit1 - 1) == Dlay160Bit2)
Dlay160Bit1 = Dlay160Bit1 + 1  ;  Shift Top up by 1
Dlay160Bit2 = -1               ;  Delete Second
 else
```

```
Dlay160Bit2 = Dlay160Bit2 + 1  ;  Shift Bottom up by
                               ;   1
  endif
  endif
Dlay5                              ;  Delay 5 msecs
  movlw   (Dlay5Value & 0x0FF00) >> 8
  movwf   Dlay
  movlw   Dlay5Value & 0x0FF
  subwf   Dlay, w
  xorlw   0x0FF
  addwf   Dlay, w
  btfsc   STATUS, Z
   decfsz Dlay, f
    goto  $ - 5
  return

LCDPORTInit                        ;  Initialize the I/O
                                   ;   Ports
  bsf     STATUS, RP0              ;  ONLY used by mid-
                                   ;   range
  movlw   0x000
  movwf   DataPort
  bcf     EPort, EPin
  bcf     RSPort, RSPin
  bcf     RWPort, RWPin
  bcf     STATUS, RP0
  bcf     EPort, EPin
  bcf     RSPort, RSPin
  bcf     RWPort, RWPin
  return

LCDIns                             ;  Send the
                                   ;   Instruction to
                                   ;   the LCD
  movwf   Dlay
  movlw   0x0FF                    ;  Read the "BF" Flag
  tris    DataPort
  bcf     RSPort, RSPin            ;  Read the
                                   ;   Instruction
                                   ;   Register
  bsf     RWPort, RWPin
  goto    $ + 1
  bsf     EPort, EPin
  nop
  movf    DataPort, w             ;  Read the Data Port
                                   ;   Value
```

```
    nop
    bcf     EPort, EPin
    andlw   0x080                   ;   Is the High Bit
                                    ;   Set?

    btfss   STATUS, Z
     goto   $ - 7
    bcf     RWPort, RWPin
    movlw   0                       ;   Put the DataPort
                                    ;    Back into Output
                                    ;    Mode
    tris    DataPort
    movf    Dlay, w                 ;   Get the Saved
                                    ;    Character
    movwf   DataPort
    if (Freq > 8000000)             ;   Make Sure Proper
                                    ;    Delay is In Place
    if (Freq < 16000000)
     nop
    else
     goto   $ + 1
    endif
    endif
    bsf     EPort, EPin
    if (Freq > 8000000)             ;   Make Sure Proper
                                    ;    Delay is In Place
    if (Freq < 16000000)
     nop
    else
     goto   $ + 1
    endif
    endif
    bcf     EPort, EPin
    return

LCDChar                             ;   Send the Character
                                    ;    to the LCD
    movwf   Dlay
    movlw   0x0FF                   ;   Read the "BF" Flag
    tris    DataPort
    bcf     RSPort, RSPin           ;   Read the
                                    ;    Instruction
                                    ;    Register
    bsf     RWPort, RWPin
    goto    $ + 1
    bsf     EPort, EPin
    nop
```

```
        movf    DataPort, w         ;  Read the Data Port
                                    ;   Value
        nop
        bcf     EPort, EPin
        andlw   0x080               ;  Is the High Bit
                                    ;   Set?
        btfss   STATUS, Z
         goto   $ - 7
        bsf     RSPort, RSPin
        bcf     RWPort, RWPin
        movlw   0                   ;  Put the DataPort
                                    ;   Back into Output
                                    ;   Mode
        tris    DataPort
        movf    Dlay, w             ;  Get the Saved
                                    ;   Character
        movwf   DataPort
        if (Freq > 8000000)         ;  Make Sure Proper
                                    ;   Delay is In Place
        if (Freq < 16000000)
        nop
        else
         goto   $ + 1
        endif
        endif
        bsf     EPort, EPin
        if (Freq > 8000000)         ;  Make Sure Proper
                                    ;   Delay is In Place
        if (Freq < 16000000)
        nop
        else
        goto    $ + 1
        endif
        endif
        bcf     EPort, EPin
        return

LCDInit                             ;  Do the 8 Bit
                                    ;   Initialization
        call    Dlay5               ;  Wait 15 msecs
        call    Dlay5
        call    Dlay5
        movlw   0x030
        movwf   DataPort
        if (Freq > 8000000)         ;  Make Sure Proper
                                    ;   Delay is In Place
```

```
      if (Freq < 16000000)
       nop
      else
       goto    $ + 1
      endif
      endif
      bsf     EPort, EPin
      if (Freq > 8000000)           ;  Make Sure Proper
                                    ;   Delay is In Place
      if (Freq < 16000000)
       nop
      else
       goto    $ + 1
      endif
      endif
      bcf     EPort, EPin           ;  Send the Reset
                                    ;   Instruction
       call    Dlay5
      if (Freq > 8000000)           ;  Make Sure Proper
                                    ;   Delay is In Place
      if (Freq < 16000000)
       nop
      else
       goto    $ + 1
      endif
      endif
      bsf     EPort, EPin
      if (Freq > 8000000)           ;  Make Sure Proper
                                    ;   Delay is In Place
      if (Freq < 16000000)
       nop
      else
       goto    $ + 1
      endif
      endif
      bcf     EPort, EPin           ;  Send the Reset
                                    ;   Instruction
      bsf     Dlay, Dlay160Bit1     ;  Delay 160 usecs
      if (Dlay160Bit2 != -1)
      bsf     Dlay, Dlay160Bit2
      endif
      decfsz  Dlay, f
       goto    $ - 1
      movlw   0x030
      call    LCDIns
```

```
        movlw   0x038                   ;   Set Interface
                                        ;     Length
        call    LCDIns
        movlw   0x010                   ;   Turn Off Display
        call    LCDIns
        movlw   0x001                   ;   Clear Display RAM
        call    LCDIns
        movlw   0x006                   ;   Set Cursor
                                        ;     Movement
        call    LCDIns
        movlw   0x00E                   ;   Turn on
                                        ;     Display/Cursor
        call    LCDIns
        return
        errorlevel 0,+224               ;   Enable "TRIS"
                                        ;     Indicators
        endm
```

The LCD should be initialized in 4-bit mode, data is written to the LCD in terms of nybbles. This is done because initially just single nybbles are sent (and not two, which make up a byte and a full instruction). When a byte is sent, the high nybble is sent before the low nybble and the "E" pin is toggled each time a nybble is sent to the LCD. To initialize in 4-bit mode:

```
1. Wait more than 15 msecs after power is applied.
2. Write 0x03 to LCD and wait 5 msecs for the
   instruction to complete.
3. Write 0x03 to LCD and wait 160 usecs for
   instruction to complete.
4. Write 0x03 AGAIN to LCD and wait 160 usecs (or
   poll the Busy Flag).
5. Set the Operating Characteristics of the LCD.
   - Write 0x02 to the LCD to Enable Four Bit Mode

All following instruction/Data Writes require two
 nybble writes:

   - Write "Set Interface Length"
   - Write 0x01/0x00 to turn off the Display
```

    - Write 0x00/0x01 to Clear the Display
    - Write "Set Cursor Move Direction" Setting
      Cursor Behavior Bits
    - Write "Enable Display/Cursor" & enable Display
      and Optional Cursor

The 4-bit LCD interfacing (the "LCD4" Macro) is modified from the "LCD8" macro. To invoke the macro, the
similar statement

```
LCD4 DataPort, DataBit, EPort, EPin, RSPort, RSPin,
  RWPort, RWPin, Freq
```

is used. The "DataBit" parameter is lowest of the four
data bits. It can only be "0" or "4". The macro requires the "LCDTemp" Variable along with "Dlay". The
Macro is

```
LCD4 Macro DataPort, DataBit, EPort, EPin, RSPort,
          RSPin, RWPort, RWPin, Freq
 variable Dlay5Value, Dlay160Value, Dlay160Bit1 =
  -1, Dlay160Bit2 = -1
 variable BitCount = 0
 variable Value = 128, Bit = 7
Dlay5Value = ((5007 * (Freq / 1000) / 4000) / 7)
          + 256
Dlay160Value = (163 * (Freq / 1000) / 4000) / 3

 if ((DataBit != 0) && (DataBit != 4))
 error "Invalid 'DataBit' Specification - Can only
  be '0' or '4'"
 endif

 while (Bit >= 0)                ;  Find the Number of
                                 ;   Bits and their
                                 ;   Positions in
                                 ;   "Dlay160Value"
 if ((Dlay160Value & Value) != 0)
 if (Dlay160Bit1 == -1)          ;  Set the Upper Bit
Dlay160Bit1 = Bit
 else
```

```
  if (Dlay160Bit2 == -1)
Dlay160Bit2 = Bit
 endif
 endif
BitCount = BitCount + 1
 endif
Value = Value >> 1
Bit = Bit - 1
 endw
 if (BitCount > 2)              ; Just Want max two
                               ;  Bits
 if ((Dlay160Bit1 - 1) == Dlay160Bit2)
Dlay160Bit1 = Dlay160Bit1 + 1 ; Shift Top up by 1
Dlay160Bit2 = -1              ; Delete Second
 else
Dlay160Bit2 = Dlay160Bit2 + 1 ; Shift Bottom up
                               ;  by 1
 endif
 endif

Dlay5                          ; Delay 5 msecs
  movlw   (Dlay5Value & 0x0FF00) >> 8
  movwf   Dlay
  movlw   Dlay5Value & 0x0FF
  subwf   Dlay, w
  xorlw   0xFF
  addwf   Dlay, w
  btfsc   STATUS, Z
   decfsz Dlay, f
   goto   $ - 5
  return

LCDPORTInit                    ; Initialize the I/O
                               ;  Ports
  bsf     STATUS, RP0          ; ONLY used by mid-
                               ;  range
 if (DataBit == 0)
 movlw   0x0F0
 else
 movlw   0x00F
 endif
 movwf   DataPort
 bcf     EPort, EPin
 bcf     RSPort, RSPin
 bcf     RWPort, RWPin
 bcf     STATUS, RP0
```

```
       bcf      EPort, EPin
       bcf      RSPort, RSPin
       bcf      RWPort, RWPin
       return

LCDIns                                    ;  Send the
                                          ;   Instruction to
                                          ;   the LCD
       movwf    LCDTemp                   ;  Save the Value
       if (DataBit == 0)
       swapf    LCDTemp, w                ;  Most Significant
                                          ;   Nybble First
       andlw    0x00F
       else
       andlw    0x0F0
       endif
       movwf    DataPort
       bcf      RSPort, RSPin
       if (Freq > 8000000)                ;  Make Sure Proper
                                          ;   Delay is In Place
       if (Freq < 16000000)
       nop
       else
       goto     $ + 1
       endif
       endif
       bsf      EPort, EPin
       if (Freq > 8000000)                ;  Make Sure Proper
                                          ;   Delay is In Place
       if (Freq < 16000000)
       nop
       else
       goto     $ + 1
       endif
       endif
       bcf      EPort, EPin
       if (DataBit == 0)
       movf     LCDTemp, w
       andlw    0x00F
       else
       swapf    LCDTemp, w                ;  Least Significant
                                          ;   Nybble Second
       andlw    0x0F0
       endif
```

```
 movwf   DataPort
 bcf     RSPin, RSPin
 if (Freq > 8000000)                    ;  Make Sure Proper
                                        ;    Delay is In Place
 if (Freq < 16000000)
 nop
 else
 goto    $ + 1
 endif
 endif
 bsf     EPort, EPin
 if (Freq > 8000000)                    ;  Make Sure Proper
                                        ;    Delay is In Place
 if (Freq < 16000000)
 nop
 else
 goto    $ + 1
 endif
 endif
 bcf     EPort, EPin
 bsf     Dlay, Dlay160Bit1              ;  Delay 160 usecs
 if (Dlay160Bit2 != -1)
 bsf     Dlay, Dlay160Bit2
 endif
 decfsz  Dlay, f
 goto    $ - 1
 movf    LCDTemp, w
 andlw   0x0FC                          ;  Have to Delay 5
                                        ;    msecs?
 btfsc   STATUS, Z
 call    Dlay5
 return

LCDChar                                 ;  Send the Character
                                        ;    to the LCD
 movwf   LCDTemp                        ;  Save the Value
 if (DataBit == 0)
 swapf   LCDTemp, w                     ;  Most Significant
                                        ;    Nybble First
 andlw   0x00F
 else
 andlw   0x0F0
 endif
 movwf   DataPort
 bsf     RSPort, RSPin
```

```
    if (Freq > 8000000)              ;   Make Sure Proper
                                     ;     Delay is In Place
    if (Freq < 16000000)
     nop
    else
     goto    $ + 1
    endif
    endif
     bsf     EPort, EPin
    if (Freq > 8000000)              ;   Make Sure Proper
                                     ;     Delay is In Place
    if (Freq < 16000000)
     nop
    else
     goto    $ + 1
    endif
    endif
     bcf     EPort, EPin
    if (DataBit == 0)
     movf    LCDTemp, w
     andlw   0x00F
    else
     swapf   LCDTemp, w             ;   Least Significant
                                    ;     Nybble Second
     andlw   0x0F0
    endif
     movwf   DataPort
     bsf     RSPort, RSPin
    if (Freq > 8000000)              ;   Make Sure Proper
                                     ;     Delay is In Place
    if (Freq < 16000000)
     nop
    else
     goto    $ + 1
    endif
    endif
     bsf     EPort, EPin
    if (Freq > 8000000)              ;   Make Sure Proper
                                     ;     Delay is In Place
    if (Freq < 16000000)
     nop
    else
     goto    $ + 1
    endif
    endif
```

```
 bcf      EPort, EPin
 bsf      Dlay, Dlay160Bit1     ;  Delay 160 usecs
if (Dlay160Bit2 != -1)
 bsf      Dlay, Dlay160Bit2
endif
 decfsz   Dlay, f
  goto    $ - 1
 return


LCDInit                         ;  Do the 8 Bit
                                ;   Initialization
 call     Dlay5                 ;  Wait 15 msecs
 call     Dlay5
 call     Dlay5
if (DataBit == 0)               ;  Send the Reset
                                ;   Instruction
 movlw    0x003
else
 movlw    0x030
endif
 movwf    DataPort
if (Freq > 8000000)             ;  Make Sure Proper
                                ;   Delay is In Place
if (Freq < 16000000)
 nop
else
 goto     $ + 1
endif
endif
 bsf      EPort, EPin
if (Freq > 8000000)             ;  Make Sure Proper
                                ;   Delay is In Place
if (Freq < 16000000)
 nop
else
 goto     $ + 1
endif
endif
 bcf      EPort, EPin
 call     Dlay5
 bsf      EPort, EPin           ;  Send Another Reset
                                ;   Instruction
if (Freq > 8000000)             ;  Make Sure Proper
                                ;   Delay is In Place
if (Freq < 16000000)
```

```
 nop
else
 goto    $ + 1
endif
endif
 bcf     EPort, EPin
 bsf     Dlay, Dlay160Bit1    ;  Delay 160 usecs
if (Dlay160Bit2 != -1)
 bsf     Dlay, Dlay160Bit2
endif
 decfsz  Dlay, f
  goto   $ - 1
 bsf     EPort, EPin          ;  Send the Third
                              ;   Reset Instruction
if (Freq > 8000000)           ;  Make Sure Proper
                              ;   Delay is In Place
if (Freq < 16000000)
 nop
else
 goto    $ + 1
endif
endif
 bcf     EPort, EPin
 bsf     Dlay, Dlay160Bit1    ;  Delay 160 usecs
if (Dlay160Bit2 != -1)
 bsf     Dlay, Dlay160Bit2
endif
 decfsz  Dlay, f
  goto   $ - 1
if (DataBit == 0)             ;  Send the Data
                              ;   Length
                              ;    Specification
 movlw   0x002
else
 movlw   0x020
endif
 movwf   DataPort
if (Freq > 8000000)           ;  Make Sure Proper
                              ;   Delay is In Place
if (Freq < 16000000)
 nop
else
 goto    $ + 1
endif
```

```
     endif
      bsf      EPort, EPin
      if (Freq > 8000000)           ;  Make Sure Proper
                                    ;   Delay is In Place
      if (Freq < 16000000)
      nop
     else
      goto     $ + 1
     endif
     endif
      bcf      EPort, EPin
      bsf      Dlay, Dlay160Bit1    ;  Delay 160 usecs
      if (Dlay160Bit2 != -1)
      bsf      Dlay, Dlay160Bit2
     endif
      decfsz   Dlay, f
       goto    $ - 1
      movlw    0x028                ;  Set Interface
                                    ;   Length
      call     LCDIns
      movlw    0x010                ;  Turn Off Display
      call     LCDIns
      movlw    0x001                ;  Clear Display RAM
      call     LCDIns
      movlw    0x006                ;  Set Cursor
                                    ;   Movement
      call     LCDIns
      movlw    0x00E                ;  Turn on
                                    ;   Display/Cursor
      call     LCDIns
      return
     endm
```

It is recommended that the I/O pins and the 4-bit "DataPort" are on the same 8-bit I/O Port. The reason for doing this is that when using this code, writes to the "DataPort" will change the output values of other I/O register bits.

The interface requirements to the PICmicro® MCU can be reduced by using the circuit shown in Fig. 8.21

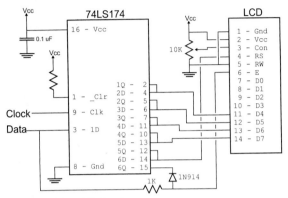

Figure 8.21 2-Wire LCD Interface

in which the serial data is combined with the contents of the shift register to produce the "E" strobe at the appropriate interval. This circuit "ANDs" (using the 1K resistor and IN914 diode) the output of the sixth "D-Flip Flop" of the 74LS174 and the "Data" bit from the device writing to the LCD to form the "E" Strobe. This method requires one less pin than a three-wire shift register interface and a few more instructions of code. The two-wire LCD interface circuit is shown in Fig. 8.21.

The 74LS174 can be wired as a shift register (as is shown in the schematic diagram) instead of a serial-in/parallel-out shift register. This circuit should work without any problems with a dedicated serial-in/parallel-out shift register chip, but the timings/clock polarities may be different. When the 74LS174 is used, note that the data is latched on the rising (from logic "low" to "high") edge of the clock signal. Figure 8.22 is a timing diagram for the two-wire interface and shows the 74LS174 being cleared, loaded, and then the "E" Strobe when the data is valid and "6Q" and incoming "Data" is high.

Before data can be written to it, loading every latch with zeros clears the shift register. Next, a "1" (to provide the "E" Gate) is written followed by the "R/S" bit and the four data bits. Once the latch is loaded in correctly, the "Data" line is pulsed to Strobe the "E" bit. The biggest difference between the three-wire and two-wire interface is that the shift register has to be cleared before it can be loaded and the two-wire operation re-

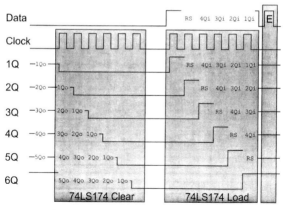

Figure 8.22 2-Wire LCD Write Waveform

quires more than twice the number of clock cycles to load 4-bits into the LCD.

One note about the LCD's "E" Strobe is that in some documentation it is specified as "high" level active while in others, it is specified as falling edge active. It *seems* to be falling edge active, which is why the two-wire LCD interface presented below works even if the line ends up being high at the end of data being shifted in. If the falling edge is used (like in the two-wire interface) then make sure that before the "E" line is output on "0", there is at least a 450 nsecs delay with no lines changing state.

The two-wire LCD interface macro uses the same parameters as the previous macros. This interface is quite a bit slower than the other ones that I have presented, but uses the fewest PICmicro® MCU I/O pins. The "LCD2" Macro only requires the "Dlay" and "LCDTemp" variables.

```
LCD2 Macro ClockPort, ClockPin, DataPort, DataPin,
 Freq
 variable Dlay5Value, Dlay160Value, Dlay160Bit1
  = -1, Dlay160Bit2 = -1
 variable BitCount = 0, i
 variable Value = 128, Bit = 7
Dlay5Value = ((5007 * (Freq / 1000) / 4000) / 7)
           + 256
Dlay160Value = (163 * (Freq / 1000) / 4000) / 3

 while (Bit >= 0)                   ;  Find the Number of
                                    ;   Bits and their
                                    ;   Positions in
                                    ;   "Dlay160Value"
 if ((Dlay160Value & Value) != 0)
 if (Dlay160Bit1 == -1)             ; Set the Upper Bit
Dlay160Bit1 = Bit
 else
 if (Dlay160Bit2 == -1)
Dlay160Bit2 = Bit
 endif
 endif
```

```
BitCount = BitCount + 1
 endif
Value = Value >> 1
Bit = Bit - 1
 endw
 if (BitCount > 2)              ;  Just Want max two
                               ;   Bits
 if ((Dlay160Bit1 - 1) == Dlay160Bit2)
Dlay160Bit1 = Dlay160Bit1 + 1  ;  Shift Top up by 1
Dlay160Bit2 = -1               ;  Delete Second
 else
Dlay160Bit2 = Dlay160Bit2 + 1  ;  Shift Bottom up
                               ;   by 1
 endif
 endif


Dlay5                          ;  Delay 5 msecs
  movlw   (Dlay5Value & 0x0FF00) >> 8
  movwf   Dlay
  movlw   Dlay5Value & 0x0FF
  subwf   Dlay, w
  xorlw   0x0FF
  addwf   Dlay, w
  btfsc   STATUS, Z
   decfsz Dlay, f
    goto  $ - 5
  return


LCDPORTInit                    ;  Initialize the I/O
                               ;   Ports
  bsf     STATUS, RP0          ;  ONLY used by mid-
                               ;   range
  bcf     ClockPort, ClockPin
  bcf     DataPort, DataPin
  bcf     STATUS, RP0
  bcf     ClockPort, ClockPin
  bcf     DataPort, DataPin
  return


LCDIns                         ;  Send the
                               ;   Instruction to
                               ;   the LCD
  movwf   LCDTemp              ;  Save the Value
  movlw   6                    ;  Clear the Shift
                               ;   Register
```

```
        movwf   Dlay
        bsf     ClockPort, ClockPin
        bcf     ClockPort, ClockPin
        decfsz  Dlay, f
         goto   $ - 3
        movwf   Dlay            ;  w still equals 6
        movf    LCDTemp, w      ;  Shift out the
                                ;    Upper 4 Bits
        swapf   LCDTemp, f
        bsf     LCDTemp, 5      ;  Make LCDTemp
                                ;    Correct for
                                ;    Shifting
        bcf     LCDTemp, 4      ;  This is "RS" Bit
        bcf     DataPort, DataPin  ;  Shift Out Each Bit
        btfsc   LCDTemp, 5      ;  5 is the Current
                                ;    MSB
         bsf    DataPort, DataPin  ;  Shift Out the Next
                                ;    Highest Bit
        bsf     ClockPort, ClockPin
        bcf     ClockPort, ClockPin
        rlf     LCDTemp, f
        decfsz  Dlay, f
         goto   $ - 7
        bsf     DataPort, DataPin  ;  Latch in the Data
        if (Freq > 8000000)     ;  Make Sure Proper
                                ;    Delay is In Place
        if (Freq < 16000000)
         nop
        else
         goto   $ + 1
        endif
        endif
        bcf     DataPort, DataPin
        bsf     Dlay, 2         ;  Dlay = 6 for Shift
                                ;    Out
        bsf     Dlay, 1
        bsf     ClockPort, ClockPin  ;  Clear the Shift
                                ;    Register
        bcf     ClockPort, ClockPin
        decfsz  Dlay, f
         goto   $ - 3
        movwf   LCDTemp         ;  Shift out the Low
                                ;    Nybble
        bsf     Dlay, 2         ;  Dlay = 6 for Shift
                                ;    Out
        bsf     Dlay, 1
```

```
    bsf      LCDTemp, 5          ;  Make LCDTemp
                                 ;   Correct for
                                 ;   Shifting
    bcf      LCDTemp, 4          ;  This is "RS" Bit
    bcf      DataPort, DataPin   ;  Shift Out Each Bit
    btfsc    LCDTemp, 5          ;  5 is the Current
                                 ;   MSB
    bsf      DataPort, DataPin   ;  Shift Out the Next
                                 ;   Highest Bit
    bsf      ClockPort, ClockPin
    bcf      ClockPort, ClockPin
    rlf      LCDTemp, f
    decfsz   Dlay, f
    goto     $ - 7
    bsf      DataPort, DataPin   ;  Latch in the Data
 if (Freq > 8000000)            ;  Make Sure Proper
                                 ;   Delay is In Place
 if (Freq < 16000000)
  nop
 else
  goto     $ + 1
 endif
 endif
    bcf      DataPort, DataPin
    bsf      Dlay, Dlay160Bit1   ;  Delay 160 usecs
 if (Dlay160Bit2 != -1)
    bsf      Dlay, Dlay160Bit2
 endif
    decfsz   Dlay, f
    goto     $ - 1
    andlw    0x0FC               ;  Have to Delay 5
                                 ;   msecs?
    btfsc    STATUS, Z
    call     Dlay5
    return

LCDChar                          ;  Send the Character
                                 ;   to the LCD
    movwf    LCDTemp             ;  Save the Value
    movlw    6                   ;  Clear the Shift
                                 ;   Register
    movwf    Dlay
    bsf      ClockPort, ClockPin
    bcf      ClockPort, ClockPin
    decfsz   Dlay, f
    goto     $ - 3
```

```
movwf    Dlay             ; w still equals 6
movf     LCDTemp, w       ; Shift out the
                          ;   Upper 4 Bits
swapf    LCDTemp, f
bsf      LCDTemp, 5       ; Make LCDTemp
                          ;   Correct for
                          ;   Shifting
bsf      LCDTemp, 4       ; This is "RS" Bit
bcf      DataPort, DataPin  ; Shift Out Each Bit
btfsc    LCDTemp, 5       ; 5 is the Current
                          ;   MSB
 bsf     DataPort, DataPin  ; Shift Out the Next
                          ;   Highest Bit
bsf      ClockPort, ClockPin
bcf      ClockPort, ClockPin
rlf      LCDTemp, f
decfsz   Dlay, f
 goto    $ - 7
bsf      DataPort, DataPin  ; Latch in the Data
if (Freq > 8000000)       ; Make Sure Proper
                          ;   Delay is In Place
if (Freq < 16000000)
 nop
else
 goto    $ + 1
endif
endif
bcf      DataPort, DataPin
bsf      Dlay, 2          ; Dlay = 6 for Shift
                          ;   Out
bsf      Dlay, 1
bsf      ClockPort, ClockPin  ; Clear the Shift
                          ;   Register
bcf      ClockPort, ClockPin
decfsz   Dlay, f
 goto    $ - 3
movwf    LCDTemp          ; Shift out the Low
                          ;   Nybble
bsf      Dlay, 2          ; Dlay = 6 for Shift
                          ;   Out
bsf      Dlay, 1
bsf      LCDTemp, 5       ; Make LCDTemp
                          ;   Correct for
                          ;   Shifting
bsf      LCDTemp, 4       ; This is "RS" Bit
bcf      DataPort, DataPin  ; Shift Out Each Bit
```

```
    btfsc   LCDTemp, 5          ;  5 is the Current
                                ;    MSB
    bsf     DataPort, DataPin   ;  Shift Out the Next
                                ;    Highest Bit
    bsf     ClockPort, ClockPin
    bcf     ClockPort, ClockPin
    rlf     LCDTemp, f
    decfsz  Dlay, f
      goto  $ - 7
    bsf     DataPort, DataPin   ;  Latch in the Data
if (Freq > 8000000)             ;  Make Sure Proper
                                ;    Delay is In Place
  if (Freq < 16000000)
  nop
  else
    goto  $ + 1
  endif
endif
    bcf     DataPort, DataPin
    bsf     Dlay, Dlay160Bit1   ;  Delay 160 usecs
if (Dlay160Bit2 != -1)
    bsf     Dlay, Dlay160Bit2
endif
    decfsz  Dlay, f
    goto    $ - 1
    return
LCDInit                         ;  Do the 8 Bit
                                ;    Initialization
    call    Dlay5               ;  Wait 15 msecs
    call    Dlay5
    call    Dlay5
    movlw   0x023               ;  Initialize the I/O
                                ;    Port
    movwf   LCDTemp             ;  Save the Value
    movlw   6                   ;  Clear the Shift
                                ;    Register
    movwf   Dlay
    bsf     Clockport, ClockPin
    bcf     Clockport, ClockPin
    decfsz  Dlay, f
      goto  $ - 3
    movwf   Dlay
    bcf     DataPort, DataPin   ;  Shift Out Each Bit
    btfsc   LCDTemp, 5          ;  5 is the Current
                                ;    MSB
```

```
    bsf     DataPort, DataPin    ;  Shift Out the Next
                                 ;   Highest Bit
  bsf     ClockPort, ClockPin
  bcf     ClockPort, ClockPin
  rlf     LCDTemp, f
  decfsz  Dlay, f
   goto   $ - 7
  bsf     DataPort, DataPin    ;  Latch in the Data
 if (Freq > 8000000)           ;  Make Sure Proper
                               ;   Delay is In Place
 if (Freq < 16000000)
 nop
 else
  goto   $ + 1
 endif
 endif
  bcf     DataPort, DataPin
  call    Dlay5
  bsf     DataPort, DataPin    ;  Send another 0x03
                               ;   to the LCD
 if (Freq > 8000000)           ;  Make Sure Proper
                               ;   Delay is In Place
 if (Freq < 16000000)
 nop
 else
 goto   $ + 1
 endif
 endif
 bcf     DataPort, DataPin
 bsf     Dlay, Dlay160Bit1    ;  Delay 160 usecs
 if (Dlay160Bit2 != -1)
 bsf     Dlay, Dlay160Bit2
 endif
 decfsz  Dlay, f
  goto   $ - 1
 bsf     DataPort, DataPin    ;  Send another 0x03
                              ;   to the LCD
 if (Freq > 8000000)          ;  Make Sure Proper
                              ;   Delay is In Place
 if (Freq < 16000000)
 nop
 else
 goto   $ + 1
 endif
 endif
```

```
 bcf      DataPort, DataPin
 bsf      Dlay, Dlay160Bit1      ;  Delay 160 usecs
if (Dlay160Bit2 != -1)
 bsf      Dlay, Dlay160Bit2
endif
 decfsz   Dlay, f
  goto    $ - 1
 movlw    0x022                  ;  Initialize the I/O
                                 ;   Port
 movwf    LCDTemp                ;  Save the Value
 movlw    6                      ;  Clear the Shift
                                 ;   Register
 movwf    Dlay
 bsf      ClockPort, ClockPin
 bcf      ClockPort, ClockPin
 decfsz   Dlay, f
  goto    $ - 3
 movwf    Dlay
 bcf      DataPort, DataPin      ;  Shift Out Each Bit
 btfsc    LCDTemp, 5             ;  5 is the Current MSB
  bsf     DataPort, DataPin      ;  Shift Out the Next
                                 ,   Highest Bit
 bsf      ClockPort, ClockPin
 bcf      ClockPort, ClockPin
 rlf      LCDTemp, f
 decfsz   Dlay, f
  goto    $ - 7
 bsf      DataPort, DataPin      ;  Latch in the Data
if (Freq > 8000000)             ;  Make Sure Proper
                                 ;    Delay is In Place
if (Freq < 16000000)
 nop
else
 goto     $ + 1
endif
endif
 bcf      DataPort, DataPin
 bsf      Dlay, Dlay160Bit1      ;  Delay 160 usecs
if (Dlay160Bit2 != -1)
 bsf      Dlay, Dlay160Bit2
endif
 decfsz   Dlay, f
  goto    $ - 1
 movlw    0x028                  ;  Set Interface
                                 ;   Length
 call     LCDIns
 movlw    0x010                  ;  Turn Off Display
```

```
      call    LCDIns
      movlw   0x001              ;   Clear Display RAM
      call    LCDIns
      movlw   0x006              ;   Set Cursor Movement
      call    LCDIns
      movlw   0x00E              ;   Turn on
                                 ;     Display/Cursor

      call    LCDIns
      return
    endm
```

I2C Bit Banging "Master" Interface

For the interface code below, make sure there is a 1K to 10K pull up on the SCL and SDA lines.

```
    I2CSetup Macro ClockPort, ClockPin, DataPort,
    DataPin, Rate, Frequency
    variable Dlay, Fraction     ;   Delay in
                                ;     Instruction
                                ;     Cycles
    Dlay = ((Frequency * 110) / (800 * Rate)) / 1000
    Fraction = ((Frequency * 110) / (800 * Rate))
            - (Dlay * 1000)
     if (Fraction > 499)
    Dlay = Dlay + 1
     endif

    I2CBitSetup                 ;   Setup I2C Lines
                                ;     for Application
      bsf     STATUS, RP0
      bcf     ClockPort, ClockPin ;  Driving Output
      bcf     DataPort, DataPin
      bcf     STATUS, RP0
      bsf     ClockPort, ClockPin ;  Everything High
                                ;     Initially
      bsf     DataPort, DataPin
     DlayMacro Dlay             ;   Make Sure Lines
                                ;     are High for
                                ;     adequate
                                ;     Period of Time

      return
```

```
I2CStart                            ;   Send a "Start"
                                    ;    Pulse to the I2C
                                    ;    Device
  bsf      ClockPort, ClockPin
  bsf      DataPort, DataPin
 DlayMacro Dlay - 2
  bcf      DataPort, DataPin     ;   Drop the Data Line
 DlayMacro Dlay
  bcf      ClockPort, ClockPin   ;   Drop the Clock
                                    ;    Line
 DlayMacro Dlay - 2              ;   Wait for the
                                    ;    Specified Period
  return                            ;   Exit with Clock
                                    ;    = Low, Data = Low


I2CStop                             ;   Pass Stop Bit to
                                    ;    I2C Device
 DlayMacro Dlay
  bsf      ClockPort, ClockPin   ;   Clock Bit High
 DlayMacro Dlay
  bsf      DataPort, DataPin
  return                            ;   Exit with Clock
                                    ;    = High, Data
                                    ;    = High


I2CRead                             ;   Read 8 Bits from
                                    ;    the Line
                                    ;    Reply with "ACK"
                                    ;    in Carry Flag
  bsf      I2CTemp, 0            ;   Put in the Carry
                                    ;    Flag
  btfsc    STATUS, C
  bcf      I2CTemp, 0            ;   If Carry Set, then
                                    ;    Send "Ack"
                                    ;    (-ative)
  bsf      STATUS, RP0          ;   Let the I2C Device
                                    ;    Drive the Data
                                    ;    Line
  bsf      DataPort, DataPin
  bcf      STATUS, RP0
  movlw    0x010 - 8

I2CRLoop
  bsf      ClockPort, ClockPin   ;   Bring the Clock
                                    ;    Line Up
 DlayMacro (Dlay / 2) - 1
```

```
      bcf     STATUS, C
      btfsc   DataPort, DataPin    ;  Sample the
                                   ;    Incoming Data
      bsf     STATUS, C
    DlayMacro (Dlay / 2) - 2
      bcf     ClockPort, ClockPin
      rlf     I2CTemp, f           ;  Shift in the Bit
      andlw   0x07F                ;  Store the Ack of
                                   ;    Bit 7 of the Data
      btfsc   STATUS, C
      iorlw   0x080                ;  If High, Set Bit 7
      addlw   0x001                ;  Finished, Do the
                                   ;    Next Bit
    DlayMacro Dlay - 9             ;  Put in "TLow"
      btfss   STATUS, DC
      goto    I2CRLoop
      bcf     DataPort, DataPin
      bsf     STATUS, RP0          ;  Send Ack Bit
      bcf     DataPort, DataPin
      bcf     STATUS, RP0
      andlw   0x080                ;  High or Low?
      btfss   STATUS, Z
      bsf     DataPort, DataPin    ;  Low, Send Ack
    DlayMacro Dlay / 18           ;  Any Reason to
                                   ;    delay?
      bsf     ClockPort, ClockPin
    DlayMacro Dlay
      bcf     ClockPort, ClockPin
      bcf     DataPort, DataPin
      movf    I2Ctemp, w           ;  Get the Received
                                   ;    Byte
      return                       ;  Return with Clock
                                   ;    = Data = Low

    I2CSend                        ;  Send the 8 Bits in
                                   ;    "w" and Return
                                   ;    Ack
      movwf   I2CTemp
      movlw   0x010 - 8
    I2CSLoop
      rlf     I2CTemp, f           ;  Shift up the Data
                                   ;    into "C"
      btfsc   STATUS, C
      goto    $ + 4
      nop
      bcf     DataPort, DataPin    ;  Low Bit
```

```
    goto     $ + 3
    bsf      DataPort, DataPin    ;  High Bit
    goto     $ + 1
    bsf      ClockPort, ClockPin  ;  Strobe Out the
                                  ;   Data
 DlayMacro Dlay
    bcf      ClockPort, ClockPin
 DlayMacro Dlay - 12
    addlw    1
    btfss    STATUS, DC
     goto    I2CSLoop
 DlayMacro 6
    bsf      STATUS, RP0          ;  Now, Get the Ack
                                  ;   Bit
    bsf      DataPort, DataPin
    bcf      STATUS, RP0
    bsf      ClockPort, ClockPin
 DlayMacro (Dlay / 2) - 1
    bcf      STATUS, C
    btfss    DataPort, DataPin
    bsf      STATUS, C            ;  Line Low, "Ack"
                                  ;   Received
 DlayMacro (Dlay / 2) - 2
    bsf      STATUS, RP0
    bcf      DataPort, DataPin
    bcf      STATUS, RP0
    bcf      ClockPort, ClockPin
    bcf      DataPort, DataPin
    return                        ;  Return with Ack in
                                  ;   Carry,
    endm                          ;  Clock = Data = Low
```

The macro is similar to

```
 I2CSetup I2CClock, I2CData, Rate, Frequency
```

where

```
 Pin       Description
 I2CClock  Port and Pin used for the "SCL" line -
           Pulled up with 1K to 10K Resistor
 I2CData   Serial Data - Pulled up with 1K to 10K
           Resistor
```

```
Rate       I2C Data Rate specified in kHz (normally
           100 or 400)
Frequency PICmicro® MCU's Clock Frequency
```

Data is sent to an I2C Device using the format:

```
idle - Start - CommandWriteA - AddressByteA - Start
 - CommandReadA - DataA - DataN - Stop - idle
```

Using the subroutines in the "I2CSetup" macro, the PICmicro® MCU code for carrying out a 16-bit read would be

```
    call    I2CStart            ;  Start the
                                ;   Transfer

    movlw   CommandWrite        ;  Send the Address
                                ;   to Read the
    call    I2CSend             ;  Sixteen Bit Word

    movlw   AddressByte
    call    I2CSend

    call    I2CStart            ;  Reset the I2C
                                ;   EEPROM to Read
                                ;   Back

    movlw   CommandRead         ;  Send the Read
                                ;   Command
    call    I2CSend

    bsf     STATUS, C           ;  Read the Byte
                                ;   with Ack
    call    I2CRead
    movwf   I2CData

    bcf     STATUS, C           ;  Read the next
                                ;   byte and stop
                                ;   the
    call    I2Cread             ;  transfer with
                                ;   the Nack
    movwf   I2CData + 1

    call    I2CStop             ;  Finished with the
                                ;   I2C Operation
```

RS-232 Interfaces

RS-232 is an older standard with somewhat unusual voltage levels. A "Mark" ("1") is actually -12 volts and a "Space" ("0") is $+12$ volts. Voltages in the "switching region" (± 3 volts) may or may not be read as a "0" or "1" depending on the device.

The "Handshaking" lines use the same logic levels as the transmit/receive lines discussed above and are used to interface between devices and control the flow of information between computers.

The "Request To Send" ("RTS") and "Clear To Send" ("CTS") lines are used to control data flow between the computer ("DCE") and the modem ("DTE" device). When the PC is ready to send data, it asserts (outputs a "Mark") on RTS. If the DTE device is capable of receiving data, it will assert the "CTS" line. If the PC is unable to receive data (i.e., the buffer is full or it is processing what it already has), it will de-assert the "RTS" line to notify the DTE device that it cannot receive any additional information.

The "Data Transmitter Ready" ("DTR") and "Data Set Ready" ("DSR") lines are used to establish communications. When the PC is ready to communicate with the DTE device, it asserts "DTR". If the DTE device is available and ready to accept data, it will assert "DSR" to notify the computer that the link is up and ready for data transmission. If there is a hardware error in the link, then the DTE device will de-assert the DSR line to notify the computer of the problem. Modems if the carrier between the receiver is lost will de-assert the DSR line.

"Data Carrier Detect" ("DCD") is asserted when the modem has connected with another device (i.e., the other device has "picked up the phone"). The "Ring Indicator" ("RI") is used to indicate to a PC whether or not the phone on the other end of the line is ringing or if it is busy. These lines (along with the other handshaking lines) are very rarely used in PICmicro® MCU applications.

There is a common ground connection between the DCE and DTE devices. This connection is critical for the RS-232 level converters to determine the actual incoming voltages. The ground pin should never be connected to a chassis or shield ground (to avoid large current flows or being shifted, preventing accurate reading of incoming voltage signals).

Most modern RS-232 connections are implemented using a "Three-Wire RS-232" set up as shown in Fig. 8.23. DTR/DSR and RTS/CTS lines are normally shorted

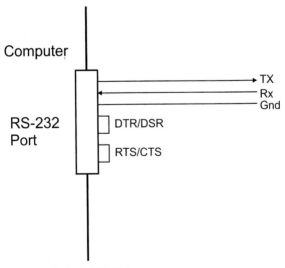

Figure 8.23 Typical RS-232 Wiring

together at the PICmicro® MCU end. The DCD and RI lines are left unconnected. With the handshaking lines shorted together, data can be sent and received without having to develop software to handle the different handshaking protocols. The 1488/1489 RS-232 Level Converter Circuits is a common method of implementing RS-232 Serial Port Interfaces if +12 and −12 volts is available to the circuit. The "#C" input is a flow control for the gates (normally RS-232 comes in the "#A" pin and is driven as TTL out of "#Y") and is normally left floating ("unconnected"). The pinout and wiring for these devices in a PC are shown in Fig. 8.24. If only a +5 volt power supply is available, the MAX232 chip can be used to provide the correct RS-232 signal levels. This circuit is wired as shown in Fig. 8.25. Another method for translating RS-232 and TTL/CMOS voltage levels is to use the transmitter's negative voltage. The circuit in Fig. 8.26 shows how this can be done using a single MOSFET transistor and two resistors.

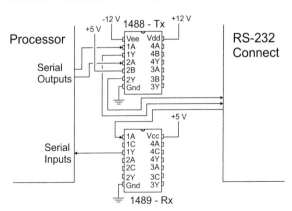

Figure 8.24 1488/1489 RS-232 Connections

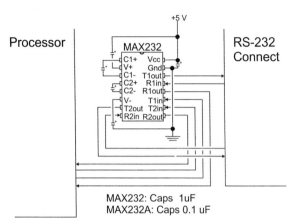

Figure 8.25 MAXIM MAX232 RS-232 Connections

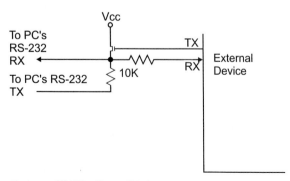

Figure 8.26 RS-232 to External Device

This circuit relies on the RS-232 communications only running in "Half-Duplex" mode (i.e., only one device can transmit at a given time). When the external device wants to transmit to the PC, it sends the data either as a "Mark" (leaving the voltage to be returned to the PC as a negative value) or as a "Space" by turning on the transistor and enabling the positive voltage output to move to the PC's receivers.

When the PICmicro® MCU transmits a byte to the external device through this circuit, it will receive the packet it is sent because this circuit connects the PICmicro® MCU's receiving pin (more or less) directly to its transmitting pin. The software running in the PICmicro® MCU (as well as the "host" device) will have to handle this.

Another issue to notice is that data out of the external device will have to be inverted to get the correct transmission voltage levels (i.e., a "0" will output a "1") to make sure the transistor turns on at the right time (i.e., a positive voltage for a space). Unfortunately, this means that the built-in serial port in the PICmicro® MCU cannot be used with this circuit because the output is "positive" and it cannot invert the data as required by the circuit.

There is a chip, the Dallas Semiconductor DS275, which basically incorporates the above-mentioned circuit (with built-in inverters) into the single package shown in Fig. 8.27. The DS1275 is an earlier version of this chip.

With the availability of low current PICmicro® MCUs, the incoming RS-232 lines can be used to power the application. This can be done using the host's RS-232 Ports to supply the current needed by the application as shown in Fig. 8.28.

When the DTR and RTS lines are outputting a "Space", a positive voltage (relative to ground) is available. This

Processor

RS-232
Connect

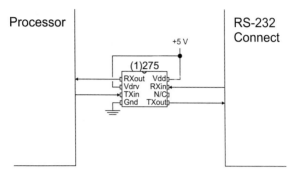

Figure 8.27 Dal Semi (1)275 RS-232 Interface

voltage can be regulated and the output used to power the devices attached to the serial port (up to about 5 mA). For extra current, the TX line can also be added into the circuit as well with a "break" being sent from the PC to output a positive voltage.

The 5 mA is enough current to power the Transistor/ Resistor type of RS-232 Transmitter and a PICmicro®

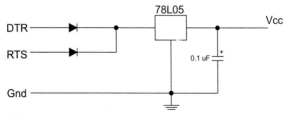

Figure 8.28 "Stealing" Power from the PC's Serial Port

MCU running at 4 MHz along with some additional hardware (such as a single LCD).

RS-485/RS-422

RS-485/RS-422 are "differential pair" serial communications electrical standards that consist of a balanced driver with positive and negative outputs that are fed into a comparator. The output from the comparator is a "1" or a "0" depending on whether or not the "positive line" is at a higher voltage than the negative. Figure 8.29 shows the normal symbols used to describe a differential pair connection.

To minimize AC transmission line effects, the two wires should be twisted around each other. "Twisted pair" wiring can either be bought commercially or made by simply twisting two wires together, twisted wires have a characteristic impedance of 75 ohms or greater.

A common standard for differential pair communications is "RS-422". This standard, which uses many commercially available chips, provides:

1. Multiple receiver operation.

2. Maximum data rate of 10 mbps.

3. Maximum cable length of 4,000 meters (with a 100 kHz signal).

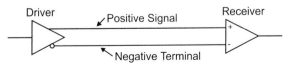

Figure 8.29 Differential Pair Serial Data Transmission

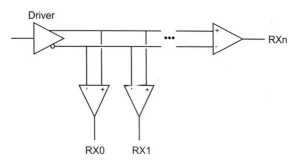

Figure 8.30 Multiple Receiver RS-422

Multiple receiver operation, shown in Fig. 8.30, allows signals to be "broadcasted" to multiple devices. The best distance and speed changes with the number of receivers of the differential pair along with its length. The 4,000 m at 100 kHz or 40 m at 10 MHz are examples of this balancing between line length and data rate. For long data lengths a few hundred ohm "terminating resistor" may be required between the positive terminal and negative terminal at the end of the lines to minimize reflections coming from the receiver and affecting other receivers.

RS-485 is very similar to RS-422, except it allows multiple drivers on the same network. The common chip is the "75176", which has the ability to drive and receive on the lines as shown in Fig. 8.31.

The only issue to be on the lookout for when creating RS-485/RS-422 connections is to keep the cable polarities correct (positive to positive and negative to negative). Reversing the connectors will result in lost signals and misread transmission values.

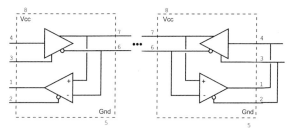

Figure 8.31 RS-485 Connection Using a 75176

Asynchronous Serial I/O Software Routines

The first method is a traditional "bit banging" interface that can be used by both the low-end and mid range PICmicro® MCUs. To set up the serial interfaces, the macro

```
NRZSerialNI Macro TXPort, TXPin, RXPort, RXPin,
   Polarity, Rate, Freq
```

is invoked, where "TXPort" and "TXPin" along with "RXPort" and "RXPin" are used to define the Transmit port and the Receive port, respectively. As I will discuss in the next section, these pairs of pins can be combined into a single define to make the definition easier. The "Polarity" of the signals is defined as "Pos" for "Positive" or positive logic and "Neg" for "Negative" or inverted logic (useful for interfacing to RS-232 directly through a current limiting resistor). "Rate" is the data rate (in bits per second) and "Freq" is the speed at which the processor is executing in Hz.

When the macro is expanded, the bit delay calculations are made internally and the bit banging serial receive and transmit subroutines are inserted into the application. The macros can be used by either low-end or mid-range PICmicro® MCUs without modification. The macro code is

```
NRZSerialNI Macro TXPort, TXPin, RXPort, RXPin,
 Polarity, Rate, Frequency
 variable BitDlay
BitDlay = Frequency / (4 * Rate)

SerialRX                      ;  Receive 8-N-1
 if (Polarity == Pos)
 btfsc   RXPort, RXPin        ;  Wait for a Bit to
                              ;   Come in
 else
 btfss   RXPort, RXPin
 endif
   goto   $ - 1

 DlayMacro BitDlay / 2        ;  Wait 1/2 a Bit to
                              ;   Confirm

 if (Polarity == Pos)
 btfsc   RXPort, RXPin        ; Confirm Data is
                              ;   Correct
 else
 btfss   RXPort, RXPin
 endif
   goto   SerialRX            ; If Just a "Glitch",
                              ;  Restart Start Bit
                              ;  Poll
 movlw   8                    ;  Wait for 8 Bits
SRXLoop
 if ((BitDlay - 10) > 770)    ;  Check to See if
                              ;   Value is Too Large
DlayMacro 770                 ;  Put in a "Double"
                              ;   Delay
 DlayMacro BitDlay - (770 + 10)
 else
DlayMacro BitDlay - 10        ;  Wait for the Middle
                              ;   of the Next Bit

 endif

  bcf      STATUS, C          ;  Check the Incoming
                                 Data
```

```
  if (Polarity == Pos)
   btfsc    RXPort, RXPin
  else
   btfss    RXPort, RXPin
  endif
   bsf      STATUS, C
   rrf      NRZTemp, f        ;  Shift in the Bit
   subwf    NRZTemp, w        ;  Decrement and End if
                             ;   == 0
   xorlw    0x0FF
   addwf    NRZTemp, w
   btfss    STATUS, Z
    goto    SRXLoop

  if ((BitDlay - 9) > 770)    ;  Check to See if
                             ;   Value is Too Large
  DlayMacro 770               ;  Put in a "Double"
                             ;   Delay
  DlayMacro BitDlay - (770 + 9)
  else
  DlayMacro BitDlay - 9       ;  Wait for the  Middle
                             ;   of the Next Bit
  endif
  if (Polarity == Pos)        ;  Is there a Stop Bit?

   btfss    RXPort, RXPin
  else
   btfsc    RXPort, RXPin
  endif
    goto    SerialRX          ;  No, Start All Over
                             ;   Again

   movf     NRZTemp, w        ;  Return the Received
                             ;   Byte
   return                    ;  Note - Zero Returned
                             ;   in Low-End
                             ;   Devices


SerialTX
   movwf    NRZTemp           ;  Save the Byte to
                             ;   Output
   movlw    10
   bcf      STATUS, C         ;  Start with Sending
                             ;   the Start Bit
```

```
STXLoop
 if (Polarity == Pos)         ;  Output the Current
                              ;   Bit
  btfsc    STATUS, C
 else
  btfss    STATUS, C
 endif
   goto    $ + 4              ;  6 Cycles Required
                              ;   Each Time
   nop
   bcf     TXPort, TXPin      ;  Output a "Low"
   goto    $ + 3
   bsf     TXPort, TXPin      ;  Output a "High"
   goto    $ + 1
 if ((BitDlay - 15) > 770)    ;  Check to See if
                              ;   Value is Too Large
 DlayMacro 770                ;  Put in a "Double"
                              ;   Delay
 DlayMacro BitDlay - (770 + 15)
 else
 DlayMacro BitDlay - 15       ;  Wait for the Middle
                              ;   of the Next Bit
 endif
   subwf   NRZTemp, w         ;  Decrement the Bit
                              ;   Counter
   xorlw   0x0FF
   addwf   NRZTemp, w
   btfsc   STATUS, Z
    return                    ;  Can Return to Caller
   bsf     STATUS, C          ;  Shift Down the Next
                              ;   Bit
   rrf     NRZTemp, f
   goto    STXLoop
 endm
```

Mid-Range "Bit Banging" NRZ Serial Interface Initialization Code

```
NRZSerialNISetup Macro TXPort, TXPin, Polarity
SerialSetup                   ;  Setup the Serial I/O
                              ;   Bits
  bsf      STATUS, RP0
  bcf      TXPort, TXPin      ;  Make TX Pin an
                              ;   Output
```

```
  bcf       STATUS, RP0
if (Polarity == Pos)
  bsf       TXPort, TXPin    ;  Transmit "idle"
else
  bcf       TXPort, TXPin
endif
  return
endm
```

The TMR0 interrupt based asynchronous serial functions shown in the macro below poll the data bit three times each data period to check for the incoming data. This method does not prevent the PICmicro® MCU from carrying out any other tasks.

```
NRZSerialI Macro TXPort, TXPin, RXPort, RXPin,
 Polarity, Rate, Frequency
 variable BitDlay, Prescalor, TMR0Reset
BitDlay = (Frequency / (3 * 4 * Rate)) - 10
TMR0Reset = BitDlay / 2    ;  Using TMR0,
                           ;    Calculate the
                           ;    Timer Reset Value
Prescaler = 0              ;  And the Prescaler
 while (TMR0Reset > 0x0FF)  ;  Find the Proper
                           ;    Reset Value
TMR0Reset = TMR0Reset / 2
Prescaler = Prescaler + 1
 endw
 if (Prescaler > 7)        ;  Can't Use TMR0
 error "Bit Delay cannot use TMR0 for Polling Clock"
 endif
TMR0Reset = 256 - TMR0Reset ;  Get the TMR0 Reset
                           ;    Value

  goto      AfterInt       ;  Jump to After
                           ;    Interrupt
 org        4
Int                        ;  Interrupt Handler
  movwf     _w
  movf      STATUS, w
  bcf       STATUS, RP0    ;  Make Sure in Bank 0
  movwf     _status
```

```
    bcf      INTCON, T0IF       ;  Reset the Timer
                                ;   Interrupt

    movlw    TMR0Reset
    movwf    TMR0

;  First, Check for a Received Character
Int_RX

    movlw    0x004              ;  Check for Bit?
    addwf    RXCount, f
    btfss    STATUS, DC         ;  DC Not Affected by
                                ;   "clrf"
    goto     _RXNo              ;  Nothing to Check for
                                ;   (Yet)

    movf     RXCount, w         ;  Everything Read
                                ;   Through?
    xorlw    0x091
    btfsc    STATUS, Z
    goto     _RXAtEnd           ;  Yes, Check for Stop
                                ;   Bit

    bcf      STATUS, C          ;  Read the Current
                                ;   State
    if (Polarity == Pos)
    btfsc    RXPort, RXPin      ;  Sample at 10 Cycles
    else
    btfss    RXPort, RXPin
    endif
    bsf      STATUS, C
    rrf      RXByte, f

    bsf      RXCount, 2         ;  Start Counting from 4

_RXEnd13
    nop
    goto     _RXEnd             ;  End 15 Cycles From
                                ;   "Int_RX" -
                                ;   Finished Receiving
                                ;   Bit

_RXEnd8                         ;  Finished - 8 Cycles
                                ;   to Here
    goto     $ + 1
```

```
      nop
      goto      _RXEnd13

_RXNo                                 ;  5 Cycles from
                                      ;   "Int_RX" - No Bit
                                      ;   to Receive

   btfsc     RXCount, 0               ;  Something Running?
    goto     _RXEnd8                  ;  End 8 Cycles from
                                      ;   "Int_RX" - Yes,
                                      ;    Skip
Over
   btfsc     RXCount, 3               ;  Checking Start Bits?
    goto     _RXStartCheck

  if (Polarity == Pos)
   btfsc     RXPort, RXPin            ;  If Line Low -
                                      ;   "Start" Bit
   else
   btfss     RXPort, RXPin
   endif
    bcf      RXCount, 2               ;  Don't Have a "Start"
                                      ;   Bit

    goto     _RXEnd13                 ;  End 18 cycles from
                                      ;   "Int_RX"

_RXStartCheck                         ;  10 Cycles to Here

  if (Polarity == Pos)
   btfsc     RXPort, RXPin
   else
   btfss     RXPort, RXPin
   endif
    movlw    0x0FF                    ;  Nothing - Clear
                                      ;   "RXCount"

    addlw    1
    movwf    RXCount

    goto     _RXEnd                   ;  16 Cycles to End

_RXAtEnd                              ;  9 Cycles from
                                      ;   "Int_RX" - Check
                                      ;    Last
                                      ;    Bit
  if (Polarity == Pos)
```

```
  btfsc    RXPort, RXPin
 else
  btfss    RXPort, RXPin
 endif
  bsf      RXFlag

  clrf     RXCount           ;   Finished - Reset
                             ;   Check - 12
                             ;   Cycles

  goto     $ + 1
  goto     _RXEnd

_RXEnd

;  Next, Check for Transmitting a Character -
Intrinsic Dlay 22 Cycles
Int_TX

  movlw    0x004             ;   Interrupt Transmit
                             ;     Increment Value
  addwf    TXCount, f
  btfss    STATUS, DC        ;   Send the Next Byte?
   goto    _TXSendDlayCheck

  bsf      TXCount, 2        ;   Want to Increment 3x
                             ;     not Four for each
                             ;     Bit
  bsf      STATUS, C
  rrf      TXByte, f
  movf     TXPort, w         ;   Send Next Bit
  andlw    0xOFF ^ (1 << TXPin)
 if (Polarity == Pos)
  btfsc    STATUS, C
 else
  btfss    STATUS, C
 endif
   iorlw 1 << TXPin
  movwf    TXPort            ;   Cycle 12 is the Bit
                             ;     Send

  goto     _TXCompletedGoOn  ;   TX Takes 14 Cycles

_TXSendDlayCheck             ;   Don't Send Bit,
                             ;   Check for Start Bit
```

```
    btfss    TXCount, 0        ;  Bit Zero Set (Byte
    goto     _TXNothingtoCheck ;  to Send)?

    movlw    0x004             ;  Setup the Timer to
    movwf    TXCount           ;   Increment 3x

    movf     TXPort, w         ;  Output Start Bit
  if (Polarity == Pos)
    andlw    0x0FF ^ (1 << TXPin)
  else
    iorlw    1 << TXPin
  endif
    movwf    TXPort

    goto     _TXCompletedGoOn  ;  TX First Bit Takes
                               ;   14 Cycles

_TXNothingtoCheck              ;  Nothing Being Sent?

    movf     TXCount, w
    xorlw    0x004             ;  Zero (Originally)
                               ;   TXCount?
    btfss    STATUS, Z
    xorlw    0x004 ^ 0x09C
    btfsc    STATUS, Z
    clrf     TXCount

_TXCompletedGoOn               ;  Finished with TX, Do
                               ;   RX

    movf     _status, w        ;  Restore the
    movwf    STATUS            ;   Interrupts
    swapf    _w, f
    swapf    _w, w
    retfie

SerialRX

    bcf      RXFlag            ;  Reset the Character
                               ;   available Flag

    btfss    RXFlag            ;  Wait for a Character
    goto     $ - 1             ;   to be Received
```

```
    movf      RXByte, w             ;   Return the Character
                                    ;   Read in

    return

SerialTX

    movf      TXCount, f            ;   Anything Being Sent?
    btfss     STATUS, Z             ;   Wait for the
                                    ;   Previous Send to
                                    ;   End
    goto      $ - 2

    movwf     TXByte                ;   Send out the
                                    ;   Character

    bsf       TXCount, 0            ;   Indicate to the
                                    ;   Interrupt Handler
                                    ;   that it can Send
                                    ;   Something

    return

AfterInt                            ;   Can Return the Value
    bsf       STATUS, RP0           ;   Setup the
                                    ;   Interrupts/TX
                                    ;   Output
    bcf       TXPort, TXPin
    movlw     0x0D0 + Prescaler
    movwf     OPTION_REG ^ 0x080;   User Prescaler with
                                    ;   TMR0
    bcf  STATUS, RP0
    if (Polarity == Pos)
    bsf       TXPort, TXPin         ;   Output "Idle" for
                                    ;   Data Transmit
    else
    bcf       TXPort, TXPin
    endif
    movlw     TMR0Reset             ;   Reset the Timer
    movwf     TMR0
    movlw     (1 << GIE) + (1 << T0IE)
    movwf     INTCON                ;   Start up the
                                    ;   Interrupts
    clrf      RXCount               ;   Make Sure No Counts
                                    ;   are Starting
    clrf      TXCount

    endm
```

Along with the macro invocation, the following variables will have to be declared for the code to work:

```
_w, _status          - Interrupt Handler Context Save
                         Registers
RXByte, TXByte       - Data Transmit and Receive Bytes
RXCount, TXCount     - Serial Transfer Data
                         Count/Status Variables
Flags                - Execution Flag Variable
```

Along with these variables the "RXFlag" bit will also have to be defined for use by the code to indicate when a valid byte has been received.

Dallas Semiconductor One-Wire Interface

Dallas Semiconductor has created a line of peripherals that are very attractive for use with microcontrollers as they only require one line for transferring data. This single-wire protocol is available in a variety of devices, but the most popular are the DS1820 and DS1821 digital thermometers. These devices can be networked together on the same bus (they have a built-in serial number to allow multiple devices to operate on the same bus) and are accurate to within one degree Fahrenheit.

The DS1820 is available in a variety of packages, with the most common being a three-pin "TO-92" package that looks like a plastic transistor package and can be wired to a PICmicro® MCU as shown in Fig. 8.32.

The DS1820 has many features that would be useful in a variety of different applications. These include the ability of sharing the single-wire bus with other devices. A unique serial number is burned into the device that allows it to be written to individually and gives it the ability to be powered by the host device. Data transfers over the "one-Wire" bus are initiated by the Host system (in the application cases, this is the PICmicro® MCU) and are carried out 8-bits at a time (with the least significant bit transmitted first). Each bit transfer takes at least 60 usec. The "one-Wire" bus is pulled up externally (as is shown in Fig. 8.32) and is pulled down by either the host or the peripheral device to transfer data. If the Bus is pulled down for a very short interval, a "1" is being transmitted. If the Bus is pulled down for more than 15 usecs, then a "0" is being transmitted. The differences in the "1" and "0" bits are shown in Fig. 8.33.

All Data Transfers are initiated by the host system. If it is transmitting data, then it holds down the line for the specified period. If it is receiving data from the DS1820, then the host pulls down the line and releases it and then polls the line to see how long it takes for it to go back up. During Data Transfers, make sure that Interrupts cannot take place (because this would affect how the data is sent or read if the interrupt takes place during the data transfer).

Before each command is set to the DS1820, a "Reset" and "Presence" Pulse is transferred. A "Reset" Pulse con-

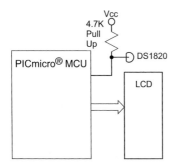

Figure 8.32 Example Thermometer Application

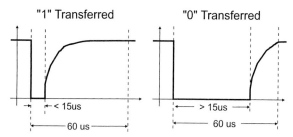

Figure 8.33 Dallas Semi. "1-Wire" Data Transfer

sists of the host pulling down the line for 480 usecs to 960 usecs. The DS1820 replies by pulling down the line for roughly 100 usecs (60 to 240 usecs is the specified value). To simplify the software, do not check for the presence pulse (because I knew in the application that I had the thermometer connected to the bus). In another application (where the thermometer can be disconnected), putting a check in for the "Presence" Pulse may be required.

To carry out a temperature "read" of a single DS1820 connected to a microcontroller, the following instruction sequence is used:

```
1. Send a "Reset" Pulse and delay while the
   "Presence" Pulse is returned.
2. Send 0x0CC, which is the "Skip ROM" command,
   which tells the DS1820 to assume that the next
   command is directed towards it.
3. Send 0x044, which is the Temperature Conversion
   Initiate instruction. The current temperature
   will be sampled and stored for later read back.
4. Wait 500+ usecs for the Temperature Conversion to
   complete.
5. Send a "Reset" Pulse and delay while the
   "Presence" Pulse is returned.
6. Send 0x0CC, "Skip ROM" Command Again.
```

7. Send 0x0BE to read the "Scratchpad" RAM that
   contains the Current Temperature (in degrees
   Celsius times two).
8. Read the nine bytes of "Scratchpad" RAM.
9. Display the Temperature (if bit 0 of the second
   byte returned from the "Scratchpad" RAM, the
   first byte is negated and a "-" is put on the
   LCD Display) by dividing the first byte by 2 and
   sending the converted value to the LCD.

The total procedure for doing a temperature measurement takes about 5 msecs. PICmicro® MCU code to access the DS1820 is given in the following listings for a PICmicro® MCU running at 4 MHz:

```
DSReset                         ;  Reset the DS1820
  bcf       DS1820              ;  Hold the DS1820 Low for
                                ;    500 usecs to reset
  movlw     125
  addlw     0xFF                ;  Add -1 until Reset is
                                ;    Equal to Zero
  btfss     STATUS, Z
   goto     $ - 2
  bsf       DS1820
  bcf       DSTRIS
  movlw     0                   ;  Wait 1 msec before
                                ;    sending a command
  addlw     0xFF
  btfss     STATUS, Z
   goto     $ - 2
  bsf       DSTRIS
  bsf       DS1820
  return
```

DS1820 Byte Send Code

```
DSSend                          ;  Send the Byte in "w" to
                                ;    the DS1820
  movwf     Temp
  movlw     8
DSSendLoop
  bcf       INTCON, GIE         ;  Make Sure Operation
                                ;    isn't interrupted
```

```
bcf        DS1820         ;  Drop the DS1820's
                          ;   Control Line
rrf        Temp, f        ;  Load Carry with Contents
                          ;   of the Buffer
btfsc      STATUS, C      ;  If "1" Sent, Restore
 bsf       DS1820         ;   After 4 Cycles
bsf        Count, 3       ;  Loop for 24 Cycles
decfsz     Count, f
 goto      $ - 1
goto       $ + 1          ;  Add 2 Cycles for a 30
                          ;   usec delay
bsf        DS1820         ;  The Line is High
bsf        INTCON, GIE    ;  Restore the Interrupts
bsf        Count, 3       ;  Loop Another 24 Cycles
                          ;   for Execution Delay
decfsz     Count, f
 goto      $ - 1
addlw      0x0FF          ;  Subtract 1 from the Bit
                          ;   Count
btfss      STATUS, Z
 goto      DSSendLoop
return                    ;  Finished, Return to Caller
```

DS1820 Byte Read Code

```
DSRead                    ;  Receive the Byte from
                          ;   the DS1820 and put in
 movlw     8              ;   "w"
DSReadLoop
 bcf       INTCON, GIE    ;  Make Sure Operation
                          ;   isn't interrupted
 bcf       DS1820         ;  Drop the DS1820's
                          ;   Control Line
 bsf       DSTRIS         ;  Turn Port into a
                          ;   simulated Open Drain
                          ;   Output
 nop
 bsf       STATUS, C      ;  What is Being Returned?
 btfss     DS1820
  bcf      STATUS, C      ;  If Line is high, a "1"
 rrf       Temp, f        ;  Shift in the Data bit
 bsf       INTCON, GIE    ;  Can Interrupt from here
 clrf      Count
 decfsz    Count, f
  goto     $ - 1
 bsf       DS1820
```

```
        bcf     DSTRIS
        bsf     DS1820
        addlw   0x0FF       ; Loop Around for another
                            ;  Bit
        btfss   STATUS, Z
        goto    DSReadLoop
        movf    FSR0L, w    ; Return the Byte Read in
        return              ; Finished, Return to
                            ;  Caller
```

The Process to read from the DS1820 is

1. Reset the DS1820
2. Send 0CCh followed by 044h to begin the
   Temperature sense and conversion.
3. Wait 480 usecs for the Temperature conversion to
   complete.
4. Send another reset to the DS1820
5. Send 0CCh and 0BEh to read the temperature
6. Wait 100 usecs before reading the first byte in
   the DS1820
7. Read the first, or "SP0" byte of the DS1820
8. Wait another 100 usecs before reading the second
   or "SP1" byte of the DS1820

Reading a Potentiometer Using Parallel I/O Pins

For measuring resistance values without an ADC, a simple RC network can be used with the PICmicro® MCU as is shown in Fig. 8.34. To measure the resistance (assuming the capacitor is of a known value), the PICmicro® MCU first charges the capacitor to 5 volts (or its nominal output) using the I/O pin in "output" mode. Once this is done, the pin changes to "input" mode and waits for the capacitor to discharge through the potentiometer. Looking at this operation on an oscilloscope, the waveform produced by the circuit looks like Fig. 8.35. In Fig. 8.35, the "Charge" cycle and "Discharge" cycle can

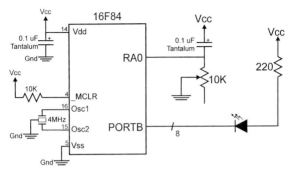

Figure 8.34 Reading a Potentiometer Position without an ADC

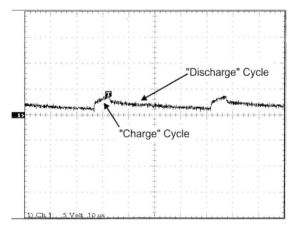

Figure 8.35 Oscilloscope Picture for ADCLess Operation

clearly be seen. From basic electronic theory, we know that the time required for the capacitor to charge is

$$time = R \times C \times ln(Vend / Vstart)$$

where the Vstart and Vend are the starting and ending voltages. Since the capacitor value, the voltages and the time it took for the capacitor to discharge, the formula above can be rearranged to find R:

$$R = time / (C \times ln(Vend / Vstart))$$

The code used to test the analog I/O uses the following logic:

```
int PotRead()          //  Read the Resistance at
                       //   the I/O Pin
{

int i;
   TRIS.Pin = Output;  //  Set the Output Mode
   Pin = 1;            //  Output a "1" to Charge
                       //   the Capacitor
   for (i = 0; i < 5usec, i++ );

   TRIS.Pin = Input;   //  Now, Time How Long it
   TMR0 = 0;           //   Takes for
   while (Pin == 1);   //   the Capacitor to
                       //   Discharge through
                       //   the Potentiometer

   return TMR0;        //  Return the TMR0 Value
                       //   for the
                       //   Discharge Time

}  //  end PotRead
```

Motor Drivers

A network of switches (transistors) can be used to control turning a motor in either direction. This is known as an "H-Bridge" and is shown in Fig. 8.36.

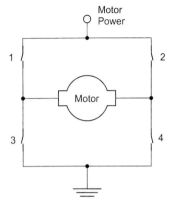

Figure 8.36 "H" Bridge Motor Driver

In this circuit, if all the switches are open, no current will flow and the motor won't turn. If switches "1" and "4" are closed, the motor will turn in one direction. If switches "2" and "3" are closed, the motor will turn in the other direction. Both switches on one side of the bridge should *never* be closed at the same time as this will cause the motor power supply to burn out or a fuse to blow because there is a short circuit directly between the motor power and ground.

Controlling a motor's speed is normally done by "pulsing" the control signals in the form of a Pulse Wave Modulated ("PWM") signal as shown in Fig. 8.37. This will control the average power delivered to the motors. The higher the ratio of the "Pulse Width" to the "Period", the more power delivered to the motor.

The frequency of the PWM signal should be greater than 20 kHz to avoid the PWM from producing an audible signal in the motors as the field is turned on and off.

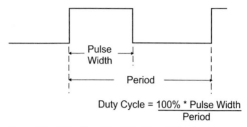

$$\text{Duty Cycle} = \frac{100\% * \text{Pulse Width}}{\text{Period}}$$

Figure 8.37 Pulse Wave Modulated Signal Waveform

The 293D chip can control two motors (one on each side of a robot), connected to the buffer outputs (pins 3, 6, 11, and 14). Pins 2, 7, 10, and 15 are used to control the voltage level (the "switches" in the H-Bridge diagram above) of the buffer outputs. Pin 1 and Pin 9 are used to control whether or not the buffers are enabled. These can be PWM inputs, which makes control of the motor speed very easy to implement. "Vs" is + 5 volts used to power the logic in the chip and "Vss" is the Power supplied to the motors and can be anywhere from 4.5 to 36 volts. A maximum of 500 mA can be supplied to the motors. The 293D contains integral shunt diodes. This means that to attach a motor to the 293D, no external shunt diodes are required as shown in Fig. 8.38. In Fig. 8.38, there is an optional "snubber" resistor and capacitor. These two components, wired across the brush contacts of the motor will help reduce electromagnetic emissions and noise "spikes" from the motor. Erratic operation from the microcontroller when the motors are running can be corrected by adding the 0.1 uF capacitor and 5 ohm (2 watt) resistor "snubber" across the motor's brushes as shown in the circuit above.

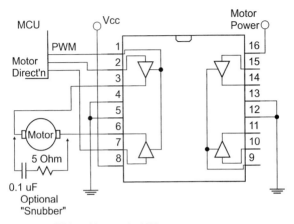

Figure 8.38 Wiring a Motor to the 293D

There is an issue with using the 293D and 298 motor controller chips which is that they are bipolar devices with a 0.7 volt drop across each driver (for 1.4–1.5 volts for a dual driver circuit as is shown in Fig. 8.38. This drop, with the significant amount of current required for a motor, results in a fairly significant amount of power dissipation within the driver. The 293D is limited to 1 Amp total output and the 298 is limited to 3 Amps. For these circuits to work best, a significant amount of heat sinking is required.

To minimize the problem of heating and power loss Power MOSFET transistors can be used to implement an H-Bridge motor control.

Stepper motors are much simpler to develop control software for than regular DC motors. This is because the motor is turned one step at a time or can turn at a specific rate (specified by the speed in which the "Steps" are executed). In terms of the hardware interface, stepper motors are a bit more complex to wire and require more current (meaning that they are less efficient), but these are offset by the advantages in software control.

A "Bipolar" Stepper motor consists of a permanent magnet on the motor's shaft that has its position specified by a pair of coils (Fig. 8.39). To move the magnet and the shafts, the coils are energized in different patterns to attract the magnet. For the example above, the following sequence would be used to turn the magnet (and the shaft) clockwise.

Commands to Move a Stepper Motor			
Step	Angle	Coil "A"	Coil "B"
1	0	S	
2	90		N
3	180	N	
4	270		S
5	360/0	S	

In this sequence, Coil "A" attracts the North Pole of the magnet to put the magnet in an initial position. Then Coil "B" attracts the South Pole, turning the magnet 90

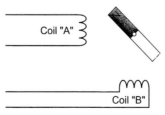

Figure 8.39 Stepper Motor

degrees. This continues on to turn the motor, 90 degrees for each "Step".

The output shaft of a stepper motor is often geared down so that each step causes a very small angular deflection (a couple of degrees at most rather than the 90 degrees in the example above). This provides more torque output from the motor and greater positional control of the output shaft.

R/C Servo Control

Servos designed for use in radio-controlled airplanes, cars, and boats can be easily interfaced to a PICmicro® MCU. They are often used for robots and applications where simple mechanical movement is required. This may be surprising to you because a positional servo is considered to be an analog device. The output of an R/C Servo is usually a wheel that can be rotated from zero to 90 degrees. (There are also servos available that can turn from zero to 180 degrees as well as servos with very high torque outputs for special applications). Typically, they only require +5 volts, ground, and an input signal.

An R/C Servo is indeed an analog device, the input is a PWM signal at digital voltage levels. This pulse is between 1.0 and 2.0 msecs long and repeats every 20 msecs (Fig. 8.40).

The length of the PWM Pulse determines the position of the servo's wheel. A 1.0 msec pulse will cause the wheel to go to zero degrees while a 2.0 msecs pulse will cause the wheel to go to 90 degrees.

For producing a PWM signal using a PICmicro® MCU, the TMR0 timer interrupt (set for every 20 msecs),

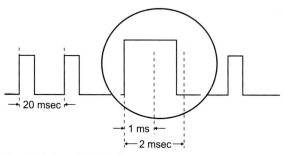

Figure 8.40 Servo PWM Waveform

which outputs a 1.0 to 2.0 msecs PWM signal can be used. The pseudo-code for the interrupt handler is

```
Interrupt() {                    //   Interrupt Handler
                                        Code
int    i = 0;

  BitOutput( Servo, 1);          //   Output the Signal

  for (i = 0; i < (1 msec + ServoDlay); i++ );

  BitOutput( Servo, 2);

  for (; i < 2 msec; i++ );      //   Delay full 2
                                        msecs
}   //  End Interrupt Handler
```

Audio Output

The PICmicro® MCU can output audio signals using a circuit like the one shown in Fig. 8.41.

Timing the output signal is generally accomplished by toggling an output pin at a set period within the TMR0 in-

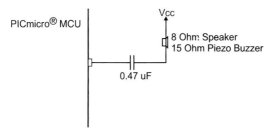

Figure 8.41 Circuit for Driving PICmicro® MCU Audio

terrupt handler. To generate a 1kHz signal shown in a
PICmicro® MCU running a 4MHz, you can use the code
below (which does not use the prescaler) for TMR0 and
the PICmicro® MCU's interrupt capability.

```
    org     4
int
  movwf   _w                  ;  Save Context
                              ;   Registers
  bcf     INTCON, TOIF        ;  Reset the
                              ;   Interrupt
  movlw   256  - (250  - 4)
  movwf   TMR0                ;  Reset TMR0 for
                              ;   another 500 usecs
  btfsc   SPKR                ;  Toggle the Speaker
   goto   $ + 2
  bsf     SPKER               ;  Speaker Output
                              ;   High
  goto    $ + 2
  bcf     SPKER               ;  Speaker Output Low
  swapf   _w, f               ;  Restore Context
                              ;   Registers
  swapf   _w, w
  retfie
```

AC Power Control

"TRIACS" come under the heading of "Thyristors" and are used to switch AC signals on and off. TRIACS do not rectify the AC voltage because they consist of two "Silicon Controlled Rectifiers" ("SCRs"), which allows the AC Current to pass without any "clipping". A typical circuit for Triacs is shown in Fig. 8.42.

TRIACS do not allow AC current to pass unless their "gates" are biased relative to the two AC contacts. To do this, a PICmicro® MCU output can pull the control to ground. The current required to "close" many of the TRIACS is 25 mA and can be provided by standard PICmicro® MCU outputs easily.

CAUTION: AC voltages and currents can damage components, start fires, burn, or even kill through electrocution. It is recommended that AC control circuits are tested with low-voltage sources before they are used in "mains" voltage circuits.

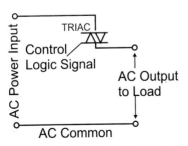

Figure 8.42 Typical TRIAC AC Control Circuit

Hall-Effect Sensors

A "Hall-effect switch" is a device in which if a current passing through a piece of silicon is deflected by a magnetic field, the output changes state as shown in Fig. 8.43.

The Hall-effect switch output can either be "Totem Pole" or "Open Collector" and can drive a PICmicro® MCU input directly. If an "Open Collector" output is used with the Hall-effect switch, then a pull up is required to ensure positive voltages will be received by the PICmicro® MCU when there is no magnetic field in place.

Sony Infrared TV Remote Control

Most (if not all) I/R TV remotes use a "Manchester" encoding scheme in which the data bits are embedded in the packet by varying the lengths of certain data levels. This can be seen in Fig. 8.44, from a theoretical Perspective and in Fig. 8.45, which shows the output from a 40-kHz Infrared Receiver receiving a signal from a "Sony" brand TV remote control. The normal signal

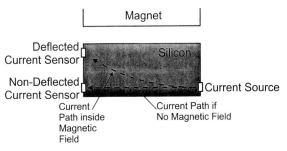

Figure 8.43 Hall-Effect Switch Operation

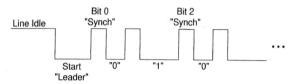

Figure 8.44 I/R TV remote data stream.

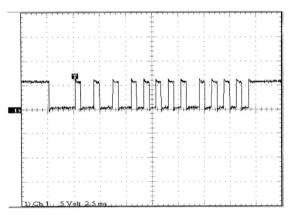

Figure 8.45 Scope View of TV I/R Remote Control Input

coming from an I/R receiver circuit is "High" when nothing is coming (line idle) and then goes "low" with a "leader" signal to indicate that data is coming in. The data consists of a bit "Synch," which when it completes the bit value is transmitted as the length of time before the next bit "Synch".

"Sony" brand TV remotes have 12 data bits and a 40-kHz carrier. The timings are as follows (and use a "Base Timing" "T" of 550 usecs).

```
Sony I/R Remote Control Timing

   Feature      "T" Timing     Actual Length
   -------      ----------     -------------
   Leader          4T            2.20 msec
   Synch            T            0.55 msec
   "0"              T            0.55 msec
   "1"             2T            1.10 msec
```

To read the incoming signal, the following code can be used in a PICmicro® MCU running at 4 MHz and a single I/R receiver can be used to pass the signal to the PICmicro® MCU:

```
Sony I/R Read Code

   clrf    IntCount          ;  Reset the
                             ;    Counters
   clrf    ReadCount

GetPack                      ;  Get the Next
                             ;    Packet Coming In

   movlw   0x088             ;  Wait for Port
                             ;    Change Interrupt
   movwf   INTCON
```

Sony I/R Read Code (*Continued*)	

```
Loop                            ;  Loop Here for
                                ;    Each Update of
                                ;    the Screen

    movlw   150                 ;  Wait fur 25 msec
                                ;    of Data from I/R
    subwf   IntCount, w
    btfss   STATUS, Z
    goto    Loop                ;  Has NOT timed out

    clrf    INTCON              ;  No more interrupts
                                ;    for a while

    movf    ReadCount, w        ;  Get the Read in
                                ;    CRC

    clrf    IntCount            ;  Reset for the
                                ;    Next Packet
    clrf    ReadCount

    call    DispHex             ;  Now, Display the
                                ;    Character

    movlw   0x08E               ;  Reset the Cursor
    call    WriteINS

    goto    GetPack             ;  Wait for the Next
                                ;    I/R Packet

Int                             ;  Interrupt, Check
                                ;    I/R Input

    movwf   _w                  ;  Save the Context
                                ;    Registers
    swapf   STATUS, w
    movwf   _status

    movlw   0x020               ;  Just wait for a
                                ;    Timer Interrupt
    movwf   INTCON

    movlw   256  -  20          ;  Reset the Timer
```

| Sony I/R Read Code (*Continued*) |

```
    movwf   TMR0

    incf    IntCount          ;   Increment the
                              ;     Count Register

    bcf     STATUS, C         ;   Now, Figure out
                              ;     what to Add to
                              ;     LSB
    btfsc   PORTB, 6          ;   Is the Incoming
                              ;     Value Set?
    goto    Int_Set

    btfsc   ReadCount, 5      ;   Do we Update the
                              ;     Value coming in?
    bsf     STATUS, C

    goto    Int_End           ;
Int_Set                       ;   Incoming Set
    btfss   ReadCount, 5      ;   Is the Current
                              ;     Bit Set?
    bsf     STATUS, C         ;   No, Turn on the
                              ;     Incoming Bit
Int_End
    rlf     ReadCount         ;   Shift Over with
                              ;     New Input Data

    swapf   _status, w        ;   Restore the
                              ;     Context
                              ;     Registers
    movwf   STATUS
    swapf   _w
    swapf   _w, w
    retfie
```

This code starts sampling the incoming data after the
Leader was received and the "1"s and "0"s were treated
as the inputs to a Linear Feedback Shift Register. For
the Code above, an 8-bit LFSR was used to produce
"Cyclical Redundancy Check" ("CRC") codes. In this

case, the Input wasn't the high bit of the shift register—instead it is the input from the I/R Receiver.

Using this code, the following CRC codes were generated from the "Sony" I/R TV Remote Control Transmitter:

```
Key     Code
Power   0x052
Vol+    0x05E
Vol-    0x0BB
Ch+     0x0DC
Ch-     0x062
"0"     0x017
"1"     0x07A
"2"     0x08D
"3"     0x033
"4"     0x01F
"5"     0x04E
"6"     0x072
"7"     0x0CC
"8"     0x0B9
"9"     0x023
```

PICmicro® MCU Programming

"Hex" File Format

The purpose of MPLAB and other assemblers and compilers is to convert PICmicro® MCU application source code into a data format that can be used by a programmer to load the application into a PICmicro® MCU. The most popular format (used by Microchip and most other programmers, including the two presented in this chapter) is the Intel 8-bit hex file format.

When source code is assembled, a hex file ("Example.hex") is generated. This file is in the format:

```
:10000000FF308600831686018312A001A101A00B98
:0A0010000728A10B07288603072824
:02400E00F13F80
:00000001FF
```

Each line consists of a starting address and data to be placed starting at this address. The different positions of each line are defined by:

"Hex File" Line Definition	
Byte	Function
First (1)	Always ":" to indicate the Start of the Line.
2-3	Two Times the Number of Instructions on the Line.
4-7	Two Times the Starting Address for the Instructions on the Line. This is in "Motorola" Format (High Byte followed by Low Byte).
8-9	The Line Type (00 - Data, 01 - End).
10-13	The First Instruction to be loaded into the PICmicro® MCU at the Specified Address. This data is in "Intel" Format (Low Byte followed by High Byte).
:	Additional Instructions to be loaded at Subsequent Addresses. These instructions are also in "Intel" Format.
End - 4	The Checksum of the Line.
End - 2	ASCII Carriage Return/Line Feed Characters

The checksum is calculated by summing each byte in a line and subtracting the least significant bits from 0x0100. For the second line in the example hex file above, the checksum is calculated as:

```
      0A
      00
      10
      00
      07
      28
      A1
      0B
      07
      28
      86
      03
      07
    + 28
    ------
     1DC
```

The least significant 8-bits (0x0DC) are subtracted from 0x0100 to get the checksum:

```
      0x0100
    - 0x00DC
    ---------
      0x0024
```

This calculated checksum value of 0x024 is the same as the last two bytes of the original line.

Low-End PICmicro® MCU Programming

The low-end PICmicro® MCUs use 17 pins for programming and are programmed using a "Parallel" protocol. The pins are defined as:

Low-End PICmicro® MCU Programming Pins	
Pins	Function
RA0–RA3	D0–D3 of the Instruction Word
RB0–RB7	D4–D11 of the Instruction Word

Low-End PICmicro® MCU Programming Pins (*Continued*)	
Pins	Function
T0CK1	Program/Verify Clock
OSC1	Program Counter Input
_MCLR/Vpp	Programming Power
Vdd	PICmicro MCU Power
Vss	Ground ("Gnd")

A programmer designed for low-end PICmicro® MCU programming generally looks like Fig. 9.1.

To program a memory location, the following procedure is used:

1. The new word is driven onto RA0-RA3 and RB0-RB7.

2. The "prog" single shot sends a 100 usec programming pulse to the PICmicro® MCU.

3. The data word driver ("driver enable") is turned off.

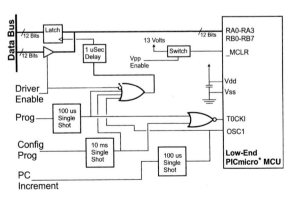

Figure 9.1 Low-End PICmicro® MCU Programmer

4. A programming pulse is driven that reads back the word address to confirm the programming was correct. In Fig. 9.1, the read-back latch is loaded on the falling edge of the "on" gate to get the data driven by the PICmicro® MCU.

5. Steps two through four are repeated a maximum of 25 times or until the data stored in the latch are correct.

6. Steps one through four are repeated three times and each time it is required to get the correct data out of the PICmicro® MCU. This is used to ensure the data is programmed in reliably.

7. "OSC1" is pulsed to increment to the next address. This operation also causes the PICmicro® MCU to drive out the data at the current address before incrementing the PICmicro® MCU's program counter (which happens on the falling edge of OSC1).

In Fig. 9.2, the programming steps one to four listed above are shown along with the latch clock signal. Note

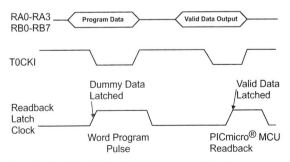

Figure 9.2 Low-End PICmicro® MCU Programming Waveform

that upon power up, the "configuration fuses" are the first address to be accessed, followed by the contents of the program memory.

Just pulsing the "OSC1" pin can be used to implement a "fast verify" as shown in Fig. 9.3. As noted above, each time "OSC1" is pulsed, data at the current address will be output and then increment the PICmicro® MCU's Program Counter. Figure 9.3 shows the fast verify right from the start with the configuration fuse value output first before the contents of the program memory are output.

The low-end PICmicro® MCU configuration fuses, while Microchip documentation indicates they are at address 0x0FFF, are the first words to be programmed. When programming a low-end PICmicro® MCU, the configuration fuses should be skipped over the first time the PICmicro® MCU is programmed. After doing this, power should be shut off and the PICmicro® MCU put back into programming mode. The reason for programming the configuration fuses last is to make sure the "code protect" bit of the configuration register is not reset (en-

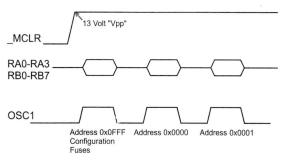

Figure 9.3 Low-End PICmicro® MCU "Fast Verify" Waveform

abled) during program memory programming. If code protection is enabled then data read back will be scrambled during programming, which makes verification of the code impossible.

Mid-Range Serial Programming

Serial programming (known as "In Circuit Serial Programming" or "ICSP") for the low-end (which implements) and mid-range parts consists of pin access:

Mid-Range PICmicro® MCU ICSP Programming Pins								
Pin		12C5xx		16C50x	18 Pin Mid	28 Pin Mid		40 Pin Mid	
1 Vpp	4	_MCLR	4	_MCLR	4 _MCLR	1	_MCLR	1 _MCLR	
2 Vdd	1	Vdd	1	Vdd	14 Vdd	26	Vdd	11,32 Vdd	
3 GND	8	Vss	14	Vss	5 Vss	8,21	Vss	12,31 Vss	
4 DATA	7	GPO	13	RB0	13 RB7	28	RB7	40	RB7
5 CLOCK	6	GP1	12	RB1	12 RB6	27	RB6	39	RB6

To program and read data, the PICmicro® MCU must be put into "programming mode" by raising the "_MCLR" pin to 13 to 14 volts, and the "data" and "clock" lines pulled low for several milliseconds. Once, the PICmicro® MCU is in programming mode, "Data" can then be shifted in and out using the "Clock" line.

Putting the PICmicro® MCU into programming mode requires the data waveform shown in Fig. 9.4. When _MCLR is driven to Vpp, the internal program counter of the PICmicro® MCU is reset. The PICmicro® MCU's program counter is used to keep track of the current program memory address in the EPROM that is being accessed.

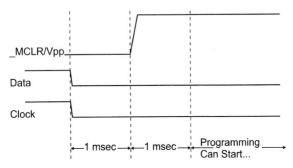

Figure 9.4 Programmer Initialization

Mid-Range PICmicro® MCU EPROM ICSP Programming Commands			
Command	Bit Pattern	Data	Comments
Load Data	0b0000010	0,14 Bits Data,0	Load word for programming
Begin Programming	0b0001000	none	Start Programming Cycle
End Programming	0b0001110	none	End Programming Cycle after 100 msec
Increment Address	0b0000110	none	Increment the PICmicro® MCU's Program Counter
Read Data	0b0000100	0, 14 Bits Data, 0	Read Program Memory at Program Counter
Load Config	0b0000000	0x07FFE	Set the PICmicro® MCU's Program Counter to 0x02000

Data is shifted in and out of the PICmicro® MCU using a synchronous protocol. Data is shifted out least significant bit (LSB) first on the falling edge of the clock line. The minimum period for the clock is 200 nsecs with the data bit centered as shown in Fig. 9.5, which is sending an "increment address" command.

When data is to be transferred, the same protocol is used, but a 16-bit transfer (LSB first) follows after one microsecond has passed since the transmission of the command. The 16 bits consist of the instruction word shifted to the left by one. This means the first and last bits of the data transfer are always "zero".

Before programming can start, the program memory should be checked to make sure it is blank. This is accomplished by simply reading the program memory ("Read Data" command listed above) and comparing the data returned to 0x07FFE. After every compare, the PICmicro® MCU's program counter is incremented (using the "increment address" command) to the size of the devices program memory. Once the program memory is checked, the PICmicro® MCU's program counter is "jumped" to 0x02000 (using the "Load Configuration" Command) and then the next eight words are checked for 0x07FFE.

For different mid-range devices, the following table of

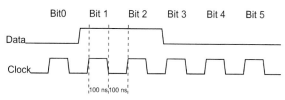

Figure 9.5 Programmer Command—6 Bits

PICmicro® MCU part numbers can be used to determine the amount of program memory available within them:

Mid-Range PICmicro® MCU EPROM Sizes by Part Number Mask	
Device	Sizes
PIC16Cx1	1k
PIC16Cxx0	0.5k
PIC16Cxx1	1k
PIC16Cxx2	2k
PIC16Cx2	2k
PIC16Cx3	4k
PIC16Cx4	2k
PIC16Cx5	4k
PIC16Cx6	8k
PIC16Cx7	8k

The process for programming an instruction word in a mid-range EPROM-based PICmicro® MCU is

1. A "Load Data" command with the word value is written to the PICmicro® MCU.

2. A "Begin Programming" command is written to the PICmicro® MCU.

3. Wait 100 msecs.

4. An "End Programming" command is written to the PICmicro® MCU.

5. A "Read Data" command is sent to the PICmicro® MCU and 14 bits (the LSB and MSB of the 16-bit data transferred are ignored) in the program memory instruction are read back.

6. Steps one through five are repeated up to 25 times until the data read back is correct. Steps one through five are known as a "Programming Cycle".

7. Steps one through four are repeated three times the number of cycles required to get a valid instruction word read. This is known as "Over-programming".

8. The PICmicro® MCU's Program Counter is incremented using the "Increment Address" command.

9. Steps one through eight are repeated for the entire application to the configuration fuses.

10. A "Load Config" command is sent to the PICmicro® MCU to set the Program Counter to 0x02000.

11. The PICmicro® MCU's configuration fuses are programmed using the "Programming Cycle" detailed in steps one through seven.

The process for programming program memory could be blocked out with the pseudo-code:

```
ICSPProgram()      //  Program to be burned in is in
                   //    an array of
{                  //    addresses and data

int PC = 0;        //  PICmicro® MCU's program counter
int i, i j k;
int retvalue = 0;

  for (i = 0; (i - PGMsize) && (retvalue == 0); I++)
  {

    if (PC ! = address[i]) {
      if ((address[I] >= 0x02000) && (PC < 0x02000))
        {
        LoadConfiguration(0x07FFE);
        PC = 0x02000;
      }

    for (; PC < address[i]; PC++)
      IncrementAddress();
```

```
for (i = 0; (i < 25) && (retvalue != data[I]);
  I++)    {
  LoadData(ins[i] << 1);  //  Programming Cycle
  BeginProgramming();
  Dlay(100usec);
  EndProgramming();

  Retvalue = ReadData();
}

if (i == 25)
  retvalue = -1;  //  Programming Error
else {
  retvalue = 0;   //  Okay, Repeat Programming
                  //  Cycle 3x
  for (k = 0; k < (j * 3); k++){
    LoadData(ins[i] << 1);
    BeginProgramming();
    Dlay(100usec);
    EndProgramming();
  } //  endif
} //  endif
} //  endfor
} //  end ICSPProgram
```

After the program memory has been loaded with the application code, Vpp should be cycled off and on and the PICmicro® MCUs program memory is read out and compared against the expected contents. When this verify is executed, Vpp should be cycled again with Vdd a minimum voltage (4.5 volts) and then repeated again with Vdd at a maximum voltage (5.0 volts) value. This second verify is used by "Production PICmicro® MCU Programmers". Programmers not having these minimum/maximum Vdd verify steps are known as "Prototype PICmicro® MCU Programmers".

The PIC12C50x and PIC16C50x low-end parts are programmed using a similar protocol as the EPROM based mid-range PICmicro® MCUs. The command first enters programming mode (with the data and clock pins pulled low followed by _MCLR driven to +13 volts) and the PICmicro® MCU's program counter is set to 0x0FFF, which is the configuration register address.

The PIC12C50x and PIC16C505 use a 12-bit instruction word. When data is passed to the PICmicro® MCU, the upper 3 bits (instead of the upper one) are "zero" and ignored by the device as it is programmed. The first bit sent is still "0", with the LED of the instruction word following.

A simple way of calculating the 16 data bits to be programmed into the PIC12C50x and PIC16C505 microcontrollers from the instruction is to save the instruction in a 16 bit variable and shift it "up" (to the "left") by one bit. The commands available for programming the PIC12C50x and PIC16C505 have a 6-bit header and optional 16-bit instruction or configuration fuse data word.

PIC12C5xx and PIC16C505 Programming Commands

Command	Bits
Load data	000010 +0, data(12), 000
Read data	000100 +0, data(12), 000
Increment PC	001000
End programming	001110

Microchip uses a modified version of the programming algorithm outlined above for the 16F8x "Flash"-based parts. Along with the programming algorithm being much simpler, the actual programming circuit is much easier to implement in hardware. Figure 9.6 shows a typical Flash programming circuit.

Electrically, the programming voltages are basically the same as what is required for the mid-range devices.

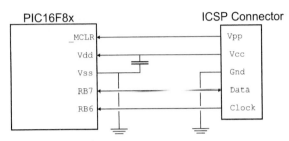

Figure 9.6 PIC16F8x ICSP Connection

There is one difference, however, and that is in the voltage and current required for Vpp. For the mid-range parts, up to 50 mA is required for EPROM programming. Because the 16F8x parts have built-in EPROM data and Flash VPP generator, this circuit will provide the actual voltages and currents to program and engage the data and program memory resulting in micro-Amperes of current required from the programmer in "Programming Mode".

The same data packet format is used for the 16F8x as was used for the mid-range EPROM parts, but the commands and how they work are slightly different. The table below lists the different commands:

Mid-Range Flash PICmicro® MCU Programming Commands		
Command	Bits	Data
Load Configuration	000000	07FFE In
Load Data for Program	000010	Word x 2 Going In Memory
Load Data for Data Memory	000011	Byte x 2 Going In
Read Data from Program	000100	Word x 2 Going Out Memory
Read Data from Data Memory	000101	Byte x 2 Going Out

Mid-Range Flash PICmicro® MCU Programming Commands (*Continued*)		
Command	Bits	Data
Increment PICmicro® MCU's PC	000110	none
Begin Programming	001000	none
Bulk Erase Program Memory	001001	none
Bulk Erase Data Memory	001011	none

The data, like in the mid-range EPROM part, is always 16 bits with the first and last bit always equal to zero. Data is always transferred LSB first using the same timings as specified earlier in the chapter for the mid-range parts. When transferring 14 bits of data from the hex file instruction word, it can be multiplied by two, leaving the LSB and MSB reset.

The programming cycle for the PIC 16F8x is as follows:

1. "Load Data for Program Memory" command with Instruction word.

2. Send "Begin Programming" command.

3. Wait 10 msecs.

4. "Read Data from Program Memory" command and verify the contents of the Program Memory.

5. Send "Increment PICmicro® MCU's Program Counter" command.

6. Steps one through five are the Flash PICmicro® MCU "Programming Cycle". These steps are repeated for every instruction in the hex file.

7. A "Load Configuration" command is sent to set the Program Counter to point to address 0x02000.

8. Steps one through four are repeated for the Configuration Information.

To erase the Flash Program Memory, use the Microchip specified erase for code protected devices. This operation will erase all Flash and EEPROM memory in the PICmicro® MCU device, even if code protection is enabled.

1. Hold RB6 and RB7 low and apply Vpp, wait at least 2 msecs.

2. Execute Load Configuration (0b0000000+ 0x07FFE).

3. Increment the PC to the Configuration Register Word (Send 0b0000110 seven times).

4. Send command 0b0000001 to the PICmicro® MCU.

5. Send command 0b0000111 to the PICmicro® MCU.

6. Send "begin programming" (0b0001000) to the PICmicro® MCU.

7. Wait 10 ms.

8. Send command 0b0000001.

9. Send command 0b0000111.

Note that there are two undocumented commands ("0b0000001" and "0b0000111") in this sequence.

PIC17Cxx Programming

The PIC17Cxx is connected to a programmer using the wiring shown in Fig. 9.7. Note that PORTB and PORTC are used for transferring data 16 bits at a time and PORTA is used for the control bits that control the operation of the programmer. The "_MCLR" pin is pulled high to 13 volts as would be expected to put the PICmicro® MCU into "Programming Mode".

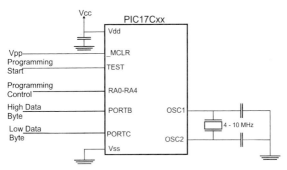

Figure 9.7 PIC17Cxx Parallel Programming Connections

While the programming of the PIC17Cxx is described as being in "parallel", a special "Boot ROM" routine executes within the PICmicro® MCU and this accepts data from the I/O ports and programs the code into the PICmicro® MCU. To help facilitate this, the "test" line, which is normally tied low, is pulled high during application execution to make sure that the programming functions can be accessed. The clock, which can be any value from 4 MHz to 10 MHz is used to execute the "Boot ROM" code for the programming operations to execute.

To put the PICmicro® MCU into programming mode, the "test" line is made active before _MCLR is pulled to Vpp and then 0x0E1 is driven on PORTB to command the boot code to enter the programmer routine. This sequence is shown in Fig. 9.8. To end programming mode, _MCLR must be pulled to ground 10 msecs or more before power is taken away from the PICmicro® MCU. "Test" should be de-asserted after _MCLR is pulled low.

When programming, the RA0 pin is pulsed high for at

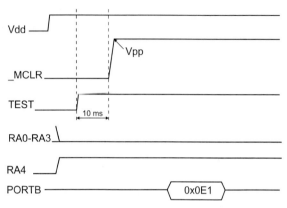

Figure 9.8 PIC17Cxx Parallel Programming Start-Up

least 10 instruction cycles (10 us for the PICmicro® MCU running at 4 MHz) to load in the instruction address followed by the PICmicro® MCU latching out the data (so that it can be verified). After the data have been verified, RA0 is pulsed high for 100 usecs to program the data. If RA1 is low during the RA0 pulse, then the PICmicro® MCU program counter will be incremented. If it goes high during the pulse, the internal program counter will not be incremented and the instruction word contents can be read back in the next RA1 cycles without having to load in a new address.

The latter operation is preferred and looks like the waveforms shown in Fig. 9.9. This waveform should be repeated until the data is loaded or up to 25 times. Once it is programmed in, then three times the number of programming cycles must be used to "lock" and "overprogram" the data in. This process is similar to that of the other EPROM parts.

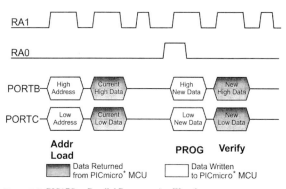

RA1

RA0

PORTB — ⟨ High Address ⟩ ⟨ Current High Data ⟩ ⟨ High New Data ⟩ ⟨ New High Data ⟩

PORTC — ⟨ Low Address ⟩ ⟨ Current Low Data ⟩ ⟨ Low New Data ⟩ ⟨ New Low Data ⟩

**Addr
Load**   **PROG**   **Verify**

▨ Data Returned
from PICmicro® MCU   ☐ Data Written
to PICmicro® MCU

Figure 9.9 PIC17Cxx Parallel Programming Waveform

Writing to the specified addresses between 0x0FE00 and 0x0FE0F programs and verifies the configuration word. To program ("make 0") one of the configuration bits, write to its register. Reading back the configuration word uses the first three RA1 cycles of Fig. 9.9 at either 0x0FE00 or 0x0FE08. Reading 0x0FE00 will return the low byte of the configuration word in PORTC (0x0FF will be in PORTB) and reading 0x0FE08 will return the high byte in PORTC.

The configuration bits for the PIC17Cxx are defined as:

PIC17Cxx Configuration Bits	
Address	Bit
F0SC0	0x0FE00
F0SC1	0x0FE01
WDTPS0	0x0FE03
WDTPS1	0x0FE04
PM0	0x0FE05
PM1	0x0FE06
PM2	0x0FE0F

where the bits are defined as:

```
PIC17Cxx Configuration Bit Definition

PM2:PM0          Processor mode
111              Microprocessor Mode
110              Microcontroller Mode
101              Extended Microcontroller Mode
000              Code Protected Microcontroller
Mode

WDTPS1:WDTPS0    Watchdog Timer and Postscaler Mode
11               WDT Enabled. Postscaler = 1:1
10               WDT Enabled, Postscaler = 256:1
01               WDT Enabled, Postscaler = 64:1
00               WDT Disabled, 16 bit Overflow
Timer

FOSC1:FOSC0      Oscillator Mode
11               External Oscillator
10               XT Oscillator
01               RC Oscillator
00               LF Oscillator
```

Note, configuration bit addresses must be written to in
ascending order. Programming the bit in nonregister ascending order can result in unpredictable programming
of the configuration word as the "Processor" Mode
changes to a "Code Protected" mode before the data is
loaded in completely.

PIC17Cxx ICSP Programming

The capability of a PIC17Cxx application to write to program memory is enabled when the _MCLR is driven by
more than 13 volts and a tablwt instruction is executed.
When tablwt is executed, the data loaded into the
TABLATH and TABLATL registers is programmed into

the memory locations. This instruction keeps executing (it does not complete after two cycles, as it would if the TBLPRH and TBLPTRL registers were pointing outside the internal EPROM) until it is terminated by an interrupt request or _MCLR reset.

To perform a word write, the following mainline process would be used:

1. Disable TMR0 interrupts.
2. Load TABPTRH and TABPTRL with the address.
3. Load TABLATH or TABLATL with the data to be stored.
4. Enable a 1,000 usec TMR0 delay interrupt (initialize TMR0 and enable TMR0 interrupt).
5. Execute tablwt instruction with the missing half of data.
6. Disable TMR0 interrupts.
7. Read back data—Check for match.
8. If no match—Return error.

The interrupt handler for this process can just be executing a "retfie" instruction. Sample code for writing to the PIC17Cxx's program memory is

```
  org      0x00010
TMR0Int                          ; Timer Interrupt
                                 ;  Request
                                 ;  Acknowledge

  retfie

   :

  movfp    SaveAddress,          ; Point to the
           TBLPTRL               ;  Memory being
```

```
movfp   SaveAddress + 1,    ; written to
        TBLPTRH

bcf     PORTA, 3            ; Turn on Programming
                           ; Voltage

movlw   HIGH ((100000 / 5) ; Delay 100 msecs
        + 256)             ; for
movwf   Dlay               ; Programming Voltage
                           ; to Stabilize
movlw   LOW ((100000 / 5)
        + 256)
addlw   0x0FF
btfsc   ALUSTA, Z
 decfsz Dlay, f
  goto  $ - 3

movlw   HIGH (65536 -      ; Delay 10 msecs for
        10000)             ; EPROM Write
movwf   TMR0H
movlw   LOW (65536 - 10000)
movwf   TMR0L

bsf     T0STA, T0CS        ; Start up the Timer

movlw   1 << T0IE          ; Enable Interrupts
movwf   INTSTA
bcf     CPUSTA, GLINTD

tlwt    0, SaveData        ; Load Table Pointer
                           ; with Data
tlwt    1, SaveData + 1
tablwt  1, 0, SaveData + 1 ; Write the Data In
nop
nop

clrf    INTSTA, f          ; Turn Off Interrupts
bsf     CPUSTA, GLINTD

movlw   2
call    SendMSG

bsf     PORTA, 3
```

To enable internal programming, _MCLR has to be "switched" from 5 volts (VDD) to 13 volts. The Microchip circuit that is recommended is shown in Fig. 9.10. This circuit will drive the PIC17Cxx's _MCLR pin at 5 volts until "RA2" is pulled low. When RA2 is pulled low, the voltage driven in to _MCLR will become 13 volts (or "Vpp"). The programming current at 13 volts is a minimum of 30 mA.

The "boot code" is a host interface application that reads data and then programs it at the specified address. This code must be burned into the PICmicro® MCU before "ICSP" can execute. The typical "boot code" for a PIC17Cxx PICmicro® MCU would be as follows:

1. Establish communication with programming host.
2. If no communication link is established jump to application code.

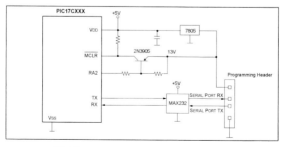

Figure 9.10 PIC17Cxxx in circuit serial programming schematic

3. Enable Vpp (RA2 = 0).

4. Wait for host to send instruction word address.

5. Program in the word.

6. Confirm word is programmed correctly.

7. Loop back to four.

PIC18Cxx Programming

Like the PIC17Cxx, the PIC18Cxx has the capability to "self program" using the table "read and write" instructions. In the PIC18Cxx, this capability is not only available within applications, but is used to program an erased device.

To program the PIC18Cxx, instructions are downloaded into the PICmicro® MCU after setting the "_MCLR" pin to Vpp (13 to 14 volts, as in the other EPROM PICmicro® MCUs) with both RB6 and RB7 low. Passing instructions (which contain the program data) to the PICmicro® MCU is accomplished by first sending a 4-bit "Special Instruction" followed by an optional 16-bit instruction. The 4-bit Special Instruction is sent most significant bit first and can either specify that an instruction follows or that it is a "mnemonic" for a "TBLRD" or "TBLWT" instruction as shown in the table below:

PIC18Cxx Programming "Mnemonics"			
Special Instruction	Mnemonic	Instruction Operation	Cycles
0000	nop	Shift in Next Instruction	1
1000	TBLRD *	Read Table	2
1001	TBLRD *+	Read Table, Increment TBLPTR	2

PIC18Cxx Programming "Mnemonics" (Continued)			
Special Instruction	Mnemonic	Instruction Operation	Cycles
1010	TBLRD *-	Read Table, Decrement TBLPTR	2
1011	TBLRD +*	Increment TBLPTR, Read Table	2
1100	TBLWT *	Write Table	2
1101	TBLWT *+	Write Table, Increment TBLPTR	2
1110	TBLWT *-	Write Table, Decrement TBLPTR	2
1111	TBLWT +*	Increment TBLPTR, Write Table	2

The data transmission looks like Fig. 9.11. The four "nop" bit code is transmitted first followed by the 16-bit instruction.

If the instruction is a table operation, then the "Special Instruction" code can be used instead of the "nop" to simplify the data transfer. At the end of Fig. 9.11, the bit pattern 0b01101 ("TBLWT *−") is sent to the PICmicro® MCU.

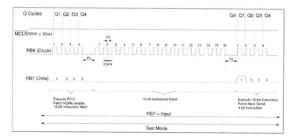

Figure 9.11 Serial Instruction Timing for 1 Cycle 16-Bit Instructions

While the table reads and writes only require 4 bits, to carry out the program operation, there are always 16 bits following the mnemonic (just as if it were a "nop") for data transfer and this avoids the need for explicitly loading and unloading the table latch registers using instructions. In Fig. 9.12, the "tblwt *" instruction (write to table and do not change TBLPTR) is shown.

After the first 20-bit sequence, a second 20-bit sequence is executed to allow the programming operation to complete (this is what is meant by the "2" in the "Cycles" in the table above). The PICmicro® MCU ignores the second sequence of 20 bits and the initial sequence is processed. Reading data from the PICmicro® MCU's program memory is accomplished in exactly the same way.

To set up a table read or write, first the TBLPTR has to be initialized. This is done using standard "movlw" and "movwf" instruction. For example, to program address

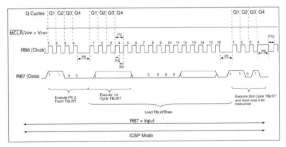

Figure 9.12 TBLWT Instruction Sequence

0x012345 with 0x06789, the data sequence is written to the PIC18Cxx:

```
Mnemonic      Instruction/Data
nop           movlw  UPPER 0x012345
nop           movwf  TBLPTRU
nop           movlw  HIGH 0x012345
nop           movwf  TBLPTRH
nop           movlw  LOW 0x012345
nop           movwf  TBLPTRL
tblwt *       0x06789
```

Microchip ICSP Programming Connector

The "ICSP" Programming Connector defined by Microchip uses the pinout shown in the table below:

Microchip "ICSP" Pin Definition}			
PIN	12C5xx	16Cxx	16Fxx
1.Vpp	-MCLR/Vpp	-MCLR/Vpp	-MCLR/Vpp
2.Vdd	Vdd	Vdd	Vdd
3.Vss	Vss	Vss	Vss
4.DATA	GP0	RB7	RB7
5.CLOCK	GP1	RB6	RB6

To connect a PICmicro® MCU, which has been put into an application circuit, the following interface shown in Fig. 9.13 should be used.

The PICmicro® MCU must be isolatable from the application circuit. The diode on the "_MCLR/Vpp" pin and the "breakable connections" on Vdd, RB7, and RB6 isolate the PICmicro® MCU. These "breaks" are best provided by unsoldered "zero ohm" resisters or uncon-

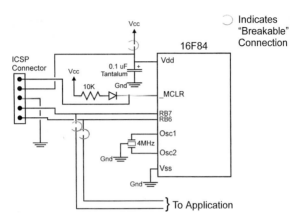

Figure 9.13 16F84 ICSP Circuit

nected jumpers in the circuit. This has to be done because the ICSP specification will only provide 50 mA for Vdd and has 1K ohm resisters in the data and clock lines to protect the driver circuits.

Third Party/Downloadable Programmers

When considering a PICmicro® MCU programmer, the following questions should be asked:

1. What are the supported PICmicro® MCU devices?

2. What is the interface and how is the application timed?

3. How are the Configuration Fuses programmed?

4. What Operating System does it run under?

PC Interfaces

Memory Map

Figure 10.1 shows a graphic of the PC's memory map.

I/O Space Map

Only the lower 10 bits of the I/O space have been specified for the basic PC/AT register operation. Some PC/XT
specific registers have been omitted from this list. It is
not obvious in the table below, but I/O port addresses
0x0000 to 0x00FF are on the motherboard while the addresses above are on adapter cards.

For some motherboards, registers are accessed at ad-

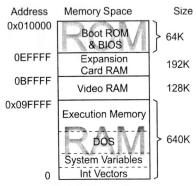

Address	Memory Space	Size
0x010000	Boot ROM & BIOS	} 64K
0EFFFF	Expansion Card RAM	192K
0BFFFF	Video RAM	128K
0x09FFFF	Execution Memory	
	DOS	} 640K
	System Variables	
0	Int Vectors	

Figure 10.1 The PC's Memory Map

dresses 0x0400 and above. To avoid problems, make sure that you only specify addresses below 0x0400.

```
Address        Register Description

0000h          DMA channel 0 address (low addressed
               first, then high)
0001h          DMA channel 0 word count (low addressed
               first, then high)
0002h          DMA channel 1 address (low addressed
               first, then high)
0003h          DMA channel 1 word count (low addressed
               first, then high)
0004h          DMA channel 2 address (low addressed
               first, then high)
0005h          DMA channel 2 word count (low addressed
               first, then high)
0006h          DMA channel 3 address (low addressed
               first, then high)
0007h          DMA channel 3 word count (low addressed
               first, then high)
0008h          Read - DMA 1 channel 0-3 status
               register
```

```
                Bit 7 = Channel 3 Request
                Bit 6 = Channel 2 Request
                Bit 5 = Channel 1 Request
                Bit 4 = Channel 0 Request
                Bit 3 = Channel Terminal Count on
                Channel 3
                Bit 2 = Channel Terminal Count on
                Channel 2
                Bit 1 = Channel Terminal Count on
                Channel 1
                Bit 0 = Channel Terminal Count on
                Channel 0
                Write - DMA 1 channel 0-3 command
                register
                Bit 7 = DACK Sense Active High
                Bit 6 = DREQ Sense Active High
                Bit 5 = Extended Write Selection
                Bit 4 = Rotating Priority
                Bit 3 = Compressed Timing
                Bit 2 = Enable Controller
0009h           DMA 1 write request register
000Ah           DMA 1 channel 0-3 mask register
                Bit 7-3 = Reserved
                Bit  2  = Mask bit
                Bit 1-0 = Channel Select
                     - 00 channel 0
                     - 01 channel 1
                     - 10 channel 2
                     - 11 channel 3
000Bh           DMA 1 channel 0-3 mode register
                Bit 7-6 = Operating Mode
                     - 00 demand mode
                     - 01 single mode
                     - 10 block mode
                     - 11 cascade mode
                Bit 5 = address increment select
                Bit 3-2 = Operation
                     - 00 verify operation
                     - 01 write to memory
                     - 10 read from memory
                     - 11 reserved
                Bit 1-0 = Channel Select
                     - 00 channel 0
                     - 01 channel 1
                     - 10 channel 2
                     - 11 channel 3
```

```
000Ch        DMA 1 clear byte pointer flip-flop
000Dh        Read - DMA 1 read temporary register
             Write - DMA 1 master clear
000Eh        DMA 1 clear mask register
000Fh        DMA 1 write mask register
0020h        Interrupt Controller 1 initialization
             command word
             Bit 7-5 = 0 - only used in 80/85 mode
             Bit  4  = ICW1 Request
             Bit  3  = Interrupt Request Mode
                     - 0 Edge triggered mode
                     - 1 Level triggered mode
             Bit  2  = Interrupt Vector Size
                     - 0 Eight Byte Interrupt Vectors
                     - 1 Four Byte Interrupt Vectors
             Bit  1  = Operating Mode
                     - 0 Cascade mode
                     - 1 Single mode
             Bit  0  = IC4 Requirements
                     - 0 not needed
                     - 1 needed
0021h        Interrupt Controller 1 Interrupt Mask
             Register
             bit 7 = 0  enable parallel printer
                        interrupt
             bit 6 = 0  enable diskette interrupt
             bit 5 = 0  enable fixed disk interrupt
             bit 4 = 0  enable serial port 1
                        interrupt
             bit 3 = 0  enable serial port 2
                        interrupt
             bit 2 = 0  enable video interrupt
             bit 1 = 0  enable keyboard, mouse, RTC
                        interrupt
             bit 0 = 0  enable timer interrupt
0040h        8254 Timer Counter 0 & Counter Divisor
             Register
0041h        8254 Timer Counter 1 & Counter Divisor
             Register
0042h        8254 Timer Counter 2 & Counter Divisor
             Register
0043h        8254 Timer Mode/Control port
             Bit 7-6 = Counter Select
                     - 00 Counter 0
                     - 01 Counter 1
                     - 10 Counter 2
```

```
              Bit 5-4 = Counter Read/Write Operation
                    - 01 Read/Write Low Counter Byte
                    - 10 Read/Write High Counter
                      Byte
                    - 11 Read/Write Low, then High
                      Counter Bytes
              Bit 3-1 = Counter Mode Select
                    - 000 mode 0
                    - 001 mode 1/Programmable One
                      Shot
                    - x10 mode 2/Rate Generator
                    - x11 mode 3/Square Wave
                      Generator
                    - 100 mode 4/Software Triggered
                      Strobe
                    - 101 mode 5/Hardware Triggered
                      Strobe
              Bit 0   = Counter Type
                    - 0 Binary Counter
                    - 1 BCD Counter
0060h         Read - Keyboard Controller
              Bit  7  = Keyboard Inhibit (Reset)
              Bit  6  = CGA (Reset)
              Bit  5  = Manufacturing Jumper Install
              Bit  4  = Reset if System RAM 512K
              Bit 3-0 = Reserved
              Write - Keyboard Controller
              Bit  7  = Keyboard Data Output
              Bit  6  = Keyboard Clock Output
              Bit  5  = Input Buffer Full (Reset)
              Bit  4  = Output Buffer Empty (Reset)
              Bit 3-2 = Reserved
              Bit  1  = Address Line 20 Gate
              Bit  0  = System Reset
0061h         Read - Keyboard Controller Port B
              control register
              Bit 7   = Parity Check
              Bit 6   = Channel Check
              Bit 5   = Current Timer 2 Output
              Bit 4   = Toggles with each Refresh
                        Request
              Bit 3   = Channel Check Status
              Bit 2   = Parity Check Status
              Bit 1   = Speaker Data Status
              Bit 0   = Timer 2 Gate to Speaker
                        Status
```

```
                   Write - 8255 Compatible Port
                    Bit 7 = Clear Keyboard
                    Bit 6 = Hold Keyboard Clock Low
                    Bit 5 = I/O Check Enable
                    Bit 4 = RAM Parity Check Enable
                    Bit 3 = Read low/high switches
                    Bit 2 = Reserved
                    Bit 1 = Speaker Clock Enable
                    Bit 0 = Timer 2 Gate to Speaker Enable
       0064h        Read - Keyboard Controller Status
                    Bit 7 = Parity Error on Keyboard
                            Transmission
                    Bit 6 = Receive Timeout
                    Bit 5 = Transmit Timeout
                    Bit 4 = Keyboard Inhibit
                    Bit 3 = Input Register Type
                        - 1 data in input register is
                          command
                        - 0 data in input register is
                          data
                    Bit 2 = System Flag Status
                    Bit 1 = Input Buffer Status
                    Bit 0 = Output Buffer Status
                   Write Keyboard Controller Input Buffer
                    20    = Read Byte Zero of Internal
                            RAM,this is the last KB
                            command send to 8042
                    21-3F = Reads the Byte Specified in
                            the Lower 5 Bits of the
                            command in the 8042's
                            internal RAM
                    60-7F = Writes the Data Byte to the
                            Address Specified in the 5
                            Lower Bits of the Command
       0065h        Address Line 20 Gate Control
                    Bit 2 = A20 gate control
                        1 - A20 enabled
                        0 - A20 disabled
       0070h        CMOS RAM index register port
                    Bit  7  = NMI Enable
                    Bit 6-0 = CMOS RAM Index
       0071h        CMOS RAM data port
                    00 = Current Second in BCD
                    01 = Alarm Second in BCD
                    02 = Current Minute in BCD
                    03 = Alarm Minute in BCD
```

```
                    04 = Current Hour in BCD
                    05 = Alarm Hour in BCD
                    06 = Day of Week in BCD
                    07 = Day of Month in BCD
                    08 = Month in BCD
                    09 = Year in BCD (00-99)
                    0A = Status Register A
                     Bit  7  = Update Progress
                     Bit 6-4 = Divider that Identifies the
                                time-based Frequency
                     Bit 3-0 = Rate Selection Output
                    0B = Status Register B
                     Bit 7 = Run/Halt Control
                     Bit 6 = Periodic Interrupt Enable
                     Bit 5 = Alarm Interrupt Enable
                     Bit 4 = Update-Ended Interrupt Enable
                     Bit 3 = Square Wave Interrupt Enable
                     Bit 2 = Calendar Format
                     Bit 1 = Hour Mode
                     Bit 0 = Daylight Savings time Enable
                    0C = Status Register C
                     Bit  7  = Interrupt Request Flag
                     Bit  6  = Periodic Interrupt Flag
                     Bit  5  = Alarm Interrupt Flag
                     Bit  4  = Update Interrupt Flag
                     Bit 3-0 = Reserved
                    0D = Status Register D
                     Bit  7  = Real-Time Clock power
0080h           "MFG_PORT" Write Address
0080h           DMA Page Register page register
                (temporary storage)
0081h           DMA Channel 2 Page Address
0082h           DMA Channel 3 Page Address
0083h           DMA Channel 1 Page Address
0084h           Extra Page Register
0085h           Extra Page Register
0086h           Extra Page Register
0087h           DMA Channel 0 Page Address
0088h           Extra Page Register
0089h           DMA Channel 6 Page Address
008Ah           DMA Channel 7 Page Address
008Bh           DMA Channel 5 Page Address
008Ch           Extra Page Register
008Dh           Extra Page Register
008Eh           Extra Page Register
008Fh           DMA Refresh Page Register
```

```
00A0h          Interrupt Controller 2 Initialization
               Command Word
               Bit 7-5 = 0 - only used in 80/85 mode
               Bit  4  = ICW1 Request
               Bit  3  = Interrupt Request Mode
                       - 0 Edge triggered mode
                       - 1 Level triggered mode
               Bit  2  = Interrupt Vector Size
                       - 0 Eight Byte Interrupt Vectors
                       - 1 Four Byte Interrupt Vectors
               Bit  1  = Operating Mode
                       - 0 Cascade mode
                       - 1 Single mode
               Bit  0  = IC4 Requirements
                       - 0 not needed
                       - 1 needed
00A1h          Interrupt Controller 2 Mask Register
               Bit 7 = Reserved
               Bit 6 = Fixed Disk
               Bit 5 = Coprocessor exception
               Bit 4 = Mouse Interrupt
               Bit 3 = Reserved
               Bit 2 = Reserved
               Bit 1 = Redirect Cascade
               Bit 0 = Real-Time Clock
00D0h          Read - DMA Controller 2 Channel 4-7
               status register
               Bit 7 = Channel 7 Request
               Bit 6 = Channel 6 Request
               Bit 5 = Channel 5 Request
               Bit 4 = Channel 4 Request
               Bit 3 = Channel 7 Terminal Count
               Bit 2 = Channel 6 Terminal Count
               Bit 1 = Channel 5 Terminal Count
               Bit 0 = Channel 4 Terminal Count
               Write DMA Controller 2 Channel 4-7
               command register
               Bit 7 = DACK Sense Active High
               Bit 6 = DREQ Sense Active High
               Bit 5 = Extended Write Selection
               Bit 4 = Rotating Priority
               Bit 3 = Compressed Timing
               Bit 2 = Enable Controller
00D2h          DMA Controller 2 Channel 4-7 Write
               Request Register
```

```
00D4h          DMA Controller 2 Channel 4-7 Write
               Single Mask Register
               Bit 7-3 = Reserved
               Bit  2  = Mask bit
               Bit 1-0 = Channel Select
                         - 00 channel 0
                         - 01 channel 1
                         - 10 channel 2
                         - 11 channel 3
00D6h          DMA Controller 2 Channel 4-7 Mode
               Register
               Bit 7-6 = Operating Mode
                         - 00 demand mode
                         - 01 single mode
                         - 10 block mode
                         - 11 cascade mode
               Bit 5 = address increment select
               Bit 3-2 = Operation
                         - 00 verify operation
                         - 01 write to memory
                         - 10 read from memory
                         - 11 reserved
               Bit 1-0 = Channel Select
                         - 00 channel 0
                         - 01 channel 1
                         - 10 channel 2
                         - 11 channel 3
00D8h          DMA Controller 2 Channel 4-7 Clear Byte
               Pointer
00Dah          Read - DMA Controller Channel 4-7 Read
               Temporary Register
               Write - DMA Controller Channel 4-7
               Master Clear
00DCh          DMA Controller 2 Channel 4-7 Clear Mask
               Register
00DEh          DMA Controller 2 Channel 4-7 Write Mask
               Register
00F0h          Math Coprocessor Clear Busy Latch
00F1h          Math Coprocessor Reset
00F8h          Opcode Transfer Register
00FAh          Opcode Transfer Register
00FCh          Opcode Transfer Register
01F0h          Hard Disk Controller Data Register
01F1h          Hard Disk Controller Error Register
               Bit  7  = Failing Drive
               Bit 6-3 = Reserved
```

```
                    Bit 2-0 = Status
                          - 001 No Error
                          - 010 Formatter Device Error
                          = 011 Sector Buffer Error
                          = 100 ECC Circuitry Error
                          = 101 Controlling Microprocessor
                                Error
01F2h           Sector Count
01F3h           Sector Number
01F4h           Cylinder Low
01F5h           Cylinder High
01F6h           Drive/Head
01F7h           Read - Hard Disk Controller Status
                Register
                bit 7 = Controller Execution Status
                bit 6 = Drive Status
                bit 5 = Write Fault
                bit 4 = Seek Complete
                bit 3 = Sector Buffer Requires
                        Servicing
                bit 2 = Disk Data Read Successfully
                        Corrected
                bit 1 = Index
                bit 0 = Previous Command Ended in
                        Error
                Write - Hard Disk Controller Command
                Register
0201h           Read - Joystick Position and Status
                Bit 7 = Status B Joystick Button 2
                Bit 6 = Status B Joystick Button 1
                Bit 5 = Status A Joystick Button 2
                Bit 4 = Status A Joystick Button 1
                Bit 3 = B joystick Y coordinate
                Bit 2 = B joystick X coordinate
                Bit 1 = A joystick Y coordinate
                Bit 0 = A joystick X coordinate
                Write - Fire Joystick's four one-shots
0220h           SoundBlaster - Left speaker
                Status/Address
                Address:
                    01 = Enable waveform control
                    02 = Timer #1 data
                    03 = Timer #2 data
                    04 = Timer control flags
                    08 = Speech synthesis mode
```

```
               20-35 = Amplitude Modulation/Vibrato
               40-55 = Level key scaling/Total level
               60-75 = Attack/Decay rate
               80-95 = Sustain/Release rate
               A0-B8 = Octave/Frequency Number
               C0-C8 = Feedback/Algorithm
               E0-F5 = Waveform Selection
0221h          SoundBlaster - Left speaker Data
0222h          SoundBlaster - Right speaker/Address
               Address:
               01 = Enable waveform control
               02 = Timer #1 data
               03 = Timer #2 data
               04 = Timer control flags
               08 = Speech synthesis mode
               20-35 = Amplitude Modulation/Vibrato
               40-55 = Level key scaling/Total level
               60-75 = Attack/Decay rate
               80-95 = Sustain/Release rate
               A0-B8 = Octave/Frequency Number
               C0-C8 = Feedback/Algorithm
               E0-F5 = Waveform Selection
0223h          Right speaker -- Data port
0278h          LPT2 data port
0279h          LPT2 Status Port
               Bit  7  = Busy
               Bit  6  = Acknowledge
               Bit  5  = Out of Paper
               Bit  4  = Printer Selected
               Bit  3  = Error
               Bit  2  = IRQ Occurred
               Bit 1-0 = Reserved
027Ah          LPT2 Control Port
               Bit 7-6 = Reserved
               Bit  5  = Data Output Control
               Bit  4  = IRQ Enable
               Bit  3  = Select Printer
               Bit  2  = Initialize
               Bit  1  = Line Feed
               Bit  0  = Strobe
02E8h          8514/A Display Status
02EBh          8514/A Horizontal Total
02EAh          8514/A DAC Mask
02EBh          8514/A DAC Read Index
02ECh          8514/A DAC Write Index
```

```
02EDh       8514/A DAC Data
02F8h       Serial Port 3 Transmitter/Receiver
              registers/Divisor Latch Low
02F9h       Serial Port 2 Interrupt Enable
              Register/Divisor Latch High
02FAh       Serial Port 2 Interrupt Identification
              Register
02FBh       Serial Port 2 Line Control Register
02FCh       Serial Port 2 Modem Control Register
02FDh       Serial Port 2 Line Status Register
02FFh       Serial Port 2 Scratchpad Register
0300h-031Fh IBM Prototype Card Addresses
0360h-036Fh Network Cards
0370h       Secondary Diskette Controller Status A
0371h       Secondary Diskette Controller Status B
0372h       Secondary Diskette Controller Digital
              Output Register
0374h       Read - Secondary Diskette Controller
              Main Status Register
            Secondary Diskette Controller Data Rate
              Select Register
0375h       Secondary Diskette Controller
              Command/Data Register
0377h       Read - Secondary Diskette Controller
              Digital Input Register
            Write - Select Register for Diskette
              Data Transfer Rate
0378h       LPT1 data port
0379h       LPT1 Status Port
              Bit  7  = Busy
              Bit  6  = Acknowledge
              Bit  5  = Out of Paper
              Bit  4  = Printer Selected
              Bit  3  = Error
              Bit  2  = IRQ Occurred
              Bit 1-0 = Reserved
037Ah       LPT1 Control Port
              Bit 7-6 = Reserved
              Bit  5  = Data Output Control
              Bit  4  = IRQ Enable
              Bit  3  = Select Printer
              Bit  2  = Initialize
              Bit  1  = Line Feed
              Bit  0  = Strobe
0380h-038Fh Secondary SDLC Registers
```

```
0390h-039Fh   IBM Cluster adapter
03A0h-03AFh   Primary SDLC Registers
03B4h         MDA CRT Index Register
03B5h         MDA CRT Data Register
              Address   Function
                00      Horizontal Total
                01      Horizontal Displayed
                02      Horizontal Sync Position
                03      Horizontal Sync Pulse Width
                04      Vertical Total
                05      Vertical Displayed
                06      Vertical Sync Position
                07      Vertical Sync Pulse Width
                08      Interlace Mode
                09      Maximum Scan Lines
                0A      Cursor Start
                0B      Cursor End
                0C      Start Address High
                0D      Start Address Low
                0E      Cursor Location High
                0F      Cursor Location Low
                10      Light Pen High
                11      Light Pen Low
03B8h         MDA Mode Control Register
              bit 7-6 = Reserved
              bit  5  = Blink Enable
              bit  4  = Reserved
              bit  3  = Video Enable
              bit 2-1 = Reserved
              bit  0  = High Resolution Mode
03B9h         EGA Color Select
03BAh         Read - EGA CRT Status Register
              Write - EGA/VGA feature control
              register
03BBh         Reserved for EGA
03BCh         LPT1 Data Port
03BDh         LPT1 Status Port
              Bit  7  = Busy
              Bit  6  = Acknowledge
              Bit  5  = Out of Paper
              Bit  4  = Printer Selected
              Bit  3  = Error
              Bit  2  = IRQ Occurred
              Bit 1-0 = Reserved
03BEh         LPT 1 Control Port
              Bit 7-5 = Reserved
```

	Bit 4 = IRQ Enable
	Bit 3 = Select Printer
	Bit 2 = Initialize
	Bit 1 = Line Feed
	Bit 0 = Strobe
03BFh	Hercules Configuration Switch Register
	Bit 7-2 = Reserved
	Bit 1 = Enable Upper 32K Graphic Buffer
	Bit 0 = Disable Graphics Mode
03C0h	EGA/VGA ATC Index/Data Register
03C1h	VGA Other Attribute Register
03C2h	Read - EGA/VGA Input Status 0 Register
	Write - VGA Miscellaneous Output Register
03C4h	VGA Sequencer Index Register
03C5h	VGA Other Sequencer Index Register
03C6h	VGA PEL Mask Register
03C7h	VGA PEL Address Read Mode/VGA DAC state register
03C8h	VGA PEL Address Write Mode
03C9h	VGA PEL Data Register
03CAh	VGA Feature Control Register
03CCh	VGA Miscellaneous Output Register
03CEh	VGA Graphics Address Register
03CFh	VGA Other Graphics Register
03D4h	CGA CRT Index Register
03D5h	CGA CRT (6845) data register
03D8h	CGA Mode Control Register
	Bit 7-6 = Reserved
	Bit 5 = Blink Enable
	Bit 4 = 640*200 Graphics Mode Select
	Bit 3 = Video Enable
	Bit 2 = Monochrome Signal Select
	Bit 1 = Text Mode Select
	Bit 0 = Text Mode Select
03D9h	CGA Palette Register
	Bit 7-6 = Reserved
	Bit 5 = Active Color Set Select
	Bit 4 = Intense Color Select
	Bit 3 = Intense Border Select
	Bit 2 = Red Border/Background/ Foreground Select
	Bit 1 = Green Border/Background/ Foreground Select

	Bit 0 = Blue Border/Background/
	Foreground Select
03DAh	CGA Status Register
	Bit 7-4 = Reserved
	Bit 3 = Vertical Retrace Status
	Bit 2 = Light Pen Status
	Bit 1 = Light Pen Trigger Set
	Bit 0 = Memory Select
03EAh	EGA/VGA Feature Control Register
03EBh	Clear Light Pen Latch
03ECh	Preset Light Pen Latch
03E8h	Serial Port 3 Transmitter/Receiver registers/Divisor Latch Low
03E9h	Serial Port 3 Interrupt Enable Register/Divisor Latch High
03EAh	Serial Port 3 Interrupt Identification Register
03EBh	Serial Port 3 Line Control Register
03ECh	Serial Port 3 Modem Control Register
03EDh	Serial Port 3 Line Status Register
03EFh	Serial Port 3 Scratchpad Register
03F0h	Primary Diskette Controller Status A
03F1h	Primary Diskette Controller Status B
03F2h	Primary Diskette Controller Digital Output Register
03F4h	Read - Primary Diskette Controller Main Status Register
	Primary Diskette Controller Data Rate Select Register
03F5h	Primary Diskette Controller Command/Data Register
03F7h	Read - Primary Diskette Controller Digital Input Register
	Write - Select Register for Diskette Data Transfer Rate
03F8h	Serial Port 3 Transmitter/Receiver registers/Divisor Latch Low
03F9h	Serial Port 1 Interrupt Enable Register/Divisor Latch High
03FAh	Serial Port 1 Interrupt Identification Register
03FBh	Serial Port 1 Line Control Register
03FCh	Serial Port 1 Modem Control Register
03FDh	Serial Port 1 Line Status Register
03FFh	Serial Port 1 Scratchpad Register

Interrupt Function by Number

Interrupt	Name and Comments
00h	Divide by Zero Error
01h	Single Step
02h	Nonmaskable
03h	Breakpoint (Instruction 0x0CC)
04h	Overflow
05h	Print Screen
06h-07h	Reserved
08h	Time of Day Services
09h	Keyboard Interrupt
0Ah	Slaved Second Interrupt Controller
0Bh	COM1/COM3 Interrupt
0Ch	COM2/COM3 Interrupt
0Dh	Hard Disk Interrupt/ LPT2 Interrupt
0Eh	Diskette Interrupt
0Fh	LPT1 Interrupt
10h	Video BIOS
11h	BIOS Equipment Check
12h	BIOS Memory Size Determine
13h	Disk I/O BIOS
14h	Serial Communications BIOS
15h	BIOS System Services
16h	Keyboard I/O BIOS
17h	Printer BIOS
18h	Resident BASIC Start Vector
19h	BootStrap Loader
1Ah	Time of Day BIOS Interrupt
1Bh	Keyboard Break Vector
1Ch	Timer Tick Vector
1Dh	Table Address of Video Parameters
1Eh	Table Address of Disk Parameters
1Fh	Table Address of Graphic Characters

```
20h       MS-DOS Program
          Terminate
21h       MS-DOS Function APIs
22h       MS-DOS Terminate
          Vector
23h       MS-DOS "Ctrl-C" Vector
24h       MS-DOS Error Handler
          Vector
25h-26h   MS-DOS Absolute Disk
          I/O
27h       MS-DOS Terminate
          Stay Resident API
28h-2Eh   MS-DOS Reserved
2Fh       MS-DOS Multiplex
          Interrupt
30h-32h   MS-DOS Reserved
33h       Mouse BIOS
34h-3Fh   MS-DOS Reserved
40h       Revectored Disk I/O
          BIOS Interrupt 13h
41h       Table Address of Hard
          Drive 0 Parameters
42h       Revectored EGA BIOS
          Interrupt 10h
43h       Table Address of EGA
          Parameters
44h-34h   Reserved
46h       Table Address of Hard
          Drive 1 Parameters
47h-49h   Reserved
4Ah       ROM BIOS Alarm
          Handler
4Bh-4Fh   Reserved
50h       PC/AT Alarm BIOS
          Interrupt
51h-59h   Reserved
5Ah       NETBIOS Function APIs
5Bh       NETBIOS Remap of
          Vector 19h
5Ch       NETBIOS Entry Point
5Dh-66h   Reserved
67h       LIM EMS Memory
          Function APIs
68h-6Fh   Reserved
```

70h	RTC Interrupt
71h	Slave Interrupt Controller Redirect
72h	IRQ10
73h	IRQ11
74h	IRQ12
75h	IRQ13
76h	IRQ14
77h	IRQ15
78h-7Fh	Not Allocated/ Available for Use
80h-85h	Reserved for Cassette BASIC
86h-F0h	Used by BASIC
F1h-FFh	Used during PC Boot as a Temporary Stack Area. Should NOT be used for Interrupts or Variables

ISA Bus

When the PC was designed, IBM designed the motherboard and specified the ISA slots in such a way that the complexity of the bus was hidden from the user (Fig. 10.2). The read/write cycle on the ISA bus is shown in Fig. 10.3. This waveform is identical for the I/O address space reads and writes.

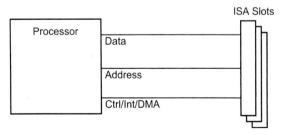

Figure 10.2 Processor/ISA Block Diagram

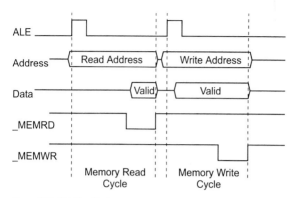

Figure 10.3 ISA Bus Timing

ISA pinouts

The 8-bit ISA bus consists of a two-sided 31-pin card
edge connector with the pins defined as:

ISA Bus Pinout		
Pin	"A" (Connector)	"B" (Solder)
1	I/O CH CHK	Ground
2	D7	Reset
3	D6	+5 V
4	D5	IRQ2
5	D4	+5 V
6	D3	DRQ2
7	D2	-12 V
8	D1	_CARD SLCTD
9	D0	+12 V
10	IO CH RDY	Ground
11	AEN	_MEMW
12	A19	_MEMR
13	A18	_IOW
14	A17	_IOR

ISA Bus Pinout (*Continued*)		
15	A16	_DACK3
16	A15	DRQ3
17	A14	_DACK1
18	A13	DRQ1
19	A12	_DACK0 (REFRESH)
20	A11	OSC
21	A10	IRQ7
22	A9	IRQ6
23	A8	IRQ5
24	A7	IRQ4
25	A6	IRQ3
26	A5	_DACK2
27	A4	T/C
28	A3	BALE
29	A2	+5 V
30	A1	CLOCK - 14.31818 MHz
31	A0	Gnd

The data and address busses are buffered to the processor. Addresses from 0x00000 to 0x0FFFFF (zero to one megabyte) can be accessed with the 8-bit connector. Memory devices can be located 0x0C0000 to 0x0DFFF, but care must be taken to avoid "contention" with other devices located within this memory space. The ISA Bus Pin Functions are given below:

Pin	Function
BALE	"Buffered ALE" and was the term used in the original PC because the ALE line was produced by the 8088's instruction sequence clock. This pin was buffered to avoid having the ISA bus directly processor driven. Today, this bit is more commonly known as "ALE" and provides essentially the same operation and timing as "BALE"
_I/O CH CHK	Pin was designed for use with parity checked memory. If a byte was read

	that did not match the saved parity, a NMI interrupt request was made of the processor. In Modern Systems, this Pin can be pulled Low (made active) to indicate a system error
I/O CH RDY	Line driven low by an adapter if it needs more time to complete an operation
_IOR/_IOW	I/O Register Read and Write Enables
_MEMR/_MEMW	Indicate the processor is reading and writing to ISA bus memory
IRQ3-IRQ7	Hardware Interrupt Request Lines. When these lines are driven high, the 8259As on the motherboard (which are known as "Programmable Interrupt Controllers" or "PICs") will process the request in a descending order of priority. These lines are driven high to request an interrupt. A PICmicro® MCU can drive these lines, but it should only be active when a "high" is driven onto the interrupt line to allow other devices to share the interrupt pins. If a PICmicro® MCU is used to drive these lines active, then there must be some way for the processor to reset the PICmicro® MCU interrupt request
CLOCK	Runs at four times NTSC "Color Burst" frequency (14.31818 MHz). The 14.31818 MHz clock was distributed to the system to provide clocking for the "MDA" and "CGA" video display cards. This clock can be useful for providing a simple clock for microcontroller and other clocked devices on adapter cards.
OSC	Pin is driven at up to 8 MHz.
DRQ#	Used to Request a DMA transfer to take place. When the corresponding "_DACK#" pin is driven high, the DMA controller is reading or writing an I/O address of an adapter card. When the DMA controllers have

	control of the bus over the processor, the "AEN" pin is active to indicate to other adapters that a DMA operation is in process. When all the DMA data has been transferred, the "T/C" bit is pulsed high to indicate the operation has completed. When the "T/C" bit becomes active, the adapter should request a hardware interrupt to indicate to software that the operation is complete.
_DACK#	Active when the DMA Channel is reading/writing the I/O device. "_DACK0" or "_REFRESH" is used with DRAM memory to request a "RAS only refresh" of the system memory
_MASTER	Driven by an adapter when it is requesting to take over the bus and drive its own signals.

Interrupts

Interrupts IRQ3, IRQ4, and IRQ7 are recommended for use in a PC system. Interrupts are driven high and should use the circuit shown in Fig. 10.4 to allow multi-

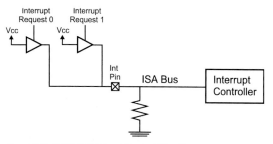

Figure 10.4 Multiple Interrupt Request Circuit

ple requests. The PC Interrupt Handler should be defined as:

1. Save the original vector using MS-DOS interrupt 021h AH = 035h API.

2. Set the new vector using MS-DOS interrupt 021h AH = 025h API.

3. Enable the Interrupt Request Mask bit in the 8259.

To enable the interrupt request mask bit in the 8259, the appropriate interrupt mask register bit has to be reset. This register is at the 8259's "Base Address" plus one. This can be done with the following statement:

```
outp(IntBase + 1, inp(IntBase + 1) & ((0x0FF ^
 (1 << Bit))));
```

To "release" the interrupt vector and the interrupt source at the end of the application, the following steps must be taken:

1. Disable the Interrupt Request mask bit in the 8259.

2. Restore the original vector using MS-DOS Interrupt 021h AH = 025h API.

Keyboard and Mouse Ports

The PCs keyboard and mouse ports operate with a synchronous serial data protocol that was first introduced with the original IBM PC. This protocol allows data to be sent from the keyboard in such a way that multiple pressed keys can be recognized within the PC without any key presses being lost. The standard was enhanced with the PC/AT as a bidirectional communication method.   Three years later, when the PS/2 was introduced, the "mouse" interface also used the keyboard's protocol, freeing up a serial port or ISA slot which, up to this point, was needed for the mouse interface. The keyboard protocol used in the PC was so successful that IBM used it for all its PC, terminal, and workstation product lines that have been developed from 1981 and it is also used by many other PC vendors.

Connector specification

The female 6-pin "Mini-DIN" keyboard connector facing out of the PC is shown in Fig. 10.5. The port can usually supply up to 100 mA over and above the keyboard requirements. The power (+5 VDC) may or may not be fused, so any hardware put on the port must not draw excessive current to prevent damage to the motherboard.

Keyboard operation with timing diagrams

Data from the keyboard looks like the waveform shown in Fig. 10.6. The parity bit is "odd", which is to say the eight data bits plus the parity is an odd number. The data line should not change for at least 5 usecs from the

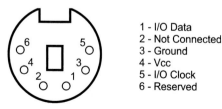

1 - I/O Data
2 - Not Connected
3 - Ground
4 - Vcc
5 - I/O Clock
6 - Reserved

Figure 10.5 PC Keyboard Connector Pinout

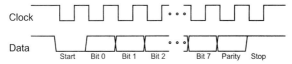

Figure 10.6 Keyboard to PC Data Protocol

change of the clock line. The clock line should be high or low for at least 30 usecs (with 40 usecs being typical).

Data that is sent from the system unit to the keyboard is similar, but with the clock inverted. The data changes while the clock is low and is latched in when the clock goes high as is shown in Fig. 10.7. When data is sent from the keyboard, the clock is pulled low and then data is sent with the keyboard accepting data when the clock is pulsed high. The bit timings are the same as data from the keyboard.

These two protocols are used to allow a device wired in parallel to monitor the communication to and from the PC.

Additional devices can be added to the keyboard/ mouse connector in parallel as is shown in Fig. 10.8.

Keyboard scan codes

In MS-DOS, the Keyboard Codes are normally a combination of the keyboard scan code and appropriate ASCII

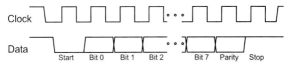

Figure 10.7 PC to Keyboard Data Protocol

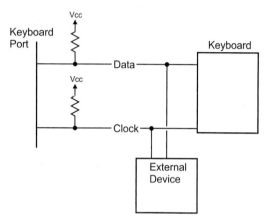

Figure 10.8 Sharing a Keyboard with Another Device

code. The table below shows the different codes returned for keystrokes by themselves, and with a "Shift", "Ctrl", or "Alt" Modifier.

The table below shows the codes in scan/ASCII configuration for the extended function keyboard characters. The standard function codes are the same except that "F11", "F12", and the keypad "Center Key" do not return any codes and for the explicit arrow and explicit "Insert", "Home", "Page Up", "Delete", "End", and "Page Down" keys, the 0x0E0 ASCII code is actually 0x000.

All values in the table below are in hex and I have put in the scan codes as they appear on my PC. I have not made allowances for upper and lower case in this table as this is processed by the PC itself. "KP" indicates the Keypad and it, or a single "A" (which indicates Alternate arrow and other keys), followed by "UA", "DA", "LA", or

"RA" indicates an Arrow. "I", "D", "H", "PU", "PD", or "E" with "KP" or "A" indicates the "Insert", "Delete", "Home", "Page Up", "Page Down", or "End" on the Keypad, respectively.

The Keypad numbers, when "Alt" is pressed is used to enter in specific ASCII codes in Decimal. For example, "Alt", "6", "5" will enter in an ASCII "A" character. These keys in the table below are marked "#".

PC Keyboard Scan Codes			
Key	Standard Codes	"Shift" Codes	"Ctrl" Codes	"Alt" Codes
Esc	01/1B	01/1B	01/1B	01/00
1	02/31	02/21	--	78/00
2	03/32	03/40	03/00	79/00
3	04/33	04/23	--	7A/00
4	05/34	05/24	--	7B/00
5	06/35	06/25	--	7C/00
6	07/36	07/5E	07/1E	7D/00
7	08/37	08/26	--	7E/00
8	09/38	09/2A	--	7F/00
9	0A/39	0A/28	--	80/00
0	0B/30	0B/29	--	81/00
-	0C/2D	0C/5F	0C/1F	82/00
=	0D/3D	9C/2B	--	83/00
BS	0E/08	0E/08	0E/7F	0E/00
Tab	0F/09	0F/00	94/00	A5/00
Q	10/71	10/51	10/11	10/00
W	11/77	11/57	11/17	11/00
E	12/65	12/45	12/05	12/00
R	13/72	13/52	13/12	13/00
T	14/74	14/54	14/14	14/00
Y	15/79	15/59	15/19	15/00
U	16/75	16/55	16/15	16/00
I	17/69	17/49	17/09	17/00
O	18/6F	18/4F	18/0F	18/00
P	19/70	19/50	19/10	19/00
[	1A/5B	1A/7B	1A/1B	1A/00
]	1B/5D	1B/7D	1B/1D	1B/00

PC Keyboard Scan Codes (*Continued*)			
Enter	1C/0D	1C/0D	1C/0A	1C/00
A	1D/61	1E/41	1E/01	1E/00
S	1F/73	1F/53	1F/13	1F/00
D	20/64	20/44	20/04	20/00
F	21/66	21/46	21/06	21/00
G	22/67	22/47	22/07	22/00
H	23/68	23/48	23/08	23/00
J	24/6A	24/4A	24/0A	24/00
K	25/6B	25/4B	25/0B	25/00
L	26/6C	26/4C	26/0C	26/00
;	27/3B	27/3A	--	27/00
`	28/27	28/22	--	28/00
`	29/60	29/7E	--	29/00
\	2B/5C	2B/7C	2B/1C	2B/00
Z	2C/7A	2C/5A	2C/1A	2C/00
X	2D/78	2D/58	2D/18	2D/00
C	2E/63	2E/43	2E/03	2E/00
V	2F/76	2F/56	2F/18	2F/00
B	30/62	30/42	30/02	30/00
N	31/6E	31/4E	31/0E	31/00
M	32/6D	32/4D	32/0D	32/00
,	33/2C	33/3C	--	33/00
.	34/2E	34/3E	--	34/00
/	35/2F	35/3F	--	35/00
KP *	37/2A	37/2A	96/00	37/00
SPACE	39/20	39/20	39/20	39/20
F1	3B/00	54/00	5E/00	68/00
F2	3C/00	55/00	5F/00	69/00
F3	3D/00	56/00	60/00	6A/00
F4	3E/00	57/00	61/00	6B/00
F5	3F/00	58/00	62/00	6C/00
F6	40/00	59/00	63/00	6D/00
F7	41/00	5A/00	64/00	6E/00
F8	42/00	5B/00	65/00	6F/00
F9	43/00	5C/00	66/00	70/00
F10	44/00	5D/00	67/00	71/00
F11	85/00	87/00	89/00	8B/00
F12	86/00	88/00	8A/00	8C/00
KP H	47/00	47/37	77/00	#
KP UA	48/00	48/38	8D/00	#
KP PU	49/00	49/39	84/00	#
KP -	4A/2D	4A/2D	8E/00	4A/00

PC Keyboard Scan Codes (*Continued*)				
KP LA	4B/00	4B/34	73/00	#
KP C	4C/00	4C/35	8F/00	#
KP RA	4D/00	4D/36	74/00	#
KP +	4E/2B	4E/2B	90/00	4E/00
KP E	4F/00	4F/31	75/00	#
KP DA	50/00	50/32	91/00	#
KP PD	51/00	51/33	76/00	#
KP I	52/00	52/30	92/00	--
KP D	53/00	53/2E	93/00	--
KP Enter	E0/0D	E0/0D	E0/0A	--
KP /	E0/2F	E0/2F	95/00	--
PAUSE	--	--	72/00	--
BREAK	--	--	00/00	--
A H	47/E0	47/E0	77/00	97/00
A UA	48/E0	48/E0	8D/E0	98/00
A PU	49/E0	49/E0	84/E0	99/00
A LA	4B/E0	4B/E0	73/E0	9B/00
A RA	4D/E0	4D/E0	74/E0	9D/00
A E	4F/E0	4F/E0	75/E0	9F/00
A DA	50/E0	50/E0	91/E0	A0/00
A PD	51/E0	51/E0	76/E0	A1/00
A I	52/E0	52/E0	92/E0	A2/00
A D	53/E0	53/E0	93/E0	A3/00

Keyboard controller commands

The PC itself has a number of commands that it can
send to the keyboard that include:

PC to Keyboard Commands	
Code	Function
0x0ED	Set Indicator LED's. The next Character out is the LED status
0x0EE	Echo - Keyboard Returns 0x0EE
0x0EF-0x0F2	Ignored by the Keyboard
0x0F3	Set Typematic rate, next character is the rate

PC to Keyboard Commands (*Continued*)	
0x0F4	Enable Key Scanning
0x0F5	Set to Default (no LEDs on, default Typematic rate) and disable Key Scanning
0x0F6	Set to Default (no LEDs on, default Typematic rate) and enable Key Scanning
0x0F7-0x0FD	Ignored by the Keyboard
0x0FE	Request Keyboard to resend the last character
0x0FF	Reset the Keyboard's Microcontroller

In all these cases (except for the "ignore" and "echo" commands), the keyboard sends back the "Acknowledge" character 0x0FA.

BIOS interfaces

When Data is transferred between the PC's processor and the keyboard controller, the following information is passed as well:

Keyboard Flags Byte	
Bit	Function
7	Set when "Insert" State Active
6	Set when "Caps Lock" Active
5	Set when "Num Lock" Active
4	Set when "Scroll Lock" Active
3	Set when a "Alt" Key Held Down
2	Set when a "Ctrl" Key Held Down
1	Set when the Left "Shift" Key Held Down
0	Set when the Right "Shift" Key Held Down

or

Extended Keyboard Flags Byte

Bit	Function
7	Set when "SysReq" Key Pressed
6	Set when "Caps Lock" Key Pressed
5	Set when "Num Lock" Key Pressed
4	Set when "Scroll Lock" Key Pressed
3	Set when Right "Alt" Key Pressed
2	Set when Right "Ctrl" Key Pressed
1	Set when Left "Alt" Key Pressed
0	Set when Left "Ctrl" Key Pressed

To access the keyboard BIOS functions, an "int 016h" instruction is executed with the registers set up as defined in the table below:

Function	Input	Output	Comments
Read Character	AH = 00h	AH = Scan Code AL = ASCII Character	This Command returns the next unread key from the buffer or waits for a Key to return.
Read Status	AH = 01h	AH = Scan Code AL = ASCII Character Zero = Set if No Character Available	Poll the Keyboard Buffer and return the next keystroke or set the Zero Flag.
Read Flags	AH = 02h	AH = 00 AL = Keyboard Flags Byte	Return the Keyboard Flags Byte

Set Typematic Rate and Delay	AH = 03h
	AL = 5
	BH = Delay
	  0 - 250ms,
	  1 - 500ms,
	  2 - 750ms,
	  3 - 1000ms
	BL = Rate
	  0 - 30 cps,
		None	Set the keyboard delay before Resending the held down Character and then the rate at which they are set. This function should

Function	Registers	Input/Output	Description
	4 – 20 cps 8 – 15 cps 12 – 10 cps 16 – 7.5 cps 20 – 5 cps 24 – 3.75 cps 28 – 2.5 cps		be set by the operating system utilities rather than from an application. I have not put in the intermediate values.
Read Typematic Rate and Delay	AH = 03h AL = 6	BH = Delay BL = Rate	Read the Current Delay and Rate set into the Keyboard.
Keyboard Write	AH = 05h BH = Scan Code BL = ASCII Character	AL = 0 if Buffer Written Successfully	This command writes a new Character into the keyboard buffer (and not to the keyboard or other external Device as the name would Imply).
Keyboard Functionality Determination	AH = 09h	AL = Function Code Bit 3 – Set If can read Delay/Rate	This API returns the capabilities of the keyboard and hardware to change the

Function	Input	Output	Comments
		Bit 2 - Set If can Set Delay/Rate Bit 1 - Set If cannot Set Delay/ Rate Bit 0 - Set If return To default Delay/Rate Supported	Typematic Rate and Delay.
Extended Keyboard Read	AH = 10h	AH = Scan Code AL = ASCII Code	Return the full keyboard code if Keyboard Buffer has an unread key or wait for a key to return.
Extended Keyboard Status	AH = 11h	AH = Scan Code AL = ASCII Code Zero = Set if No character To return	Check the Keyboard buffer and return the next key to process or set the Zero Flag.
Extended Shift Status	AH = 12h	AH = Extended Keyboard Flags AL = Keyboard Flags Byte	Return the Extended Keyboard Shift/Ctrl/Alt Status.

Keyboard commands

To simply process a keystroke in "C", the following "switch" code could be used:

```
switch ((KeySave = KEYREAD()) & 0xFF) {//  Process
                                        //     the Key
  case 0x000:                     //  Special
                                  //     Function
                                  //     Keys

  case 0x0E0:
    KeySave = (KeySave >> 8) & 0xFF;  //  Process
                                      //     the Scan
                                      //     Code
    :
  break;
  case 0x00D:                     //  Handle
                                  //     "Enter"
    :
  break;
    :                             //  Handle
                                  //     Other
                                  //     Special
                                  //     Keys
  default:                        //  Other,
                                  //     Unneeded
                                  //     Keys

    :
} //  endswitch
```

Serial Port

The PC's serial port's design has not changed since the PC was introduced in 1981. Since that time, a 9-pin connector has been specified for the port (in the PC/AT) and the ability to buffer data within the serial port has been added.

Connector pinouts

DB-9 and D-9 pin RS-232 connectors are shown in Fig. 10.9. These connectors are wired as:

PC RS-232 Pinout			
Pin Name	25 Pin	9 Pin	I/O Direction
TxD	2	3	Output ("O")
RxD	3	2	Input ("I")
Gnd	7	5	
RTS	4	7	O
CTS	5	8	I
DTR	20	4	O
DSR	6	6	I
RI	22	9	I
DCD	8	1	I

8250 block diagram

The "8250" UART is the basis for serial communications within the PC (Fig.10.10).

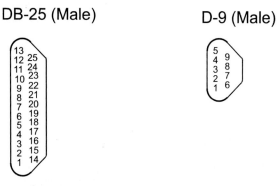

Figure 10.9 IBM PC DB-25 and D-9 Pin RS-232 Connectors

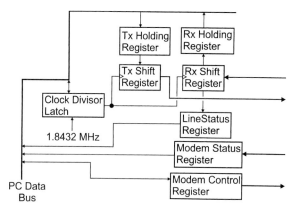

Flyure 10.10 8250 Block Diagram

Serial port base addresses

The Serial Port "Base Addresses" are as follows:

PC Serial Port Base Addresses		
Port	Base Address	Interrupt Number
COM1	0x03F8	0x00C
COM2	0x02F8	0x00B
COM3	0x03E8	0x00C
COM4	0x02E8	0x00B

Each "Base Address" is used as an initial offset to eight registers that are used by the Serial Port Controller (The "8250"). The "Interrupt Number" is the interrupt vector requested when an interrupt condition is encountered. Note that "COM4" has conflicting addresses with the 8514/A ("SuperVGA") Graphics Adapter.

8250 registers

The 8250 consists of eight registers offset from the "base address".

PC Serial Port Register Offsets	
Base Address Offset	Register Name
0	Transmitter Holding Register/Receiver Character Buffer/LSB Divisor Latch
1	Interrupt Enable Register/MSB Divisor Latch
2	Interrupt Identification Register
3	Line Control Register
4	Modem Control Register
5	Line Status Register
6	Modem Status Register
7	Scratchpad Register

Data Speed is specified by loading a 16-bit divisor value into the Rx/Tx Holding Register and Interrupt Enable Register addresses after bit 7 of the Line Control Register is set. The value loaded into the register is multiplied by 16 and divided into 1.8432 MHz to get the actual data rate.

$$\text{Data Rate} = 1.8432 \text{ MHz} / (16 \text{ X Divisor})$$

The divisors for different standard data rates are

PC Serial Port Speed Divisor Table	
Data Rate	Divisor
110 bps	0x0417
300 bps	0x0180

PC Serial Port Speed Divisor Table (*Continued*)

Data Rate	Divisor
600 bps	0x00C0
1200 bps	0x0060
2400 bps	0x0030
9600 bps	0x000C
19200 bps	0x0006
115200 bps	0x0001

After a character is received, it will set a number of conditions (including error conditions) that can only be reset by reading the character in the Receive Holding Register. For this reason it is always a good idea to read the serial port at the start of an application. By reading the port, the status and left over characters are "cleared" out.

Writing to the base address (with no offset added) loads a character into the "Transmit Holding Register", which will be loaded as soon as the shift out register has completed sending the previous character. Often, when starting transmission, nothing will be in the shift register so the character is loaded immediately into the shift register, freeing up the holding register for the next character.

When any interrupts are enabled in the 8250, they will output an interrupt request (Fig. 10.11). This may not be desirable, so in the PC, some hardware was added to globally mask the interrupt.

"_Out2" is controlled within the "Modem Control Register".

PC Serial Port Interrupt Enable Register (Base + 1)

Bit	Description
4-7	Unused, normally set to zero.

PC Serial Port Interrupt Enable Register (Base + 1)
(Continued)

Bit	Description
3	When set an interrupt request on change of state for modem interface lines.
2	Request interrupt for change in receiver holding register status
1	Request interrupt if the holding register is empty
0	Request interrupt for received character

PC Serial Port Interrupt Identification Register (Base + 2)

Bits	Description
3-7	Unused, Normally set to zero
1-2	Interrupt ID Bits

B2	B1	Priority	Request Type
0	0	Lowest	Change in Modem Status Lines
0	1	Third	Transmitter Holding Register Empty
1	0	Second	Data Received
1	1	Highest	Receive Line Status Change

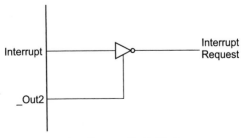

Figure 10.11 IBM PC Serial Interrupt Enable Hardware

```
PC Serial Port Line Control Register (Base + 3)

Bit     Description
7       When set, the Transmitter Holding and
          Interrupt Enable Registers are used for
          loading the data speed divisor
6       When set, the 8250 outputs a "Break
          Conditions" (sending a space) until this
          bit is reset
3-5     Parity Type Specification
        B5 B4 B3
         0  0  0 - No Parity
         0  0  1 - Odd Parity
         0  1  0 - No Parity
         0  1  1 - Even Parity
         1  0  0 - No Parity
         1  0  1 - "Mark" Parity
         1  1  0 - No Parity
         1  1  1 - "Space" Parity
?       When set, two stop bits are sent in the
          Packet, otherwise one
0-1     Number of Data Bits sent in a Packet
        B1 B1
         0  0 - 5 Bits
         0  1 - 6 Bits
         1  0 - 7 Bits
         1  1 - 8 Bits
```

```
PC Serial Port Modem Control Register (Base + 4)

Bit     Pin     Description
5-7             Unused, normally set to zero
4       Loop    When Set, Data from the transmitter
                  is looped internally to the receiver
3       Out2    When Set, Interrupt Requests from the
                  8250 are unmasked
2       Out1    This bit/pin is not controlling any
                  hardware features in the serial port
1       _RTS    When this bit is Reset, the RTS line
                  is at "Mark" State
0       _DTR    When this bit is Reset, the DTR line
                  is at "Mark" State
```

PC Serial Port Line Status Register (Base + 5)

Bit	Description
7	Unused, Normally set to zero
6	Set when the transmitter shift register is empty
5	Set when the transmitter holding register is empty
4	Set when the receive line is held at a space value for longer than the current packet size
3	This bit is Set when the last character had a framing error (ie stop bit set to "Space")
2	Set when the last character had a parity error
1	Set when the latest character has overrun the receiver holding register
0	Set when a character has been received but not read

PC Serial Modem Status Register (Base + 6)

Bit	Pin	Description
7	DCD	When Set, an asserted DCD signal is being received
6	RI	When Set, the modem is detecting a ring on the device it is connected to
5	DSR	When Set, a DSR "Mark" is being received
4	CTS	When Set, a CTS "Mark" is being received
3	DCD	When this bit is set, the DCD line has changed state since the last check
2	RI	When set, this bit indicates that the Ring Indicator line has changed from a Mark to a Space

PC Serial Modem Status Register (Base + 6) (*Continued*)

Bit	Pin	Description
1	DSR	When this bit is set, the DSR line has changed state since the last check
0	CTS	When this bit is set, the CTS line has changed state since the last check

Interrupts

To enable Interrupts for COM1/COM3 (at Interrupt 0x00C), the following code is used:

```
SetInt( 0x0C, SerIntHndlr );      // Point the
                                  //   Interrupt
                                  //   Handler to
                                  //   the Correct
                                  //   Handler
Dummy = inp( RxHoldingRegister );//  Turn Off any
                                  //   Pending
                                  //   Interrupts
outp( IntMaskRegister, inp( IntMaskRegister ) &
  0x0FB );
                                  //   Enable
                                  //   COM1/COM3
                                  //   Interrupts in
                                  //   Controller
outp( InterruptEnableRegister, 0x003 );
                                  //   Request
                                  //   Interrupts
                                  //   on TxHolding
                                  //   Register
                                  //   Empty and Rx
                                  //   Holding
                                  //   Register Full
outp( ModemControlRegister, inp
  ( ModemControlRegister ) | Out2);
                                  //   Unmask
                                  //   Interrupt
                                  //   Requests
                                  //   from
                                  //   8250
```

Once an interrupt request is made by the hardware, control is passed to the service routine:

```
SerIntHndlr:                    //  Serial Interrupt
                                //   Handler


Assume that the Interrupting COM port is identified

switch ( InterruptIDRegister ) { //  Handle the
                                 //   Interrupt
                                 //   Request
  case 4:                       //  Received Character
    InString[ i++ ] = RxHoldingRegister;
    break;
  case 2:                       //  TxHolding Register
                                //   Empty
    TxHoldingRegister = OutString[ j++ ]; //  Send
                                             the
                                             Next
                                //  Character
    break;
  default:                      //  Some other kind of
                                //   Interrupt
    Dummy = RxHoldingRegister;  //  Clear the
                                    Receiving Data
} //  endswitch
InterruptControlRegister = EOI; //  Reset the
                                //   Interrupt
                                //   Controller
returnFromInterrupt;            //   Return from the
                                //    Interrupt.
```

Interrupt 14h–RS-232 communications APIs

The following APIs are available within the PC—to access and load registers as specified and execute an "int 014h" instruction.

Function	Input	Output	Comments
Initialize Communications Port	AH = 00h AL = Init Parameter DX = Port Number	AH = Line Status AL = Modem Status	Initialize the Serial port. Note, AH = 004h Provides Extended Capabilities
Write Character	AH = 01h AL = Character DX = Port	AH = Line Status AL = Modem Status	Send the Character when the Modem Handshake Allows or Time Out.
Read Character	AH = 02h DX = Port	AH = Line Status AL = Character	Wait for the Character to be Received when the Modem Handshake allows or Time Out.
Status Request	AH = 03h DX = Port	AH = Line Status AL = Modem Status	Return the Current Serial Port Status.

Extended
Port
Initialize

AH = 04h
AL = 0 for
no Break, 1
For Sending
Break
BH = Parity
0 - No Parity
1 - Odd
Parity
2 - Even
Parity
3 - Odd
Stick
Parity
4 - Even
Stick
Parity
BL = Stop
Bits
0 - One
1 - Two
CH = Word
Length
0 - 5 Bits
1 - 6 Bits
2 - 7 Bits
3 - 8 Bits

AH = Line
Status
AL = Modem
Status

This is a more
Complete Serial
Port Initialize.

Function	Input	Output	Comments
	CL = Data Rate 0 - 110 bps 1 - 150 bps 2 - 300 bps 3 - 600 bps 4 - 1200 bps 5 - 2400 bps 6 - 4800 bps 7 - 9600 bps 8 - 19200 bps DX = Port		
Read Modem Control Register	AH = 05h AL = 0 DX = Port	BL = Modem Control Register	Return the Contents of the Modem Control Register
Write to Modem Control Register	AH = 05h AL = 1 BL = New Modem Control Register Value	AH = Line Status AL = Modem Status	Set the Modem Control Register to a New State.

Parallel Port

The parallel port is the first device that most people look to when simple I/O expansion must be implemented in the PC. The parallel port itself is very simple; the design used in the PC/AT consists of just seven TTL chips and provides a simple, byte-wide parallel bidirectional interface into the PC.

Block diagram/connector

PC Parallel Port can be blocked out as shown in Fig. 10.12. The Parallel Port Connector is shown in Fig. 10.13. The Pinout for the Connector is

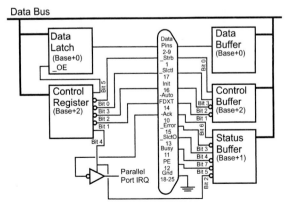

Figure 10.12 Parallel Port Block Diagram

DB-25 (Female)

Figure 10.13 IBM PC DB-25 Parallel Port Connector

Pin	Label	Direction	Function
1	_STROBE	Output	Negative Active Data Strobe
2	D0	Bi-Directional	Parallel Data Bit 0
3	D1	Bi-Directional	Parallel Data Bit 1
4	D2	Bi-Directional	Parallel Data Bit 2
5	D3	Bi-Directional	Parallel Data Bit 3
6	D4	Bi-Directional	Parallel Data Bit 4
7	D5	Bi-Directional	Parallel Data Bit 5
8	D6	Bi-Directional	Parallel Data Bit 6
9	D7	Bi-Directional	Parallel Data Bit 7
10	_ACK	Input	Pulsed Low When Data Accepted
11	BUSY	Input	High while Printer cannot accept another Character
12	NOPAPER	Input	High Indicates that Printer has run out of Paper
13	SELECTED	Input	High Indicates Printer is Active and Selected

14	_AUTOFEED	OC/Output	Forces Printer to Eject the Current Page when pulled low
15	_ERROR	Input	Low Indicates Printer cannot Print any more Characters
16	_INIT	OC/Output	Low Resets the Printer
17	_SELECT	OC/Output	Low Indicates Printer is about to be Written to
18-25	Ground	N/A	Signal Ground

Base registers

The installed Parallel Ports can be read at address 0x00040:0x00008 and 0x040:0x0000C. The common Parallel Port addresses are

```
Port    Base Address       Interrupt Number
LPT1    0x0378/0x03BC      0x00F/0x00D
LPT2    0x0378             0x00F
LPT3    0x0278             0x00D
```

Registers

Printer Port Data Register (Base Offset + 0)
Bit   Function
7-0   Data Bits. Normally Output, can be set to Input for Bi-Directional Operation by setting bit 5 of the "Control Register"

Printer Port Status Register (Base Offset + 1)

Bit	Function
7	BUSY Pin Data Passed to Parallel Port
6	_ACK. When Low, "_ACK" is active
5	NOPAPER. When High, Printer is out of Paper
4	SELECTED. When High, Printer is responding that it is Selected
3	_ERROR. When Low, "_ERROR" is active
2-0	Undefined

Printer Port Control Register (Base Offset + 2)

Bit	Function
7-6	Undefined
5	Set to Put data pins in "Input Mode"
4	Set to enable Printer Interrupt Requests from "_ACK" Pin. Can be Read back
3	_SELECT. Set to make "_SELECT" Pin Active (Low). Can be Read back
2	_INIT. Reset to make "_INIT" Pin Active (Low - Initialize Printer). Pin is NOT Inverted. Can be Read back
1	_AUTOFEED. Set to make "_AUTOFEED" Pin Active (Low) and current page ejected. Can be Read back
0	_STROBE. Set to make "_STROBE" Pin Active (Low). Can be Read back

Data output waveform

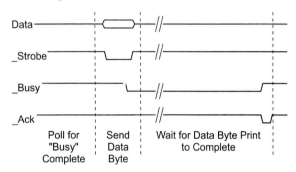

Figure 10.14 Parallel Port Printer Byte Write Waveform

BIOS interfaces

The Printer Status Byte passes back information from the printer port (along with program status information) via the "AH" register during Parallel Port BIOS Calls:

```
Bit      Function
 7       Not Busy
 6       Acknowledge
 5       Out of Paper
 4       Selected
 3       Error
 0       Time-Out
```

To Enable one of the Parallel Port BIOS Requests, an "int 017h" instruction is executed with the following Register specifications:

Function	Input	Output	Comments
Write Character	AH = 00h AL = Character DX = Printer Number	AH = Status	Send the Specified Character. If the Printer is not present or not working, the "Time-Out" Bit will be set.
Initialize Printer Port	AH = 01h DX = Printer Number	AH = Status	Initialize the Printer Port and Printer connected to it.
Status Request	AH = 02h DX = Printer Number	AH = Status	Return the Current Printer Status.

Useful Code "Snippets"

Jumping outside the Current Page

The general case, low-end PICmicro® MCU interpage "goto" is

```
movf   STATUS, w    ; Going to Modify the High
                          Three Bits
andlw  0x01F        ;  of the STATUS Word
iorlw  HIGH ((Label << 4) & 0x0E0)
movwf  STATUS
goto   (Label & 0x01FF) | ($ & 0x0E00)
```

The mid-range and PIC17Cxx interpage "goto" is

```
Mid-Range/PIC17Cxx General Case Table Jump Code

  movlw   HIGH Label        ;  Get the Current 256
                               Instruction Block
  movwf   PCLATH            ;  Store it so the Next Jump
                               is Correct
  goto    (Label & 0x07FF) | ($ & 0x01800)
```

The PIC18Cxx interpage "goto" (and "call") can jump to anywhere within the PICmicro® MCU program memory space. If a "branch always" ("BRA") instruction is to be used, the PIC8Cxx code would be

```
PIC18Cxx General Case Table Jump Code

  movlw   UPPER Label   ;  Get the Current 256
                           Instruction Block
  movwf   PCLATU        ;  Store it so the Next Jump
                           is Correct
  movlw   HIGH Label
  movwf   PCLATH
  bra     (Label & 0x07FF) | ($ & 0x01E000)
```

If a "call" to a subroutine in another page is implemented, make sure that PCLATH (and PCLATU) is restored upon return from the call.

Tables

The general case low-end PICmicro® MCU table code is

```
Table1                          ;  Return Table Value for
                                   Contents of "w"
                                ;  Anywhere in PICmicro®
                                   MCU Memory
  movwf  Temp                   ;  Save the Table Index
  movf   STATUS, w              ;  Going to Modify the
                                   High Three Bits
  andlw  0x01F                  ;   of the STATUS Word
  iorlw  HIGH ((TableEntries << 4) & 0x0E0)
  movwf  STATUS
  movlw  LOW TableEntries ;   Instruction Block
  addwf  Temp, w                ;  Compute the Offset
                                   within the 256
  movwf  PCL                    ;  Write the correct
                                   address to the
                                ;   Program Counter
TableEntries
  dt     "Table", 0
```

Note that in the low-end PICmicro® MCU case, the index to "TableEntries" should never be in the second 256 instructions of a page.

The general case mid-range and PIC17Cxx table code is

```
Table2                          ;  Return Table Value for
                                   Contents of "w"
                                ;  Anywhere in PICmicro®
                                   MCU Memory
  movwf  Temp                   ;  Save the Table Index
  movlw  HIGH TableEntries;  Get the Current 256
                                   Instruction Block
  movwf  PCLATH                 ;  Store it so the Next
                                   Jump is Correct
  movf   Temp, w                ;  ("movfp Temp, WREG" in
                                   PIC17Cxx)
```

```
    addlw  LOW TableEntries ;  Instruction Block
    btfsc  STATUS, C
     incf   PCLATH, f        ;  If in next, increment
                             ;     PCLATH
    movwf  PCL               ;  Write the correct
                             ;     address to the
                             ;  Program Counter
TableEntries
   dt      "Table", 0
```

The PIC18Cxx requires that the index be multiplied by two before PCL is changed and requires both the "PCLATU" and "PCLATH" registers to be updated:

```
Table3                     ;  Return Table Value for
                           ;     Contents of "w"
                           ;  Anywhere in PICmicro® MCU
                           ;     Memory
   movwf  Temp             ;  Save the Table Index
   movlw  UPPER            ;  Get the Current 256
    TableEntries           ;     Instruction Block
   movwf  PCLATU           ;  Store it so the Next Jump is
                           ;     Correct
   movlw  HIGH TableEntries
   movwf  PCLATH
   bcf    STATUS, C
   rlcf   Temp, w          ;  Multiply Index by 2
   btfss  STATUS, C        ;  If Carry Set, Increment
                           ;     PCLATH/PCLATU
    goto  TableSkip1
   infsnz PCLATH, f
    incf  PCLATU, f
TableSkip1
   addlw  LOW              ;  Get the Offset into the Table
    TableEntries
   btfss  STATUS, C        ;  Increment PCLATH/PCLATU if
                           ;     necessary
    goto  TableSkip2
   infsnz PCLATH, f
    incf  PCLATU, f
```

```
   movwf   PCL             ;  Write the correct address to
                                 the
                           ;  Program Counter
 TableEntries
   dt      "Table", 0
```

Conditional Branching

The following table shows the code used for different comparisons and jumping on Specific Conditions. Note that both variable and constant values are included.

11.7 Condition to Subtraction Instruction Cross Reference				
Jump "if"	Condition to Check	Code		
A == B	A - B = 0	movf	A, w/movlw	A
		subwf	B, w/sublw	B
		btfsc	STATUS, Z	
		goto	Label	; Jump if Z = 1
A != B	A - B = 0	movf	A, w/movlw	A
		subwf	B, w/sublw	B
		btfss	STATUS, Z	
		goto	Label	; Jump if Z = 0
A > B	B - A < 0	movf	B, w/movlw	A
		subwf	B, w/sublw	B
		btfss	STATUS, C	
		goto	Label	; Jump if C = 0
A >= B	A - B >= 0	movf	B, w/movlw	B
		subwf	A, w/sublw	B
		btfsc	STATUS, C	
		goto	Label	; Jump if C = 1

Condition to Subtraction Instruction Cross Reference (*Continued*)				
Jump "if"	Condition to Check	Code		
A < B	A - B < 0	movf	B, w/movlw	B
		subwf	A, w/sublw	A
		btfss	STATUS, C	
		goto	Label	; Jump if C = 0
A <= B	B - A > 0	movf	A, w/movlw	A
		subwf	B, w/movlw	B
		btfsc	STATUS, C	
		goto	Label	; Jump if C = 1

Time Delays

Here is a simple, generic delay of zero to 777 cycles as a macro.

```
DlayMacro Macro Cycles   ;  Delay Macro for Edges
 variable i, TCycles, Value, TFlag
TCycles = Cycles
Value = 1 << 7
i = 7
TFlag = 0
 if (TCycles > 5)
 while (i >= 0)
 if ((TFlag == 0) && ((Value * 3) <= TCycles))
  bsf    DlayCount, i
TFlag = 1
TCycles = TCycles - (Value * 3)
 else
 if ((TFlag != 0) && (((Value * 3) + 1) <= TCycles))
  bsf    DlayCount, i
TCycles = TCycles - ((Value * 3) + 1)
 endif
 endif
Value = Value >> 1
i = i - 1
 endw
 if (TCycles > 3)
 Error "Delay Cycles too Large for Macro"
 endif
```

```
    decfsz  DlayCount, f
     goto   $ - 1
    endif
    while (TCycles > 1)
    goto    $ + 1
 TCycles = TCycles - 2
    endw
    if (TCycles == 1)
     nop                    ;  Delay the Last Cycle
    endif
    endm
```

Below is a 16-bit Delay. Each loop Iteration requires five instruction cycles and the delay can be defined as:

$$\text{Delay} = (\text{InstructionCycleDelay} / 5)$$

Note that in the variable initialization, 256 is added to the "InstructionCycleDelay" to take into account the loop when the low byte is initially set.

```
    movlw   HIGH ((InstructionCycleDelay / 5) + 256)
    movwf   HiCount
    movlw   LOW ((InstructionCycleDelay / 5) + 256)
 Dlay:
    addlw   0x0FF         ;  Decrement the Counter by 1
    btfsc   STATUS, Z
     decfsz HiCount, f    ;  Decrement the High Byte
                             Counter

     goto   Dlay
```

Negating the Contents of a Register

Converting the contents of a File Register to its 2's complement value without affecting "w" is simply accomplished by:

```
comf  Reg, f          ;  Invert the bits in the
                          Register
incf  Reg, f          ;  Add One to them to turn
                          into 2's
                      ;  Complement
```

This code should not be used on any special hardware control registers.

The "w" register can be negated in the low-end PICmicro® MCU using the instructions:

```
addwf  Reg, w  ;   w = w + Reg
subwf  Reg, w  ;   w = Reg - w
               ;   w = Reg - ( w + Reg )
               ;   w = -w
```

Any file register can be used for this code because its contents are never changed.

In mid-range PICmicro® MCUs, the single instruction:

```
sublw  0      ;   w = 0 - w
```

could be used.

Incrementing/Decrementing "w"

The following assembly language code can be used to increment/decrement "w" in low-end PICmicro® MCUs that do not have "addlw" and "sublw" instructions.

"Reg" can be any register that does not change during the execution of the three instructions. For the low-end parts, any file register can be used because there is no danger of them being updated by an interrupt handler.

To Increment:

```
xorlw 0x0FF   ;  Get 1s Complement of Number
addwf Reg, w  ;  w = Reg + (w^0x0FF)
subwf Reg, w  ;  w = Reg + ((Reg + (w^0x0FF))^0x0FF)
              ;      + 1
              ;  w = w + 1
```

To decrement, the instructions are rearranged:

```
subwf Reg, w  ;  w = Reg + (2^0x0FF) + 1
xorlw 0x0FF   ;  Get 1s Complement of Result
addwf Reg, w  ;  w = w - 1
```

Rotating a Byte in Place

These two lines will rotate the contents of a file register without losing data in the "Carry Flag". Rotates right and left can be implemented with this snippet. Note that the carry flag is changed.

```
rlf  Register, w  ;  Load Carry with the high bit
rlf  Register, f  ;  Shift over with high bit
                       going low
```

Copy Bits from One Register to Another

Here is a fast way to save specific bits from one register into another.

```
movf    Source, w
xorwf   Destination, w
andlw   B'xxxxxxxx'   ;  Replace "x" with "1" to
                            Copy  the Bit
xorwf Destination, f
```

Converting a Nybble to ASCII

The most obvious way of doing this is

```
NybbletoASCII

  addwf  PCL, f                ;  Add the Contents of
                                  the Nybble to PCL/
  dt     "0123456789ABCDEF" ;  return the ASCII as a
                                  Table Offset
```

Another way is

```
NybbletoASCII            ;  Convert a Nybble in "w" to
                            ASCII
  addlw  0x036           ;  Add '0' + 6 to Value
  btfsc  STATUS, DC      ;  If Digit Carry Set, then
                            'A' - 'F'
  addlw  7               ;  Add Difference Between '9'
                            and 'A'
```

```
    addlw    0-6

    return              ;   Return the ASCII of Digit in
                            "w"
```

Converting an ASCII Byte to a Hex Nybble

Using the aspect that the high nybble of ASCII "A" to "F"
is 16 greater than the high nybble of "0" to "9", a value is
conditionally added to make the result 0x000 to 0x00F.

```
ASCIItoNybble
    addlw    0x0C0      ;   If "A" to "F", Set the Carry
                            Flag
    btfss    STATUS, C  ;   If Carry Set, then 'A' - 'F'
     addlw   7          ;   Add Difference Between '9'
                            and 'A'

    addlw    9

    return              ;   Return the ASCII of Digit in
                            "w"
```

Note that ASCII characters other than "0" to "9" and "A"
to "F" will result in an incorrect result.

Using T0CKI as an Interrupt Source Pin

The following code will reset TMR0 when rising edge is
received.

```
movlw  B'11000000' ; First Setup with Instruction Clock
option             ; as TMR0 Source

movlw  B'11100000' ; Option Setup for TOCK1 TMR0 Source
```

```
clrf    TMR0        ; Set TMR0 to 0x0FF
decf    TMR0, f

option              ; Enable Timer on Outside Interrupt
                    ; Edge
                    ; NOTE - Executing this Instruction
                    ;  after "decf" will Load the
                    ;  Synchronizer with a "1"

btfsc   TMR0, 1     ; Wait for incoming Rising Edge
 goto $ - 1

; When Execution Here, the Input has toggled
```

This code can also be used on a low-end PICmicro® MCU to monitor when an input changes instead of continuously polling the input pin.

Dividing by Three

Here is an algorithm from Andy Warren for dividing a positive value by three; by knowing that "divide by three" can be represented by the series:

$$x/3 = x/2 - x/4 + x/8 - x/16 + x/32 - x/64 \ldots$$

it can be implemented in the PICmicro® MCU as:

```
Div3:               ; Divide Contents of "w" by 3

  movwf Dividend
  clrf  Quotient

Div3_Loop           ; Loop Until the Dividend == 0
```

```
bcf    STATUS, C
rrf    Dividend, f  ; Dividend /2 (ie "x/2" in Series)
movf   Dividend, w  ;  Is it Equal to Zero?
btfsc  STATUS, Z
 goto  Div3_Done    ; If it is, then Stop

addwf  Quotient     ; Add the Value to the Quotient

rrf    Dividend, f  ; Dividend /2 (ie "x/4" in Series)
movf   Dividend, w
btfsc  STATUS, Z
 goto  Div3_Done

subwf  Quotient, f ; Quotient = Quotient-(Dividend  4)

goto   Div3_Loop

Div3_Done

movf   Quotient, w  ; Return the Quotient
return
```

Sixteen-Bit Pulse Measurement with 5-Cycle Delay

The code that measures the pulse width for a "high" pulse is

```
clrf    PulseWidth       ;  Reset the Timer
clrf    PulseWidth + 1

btfss   PORTn, Bit       ;  Wait for the Pulse to
                             go high
 goto   $ — 1

incfsz  PulseWidth, f    ;  Increment the Counter
 decf   PulseWidth + 1, f
btfsc   PORTn, Bit       ;  Loop while Still High
 goto   $ — 3
```

```
movf    PulseWidth, w      ;  Make 16 Bit Result
                                      Valid
addwf   PulseWidth + 1, f
```

Detect a Change in a Register

This code can be used to detect changes in the I/O ports, timers, or other registers that can be updated externally to the software execution.

```
movf   Reg, w
andlw  Mask        ;  Mask out unused bits
xorwf  old, w      ;  Compare to previous value
btfsc  STATUS, Z   ;  If Zero set, bits are the Same
 goto  no_change
xorwf  old         ;  Bits are different, Store New
                   ;  pattern in "old"
```

Test a Byte within a Range

Code that Tests "Num" to be within a specific byte range and jumps to the "in_range" label if true.

```
movf   Num, w
addlw  255 - hi_lim        ; "Num" is equal to -hi_lim
addlw  hi_lim - lo_lim + 1 ; "Num" is > 255 if it is
                                       above
btfsc  STATUS, C           ;  the lo-lim
 goto  in_range
```

Convert ASCII to Upper Case

This is a practical application of the previous snippet.

```
ToUpper:
    addlw   255 - 'z'       ; Get the High limit
    addlw   'z' - 'a' + 1   ; Add Lower  Limit to Set Carry
    btfss   STATUS, C       ; If Carry Set, then Lower Case
    addlw   h'20'           ;  Carry NOT Set, Restore
                                Character
    addlw   'A'             ; Add 'A' to restore the
                                Character
    return
```

Swap the Contents of "w" with a Register

Fast method of exchanging "w" with a register without requiring a third "temporary" file register.

```
    xorwf Reg, f    ;  w = w, Reg = Reg ^ w
    xorwf Reg, w    ;  w = w ^ (Reg ^ w), Reg = Reg ^ w
                    ;  w = Reg, Reg = Reg ^ w
    xorwf Reg, f    ;  w = Reg, Reg = Reg ^ w ^ Reg
                    ;  w = Reg, Reg = w
```

Swap the Contents of Two Registers

Here is a fast snippet to swap the contents of two file registers:

```
    movf    X, w
    subwf   Y, w     ;  W = Y - X
```

```
addwf   X, f    ;   X = X + (Y - X)
subwf   Y, f    ;   Y = Y - (Y - X)
```

Compare and Swap if Y < X

This snippet is useful for "Bubble" Sort Routines.

```
movf    X, w
subwf   Y, w        ;   Is Y >= X?
btfsc   STATUS, C   ;   If Carry Set, Yes
 goto   $ + 2       ;   Don't Swap
addwf   X, f        ;   Else, X = X + (Y - X)
subwf   Y, f        ;           Y = Y - (Y - X)
```

Counting the Number of "1"s in a Byte

The code below is Dmitry Kirashov's optimization of the classic problem of counting the number of "1"s in a byte in 12 instructions/12 cycles.

```
                    ;   (c) 1998 by Dmitry Kirashov

rrf     X, w        ;   "X" Contains Byte
andlw   0x55        ;   -a-c-e-g
subwf   X, f        ;   ABCDEFGH
                    ;   where AB=a+b, etc.
                    ;   the same trick as in example_1
movwf   X
andlw   0x33        ;   --CD--GH
addwf   X, f
rrf     X, f        ;   0AB00EF0
                    ;   00CD00GH
```

```
addwf   X, f      ;   0AB00EF0
                  ;   0CD00GH0
rrf     X, f      ;   0ABCD.0EFGH

swapf   X, w
addwf   X, w
andlw   0x0F      ;   Bit Count in "w"
```

Generating Parity for a Byte

At the end of the routine, bit 0 of "X" will have the "Even" Parity bit of the original number. "Even" Parity means that if all the "1"s in the byte are summed along with Parity Bit, an even number will be produced.

```
swapf    X, w
xorwf    X, f
rrf      X, w
xorwf    X, f

btfsc    X, 2
 incf    X, f
```

Keeping a Variable within a Range

Sometimes when handling data, you will have to keep integers within a range. The four instructions below will make sure that the variable "Temp" will always be in the range of Zero to "Constant".

```
movlw    Constant  ;   0 <= Temp <= Constant
subwf    Temp, w
```

```
btfsc    STATUS, C
subwf    Temp, f
```

Swapping Bit Pairs

```
              ;  (c) 1998 by Dmitry Kirashov

movwf    X         ;  Save the Incoming Byte in
                   ;   a temporary register
                   ;  w = X = ABCDEFGH
andlw    0x055     ;  w = 0B0D0F0H
addwf    X, f      ;  X = ABCDEFGH + 0B0D0F0H

rrf      X, f      ;  X = (ABCDEFGH + 0B0D0F0h) >> 1
addwf    X, w      ;  w = BADCFEHG
```

Bitwise Operations

Setting a bit by "ANDing" two others together is accomplished by:

```
bsf      Result    ;  Assume the result is True
btfsc    BitA      ;  If BitA != 1 then result is False
btfss    BitB      ;  If BitB == 0 then result is False
bcf      Result    ;  Result is False, Reset the Bit
```

"ORing" two bits together is similar to the "AND" operation, except the result is expected to be false and when either bit is set, the result is true:

```
bcf     Result  ;  Assume the result is False
btfss   BitA    ;  If BitA != 0 then result is True
 btfsc  BitB    ;  If BitB == 0 then result is False
 bsf    Result  ;  Result is True, Set the Bit
```

There are two ways of implementing the "NOT" operation based on where the input value is relative to the
output value. If they are the same (i.e., the operation is
to complement a specific bit), the code to be used is
simply:

```
movlw    1 << BitNumber    ;  Complement Specific
                               Bit for "NOT"
xorwf    BitRegister, f
```

If the bit is in another register, then the value stored is
the complement of it:

```
bcf     Result  ; Assume that the Input Bit is Set
btfss   Bit     ;  - If it is Set, then Result Correct
 bsf    Result  ; Input Bit Reset, Set the Result
```

Constant Multiplication

The following macro will insert 8-bit multiplication by a
constant code:

```
multiply macro Register, ; Multiply 8 bit value by a
 Value variable i = 0,    ;  constant
 TValue
```

```
TValue = Value                ; Save the Constant Multiplier
  movf    Register, w
  movwf   Temporary           ; Use "Temporary" as Shifted
                                  Value

  clrf    Temporary + 1
  clrf    Product
  clrf    Product + 1
 while (i < 8)
 if ((TValue & 1) != 0)    ; If LSB Set, Add the Value
  movf    Temporary + 1, w
  addwf   Product + 1, f
  movf    Temporary, w
  addwf   Product, f
  btfsc   STATUS, C
   incf   Product + 1, f
 endif
  bcf     STATUS, C           ; Shift Up Temporary
                                 multicand
  rlf     Temporary, f
  rlf     Temporary + 1, f
TValue = TValue >> 1       ; Shift down to check the Next
                              Bit
i = i + 1
 endw
 endm
```

Constant Division

The following code will return a rounded quotient for a
variable divided by a constant:

```
divide macro Register, Value ; Divide 8 bit value
 variable i = 0, TValue          by a constant
TValue = 0x010000 / Value    ; Get the Constant Divider
  movf    Register, w
  movwf   Temporary + 1       ; Use "Temporary" as the
                                 Shifted Value
  clrf    Temporary
  clrf    Quotient
  clrf    Quotient + 1
```

```
  while (i < 8)
   bcf      STATUS, C            ; Shift Down the Temporary
   rrf      Temporary + 1, f
   rrf      Temporary, f
   if ((TValue & 0x08000) != 0); If LSB Set, Add the
                                     Value
    movf     Temporary + 1, w
    addwf    Quotient + 1, f
    movf     Temporary, w
    addwf    Quotient, f
    btfsc    STATUS, C
     incf    Quotient + 1, f
   endif
TValue = TValue << 1          ; Shift up to check the
                                Next Bit
i = i + 1
  endw
   movf     Quotient + 1, w     ; Provide Result Rounding
   btfsc    Quotient, 7
    incf    Quotient + 1, w
   movwf    Quotient
  endm
```

12

16-Bit Numbers

Defining 16 Bit Numbers

16-bit numbers can have their addresses declared specifically, as in the example below:

```
RAM       equ    12           ;  Start of RAM for the
                              ;  PIC16C71

Reg_8     equ    RAM          ;  Define the 8 Bit
                              ;    Register
Reg_16    equ    RAM + 1      ;  Define the first 16
                              ;    Bit Register
Reg2_16   equ    RAM + 3      ;  Define the 2nd 16 Bit
                              ;    Register
```

or, using the "CBLOCK" Command in MPASM with the number of bytes in the variable specified:

```
CBLOCK 12                 ;  Start of RAM for the
                          ;   16C71
Reg_8                     ;  Define the 8 Bit
                          ;   Register
Reg_16:2                  ;  Define the first 16
                          ;   Bit Register
Reg2_16:2                 ;  Define the 2nd 16 Bit
                          ;   Register
ENDC
```

Increments and Decrements

Incrementing a 16-bit value in the low-end or mid-range is accomplished by:

```
incf    Reg, f            ;  Increment the Low byte
btfsc   STATUS, Z         ;  Do we have Zero
                          ;   (Multiple of 256)?
 incf   Reg + 1, f        ;  Increment High byte
                          ;   (if necessary)
```

For the PIC17Cxx or a PIC18Cxx, the "infsnz" instruction is used to simplify the 16-bit increment by one instruction:

```
infsnz  Reg, f            ;  Increment "Reg's" Low
                          ;   Byte and Skip
 incf   Reg + 1, f        ;  High Byte Increment
                          ;   if Result is Not
                          ;   Equal to Zero
```

The decrement of a 16-bit value for the PICmicro® MCUs is a four instruction (instruction cycle) process:

```
movf    Reg, f            ;  Set "Z" if LOW "Reg"
                          ;   == 0
btfsc   STATUS, Z
 decf   Reg + 1, f        ;  If Low byte is Zero,
                          ;   Decrement High
decf    Reg, f
```

Addition/Subtraction

Adding a Constant to a value in the low-end and mid-range PICmicro® MCUs, that is,

```
Reg = Reg + 0x01234
```

is accomplished by:

```
movlw   HIGH 0x01234      ;  Add the high byte
                          ;   first
addwf   Reg + 1, f
movlw   LOW 0x01234       ;  Add the Low Byte Next
addwf   Reg, f
btfsc   STATUS, C         ;  Don't Inc high byte if
                          ;   carry Reset
  incf  Reg + 1, f
```

In the PIC17Cxx and PIC18Cxx, the "addwfc" instructions can be used to simplify the operation:

```
movlw   LOW 0x01234       ;  Add Low Byte First
addwf   Reg, f
movlw   HIGH 0x01234      ;  Add High Byte Next
addwfc  Reg + 1, f
```

The corresponding subtraction, that is,

```
Reg = Reg - 0x01234
```

looks like the following code for the low-end and mid-range PICmicro® MCUs:

```
movlw   HIGH 0x01234      ;  Subtract the High Byte
                          ;   First
subwf   Reg + 1, f
movlw   LOW 0x01234       ;  Subtract the Low Byte
                          ;   Next
subwf   Reg, f
btfss   STATUS, C         ;  Don't Dec high byte if
                          ;   carry Set
  decf  Reg + 1, f
```

For the PIC17Cxx and PIC18Cxx, the "subwfb" instruction is used:

```
movlw   LOW 0x01234      ; Subtract the Low Byte
                         ;   First
bsf     STATUS, C        ; Don't pass any
                         ;   "Borrow"
subwfb  Reg, f           ; Reg = Reg - w - !C
movlw   HIGH 0x01234
subwfb  Reg + 1, f       ; Reg + 1 = Reg + 1 - w
                         ;   - !C
```

The "addwfc" and "subwfb" enhancements can be used in all the 16-bit addition and subtraction operations given below. When using these instructions follow the same format of finding the least significant byte's result followed by the most significant byte's result, which is opposite to how the operations are carried out in the low-end and mid-range PICmicro® MCUs.

When adding to and subtracting from a 16-bit variable and storing the result in another variable in the low-end and mid-range PICmicro® MCUs:

```
Destination = Source + 0x05678
```

the assembly code will look like:

```
movlw   HIGH 0x05678       ; Add High Byte First
addwf   Source + 1, w
movwf   Destination + 1, f ; Store Result in
                           ;   Destination
movlw   LOW 0x05678        ; Add Low Byte Next
addwf   Source, w
movwf   Destination, f     ; Store Result
btfsc   STATUS, C          ; Is the Carry Flag Set?
 incf   Destination + 1, f ; Yes, Increment High
                           ;   Byte
```

Addition of a 16-bit variable to another 16-bit variable is similar to that of adding a Constant to a 16-bit variable.

If the destination is the same as one of the values, for instance:

$$a = a + b$$

the low-end and mid-range assembly language code looks like:

```
movf    b + 1, w              ;  Add the High Bytes
addwf   a + 1, f
movf    b, w                  ;  Add the Low Bytes
addwf   a, f
btfsc   STATUS, C             ;  Add the Carry to High
                             ;    Byte
 incf   a + 1, f
```

If the Destination is different from both values to be added, for instance,

$$c = a + b$$

the code is changed to save the sums in "w" and then store them in "c":

```
movf    a + 1, w              ;  Add the High Bytes
addwf   b + 1, w
movwf   c + 1
movf    a, w                  ;  Add the Low Bytes
addwf   b, w
movwf   c
btfsc   STATUS, C             ;  Increment due to Carry
 incf   c + 1
```

Subtraction is carried out in the same way, but care must be taken to ensure that the subtracting Register is kept straight. To implement

$$c = a - b$$

in assembly language, the following code would be used in the low-end and mid-range PICmicro® MCUs:

```
movf    b + 1, w        ; Get Value to be
                        ;   subtracted
subwf   a + 1, w        ; Do the High Byte
movwf   c + 1
movf    b, w            ; Get the Value to be
                        ;   Subbed

subwf   a, w
movwf   c
btfss   STATUS, C       ; Look for the Carry
  decf  c + 1
```

Bitwise Operations on Constants and Variables

ANDing a 16-bit variable with 0x0A55A would be implemented in this way:

```
movlw   HIGH 0x0A55A    ; Get Value for ANDING
andwf   Reg + 1, f      ; Do the High Byte
movlw   LOW 0x0A55A     ; Get Value for ANDING
andwf   Reg, f          ; Do the Low Byte

bcf     STATUS, C       ; Clear the Carry Flag
                        ;   for new bit
rlf     Reg, f          ; Shift the Low Byte
rlf     Reg + 1, f      ; Shift High Byte with
                        ;   Low Carry
```

and to shift right:

```
bcf     STATUS, C       ; Clear Carry Flag for
                        ;   the New bit
rrf     Reg + 1, f      ; Shift down the High
                        ;   Byte
rrf     Reg, f          ; Shift Low Byte with
                        ;   Valid Carry
```

Comparisons with 16-Bit Variables

12.3

```
movf    Reg2 + 1, w      ; Get the High Byte of
                         ;   the Result
subwf   Reg1 + 1, w
movwf   _2               ; Store in a Temporary
                         ;   Register
movf    Reg2, w          ; Get the Low Byte
subwf   Reg1, w
btfss   STATUS, C        ; Decrement High if
                         ;   Necessary
 decf   _2
```

At the end of this series of instructions, "w" contains Reg2 − Reg1 and "_2" contains Reg2HI − Reg1HI with the borrow result of Reg2 − Reg1.

There are six basic conditions that you can look for: Equals, Not Equals, Greater Than, Greater Than or Equal To, Less Than, Less Than or Equal To. So, to discover whether or not I have any of these conditions, the following code can be added.

For Equals and Not Equals, the value in "w" is ORed with "_2" to see if the Result is equal to zero.

```
iorwf   _2, w            ; Is the Result == 0?
```

for Equals add the lines:

12.4

```
btfss   STATUS, Z        ; Execute following Code
                         ;   if == 0
 goto   Zero_Skip        ; Else, Code != 0, Skip
                         ;   Over
```

for Not Equals, append:

```
btfsc    STATUS, Z        ; Execute following if
                          ;   != 0
goto     NotZero_Skip     ; Else, Code == 0, Skip
                          ;   Over
```

If Greater Than (the 16-bit variable is greater than the comparison value), then the result will not be less than Zero. Actually, the same code (just with a different Bit Skip) can be used to test.

For Greater Than:

```
btfsc    _2, 7            ; Not Negative, 16 Bit
                          ;   is Greater
goto     NotGreater_Skip  ; Else, Skip if Not
                          ;   Greater than
iorwf    _2, w            ; Is it Equal to Zero?
btfsc    STATUS, z        ; No, It is Greater
                          ;   than
Goto     NotGreater_Skip  ; Else, if Zero, Not
                          ;   Greater than
```

Note that just the most significant bit of the 16-bit difference is checked. If this bit is set (= 1), then the 16-bit variable is less than the Comparison. If it is reset (= 0), then it is greater than and you should check to see if the result is not equal to zero (or else it is equal).

For Less Than:

```
btfss   _2, 7                ;  Negative, 16 Bit is
                             ;    Less Than
goto    NotLess_Skip         ;  Else, Skip because Not
                             ;    Less Than
```

To check for Greater Than or Equal To, the last three lines of the code checking for Greater Than are simply erased. To check for Less Than or Equal To, the three lines from Not Equals are added before the check for less than.

Here is the complete code for compare and skip on Reg1 less than or equal to Reg2:

```
movf    Reg2 + 1, w          ;  Get the High Byte of
                             ;    the Result
subwf   Reg1 + 1, w
movwf   _2                   ;  Store in a Temporary
                             ;    Register
movf    Reg2, w              ;  Get the Low Byte
subwf   Reg1, w
btfss   STATUS, C            ;  Decrement High if
                             ;    Necessary
decf    _2
iorwf   _2, w                ;  Check for Equal to
                             ;    Zero
btfsc   STATUS, Z            ;  If Not Zero, Jump Over
goto    EqualLess_Skip       ;  Equals, Jump to the
                             ;    Code
btfsc   _2, 7                ;  If Number is Negative,
                             ;    execute
goto    EqualLess_Skip       ;  Else, Jump Over
```

Multiplication

Here is multiplication that requires a separate byte for counting the iterations through "Loop":

```
    clrf    Product
    clrf    Product + 1

    movlw   16                      ; Operating on 16 Bits
    movwf   BitCount

Loop                                ; Loop Here for Each Bit

    rrf     Multiplier + 1, f       ; Shift the Multiplier
                                    ;   down
    rrf     Multiplier, f           ; by one
    btfss   STATUS, C               ; If the bit is set, add
     goto   Skip                    ;   the Multiplicand to
                                    ;   the "Product"

    movf    Multiplicand + 1, w
    addwf   Product + 1, f
    movf    Multiplicand, w
    addwf   Product, f
    btfsc   STATUS, C
     incf   Product + 1, f

Skip                                ; Shift up Multiplicand
                                    ;   and
    bcf     STATUS, C               ;   Loop Around
    rlf     Multiplicand, f
    rlf     Multiplicand + 1, f

    decfsz  BitCount
     goto   Loop
```

The code given below is the most efficient way of doing a 16-bit multiply with a 32-bit result. It is not immediately obvious, but it is very clever. Rather than use a

32-bit add each time the shifted data is detected, it provides a 16-bit (with valid carry) add and then shifts the data down. This Code does not change "Multiplicand", but does change "Multiplier".

Note that in the code, a 32-bit value for "Product" (using a "Product:5" line in the "CBLOCK" variable declare statement) is used.

```
clrf    Product + 2          ;  "Product" will be the
clrf    Product + 3          ;  Result of the
                             ;   Operation

movlw   16                   ;  Operating on 16 Bits
movwf   BitCount

Loop                         ;  Loop Here for Each Bit

  rrf   Multiplier + 1, f    ;  Shift the Multiplier
  rrf   Multiplier, f        ;  down by one

  btfss STATUS, C            ;  If the bit is set, add
   goto  Skip                ;   the Multiplicand to
                             ;   the Product"

  clrf  Product + 4
  movf  Multiplicand + 1, w
  addwf Product + 3, f
  btfsc STATUS, C            ;  Make Sure the Carry is
                             ;   Passed
   incf Product + 4, f       ;   to the Next Byte
  movf  Multiplicand, w
  addwf Product + 2, f
  btfsc STATUS, C
   incfsz Product + 3, f     ;  Make Sure Carry is
                             ;   Passed with
    goto $ + 2               ;   the Shift
  incf  Product + 4, f

Skip                         ;  Shift "Product" Down
                             ;   with
```

```
bcf      STATUS, C
rrf      Product + 4, f
rrf      Product + 3, f      ;   the Reset Carry from
                             ;     the
rrf      Product + 2, f      ;   Multiplier shift down
                             ;     or
rrf      Product + 1, f      ;   the result of the
                             ;     sixteen
rrf      Product, f          ;   bit addition.

decfsz   BitCount
 goto    Loop
```

Both of the Multiplication routines shown here will work with positive and negative numbers.

For the PICmicro® MCUs that have built-in eight by eight multipliers (PIC17Cxx and PIC18Cxx), the code for 16-bit multiplication can be accomplished using the code:

```
clrf     Product + 2         ;   Clear the High-Order
                             ;     Bits
clrf     Product + 3
movf     Al, w               ;   Do the "L"
                             ;     Multiplication first
mulwf    Bl
movf     PRODL, w            ;   Save result
movwf    Product
movf     PRODH, w
movwf    Product + 1
movf     Al, w               ;   Do the "I"
                             ;     Multiplication
mulwf    Bh
movf     PRODL, w            ;   Save the Most
                             ;     Significant Byte
                             ;     First
addwf    Product + 1, f
movf     PRODH, w
```

```
addwfc   Product + 2, f     ;  Add to the Last Result
movf     Ah, w              ;  Do the "O"
                            ;    Multiplication

mulwf    Bl
movf     PRODL, w           ;  Add the Lower Byte
                            ;    Next
addwf    Product + 1, f
movf     PRODH, w           ;  Add the High Byte
                            ;    First
addwfc   Product + 2, f
btfsc    STATUS, C          ;  Add the Carry
 incf    Product + 3, f
movf     Ah, w              ;  Do the "F"
                            ;    Multiplication

mulwf    Bh
movf     PORDL, w
addwf    Product + 2, f
movf     PRODH, w
addwfc   Product + 3, f
```

Division

The division routine provided here first finds how far
the divisor can be shifted up before comparing to
the quotient. The "Count" variable in this routine is a
16-bit variable that is used both to count the bits
and add to the quotient. "Temp" is an 8-bit temporary
Storage Variable. At the end of the division routine,
"Dividend" will contain the remainder of the operation.

```
clrf     Quotient
clrf     Quotient + 1

movlw    1                  ;  Initialize Count
movwf    Count
clrf     Count + 1
```

```
StartLoop                       ;  Find How Large
                                ;    "Divisor" can
                                ;    be
  btfsc    Divisor + 1, 7       ;  If at the "top", then
                                ;    do
   goto    Loop                 ;    the Division

  bcf      STATUS, C            ;  Shift Count and
                                ;    Divisor Up
  rlf      Count, f
  rlf      Count + 1, f

  rlf      Divisor, f
  rlf      Divisor + 1, f

  goto     StartLoop

Loop                            ;  Now, Take Away
                                ;    "Divisor"
                                ;    from "Dividend"
  movf     Divisor + 1, w       ;  If Divisor < Dividend
                                ;    then
  subwf    Dividend + 1, w      ;    Don't Take Away
  movwf    Temp
  movf     Divisor, w
  subwf    Dividend, w
  btfss    STATUS, C
   decf    Temp, f
  btfsc    Temp, 7              ;  If "Temp" Negative
                                ;    then
   goto    Skip                 ;  Divisor < Dividend

  movwf    Dividend             ;  Save the New Dividend
  movf     Temp, w
  movwf    Dividend + 1

  movf     Count, w             ;  Add Count to the
                                ;    Quotient
  addwf    Quotient + 1, f
  movf     Count, w
  addwf    Quotient + 1, f      ;  No Opportunity for
                                ;    Carry

Skip                            ;  Shift Divisor/Count
                                ;    Down
```

```
bcf     STATUS, C
rrf     Divisor + 1, f
rrf     Divisor, f

rrf     Count + 1, f        ;  If Carry Set after
                            ;    Count
rrf     Count, f            ;  Shift, Finished

btfss   STATUS, C           ;  If Carry NOT Set, then
 goto   Loop                ;    Process next Bit
```

This division routine is designed to only handle positive numbers—there is not a general algorithm that handles both positive and negative numbers and passes back both the quotient and remainder with the correct polarity efficiently.

A general form for a division routine (using the algorithm shown above) could be the division of the core of the pseudo-code in a bit-shift analogous algorithm to multiplication that can handle positive and negative numbers.

```
if (Dividend < 0) {         //  Change dividend to
                            //    positive number
 Dividend = 0 - Dividend;
 dividendneg = 1;           //  Mark we have to
                            //    change it back
} else
 dividendneg = 0;
if (Divisor < 0) {          //  Repeat with the
                            //    Divisor
 Divisor = 0 - Divisor;
 divisorneg = 1;
} else
 divisorneg = 0;

Count = 0;                  //  Going to Count where
                            //    division starts
```

```
Quotient = 0;                    //  Store the Quotient
while (( Divisor & 0x0400 ) != 0) {
                                 //  Find the Start of the
                                 //    Division
 Count = Count + 1;              //  Increment the Number
                                 //    of Bits Shifted
 Divisor = Divisor << 1;
 }

while (Count != 0) {             //  Now, do the Division
 if (Dividend >= Divisor)  {//  A subtract can take
                                 //    place
    Quotient = Quotient + 2 ^ Count;
    Dividend = Dividend - Divisor;
 }
Count = Count - 1;
Divisor = Divisor >> 1;
 }

if (Dividendneg == 1)            //  Now, change the
                                 //    values
 if (Divisorneg == 1) {
  Quotient = Quotient;
  Remainder = 0 - Dividend;
 } else {
   Quotient = 0 - Quotient;
   Remainder = 0 - Dividend;
else                             //  The Dividend was
                                 //    Positive
 if (Divisorneg == 1) {
   Quotient = 0 - Quotient;
   Remainder = Dividend;
 } else {
   Quotient = Quotient;
   Remainder = Dividend;
  }
```

13

PICmicro® MCU Operations Tables

The following information is based on the datasheets available at the time of printing and are meant to be used for providing a basic operating reference. Some data is not complete due to "Advanced" copies of the datasheets. "Idd", or "intrinsic" current requirements, is the amount of current required for the base PICmicro® MCU to operate and does not include current required for peripheral functions.

I/O Pin Current Capabilities

Current Source/Sink requirements are in milli-Amperes ("mA").

I/O Pin Current Source/Sink Capabilities

Device	Pin Source/Sink	Port Source/Sink	Device Source/Sink	Comments
PIC12C5xx	25/25	25/25	100/100	GPIO used for Data I/O Pins
PIC14C000	25/25	200/200	250/300	Also Available for LCD driving
PIC16C5x	20/25	40/50	50/150	
PIC16C55x	25/25	200/200	250/300	
PIC16C6x	20/25-50/80	100/150	100/150	
PIC16C62x	25/25	200/200	250/300	
PIC16C7x	25/25	200/200	250/300	
PIC16F84	20/25	50/80-100/150	100/150	
PIC16F87x	25/25	200/200	250/300	
PIC17C4x	20/35	100/150	200/250	RA2/RA3 able to Sink 60 mA
PIC18Cxx(x)	25/25	200/200	50/300	

RC Oscillator Component Values

The following table and chart outline different Resistor/Capacitor values and current requirements for the low-end PICmicro® MCUs and the PIC16F84 using an RC oscillator. Note that RC oscillator operation can have variances up to 30% according to Microchip documentation and are only recommended for time-insensitive applications.

For the low-end PICmicro® MCUs, Table 13.1 shows different capacitor values and Fig. 13.1 shows current consumption for different operating speeds.

The PIC16F84 is a very common beginning user PICmicro® MCU. Table 13.2 shows different frequencies for different resistor/capacitor combinations and Fig. 13.2 shows the varying Idd current required for different frequencies.

TABLE 13.1

C$_{EXT}$	R$_{EXT}$	Average Fosc @ 5 V, 25°C	
20 pF	3.3 k	4.973 MHz	± 27%
	5 k	3.82 MHz	± 21%
	10 k	2.22 MHz	± 21%
	100 k	262.15 kHz	± 31%
100 pF	3.3 k	1.63 MHz	± 13%
	5 k	1.19 MHz	± 13%
	10 k	684.64 kHz	± 18%
	100 k	71.56 kHz	± 25%
300 pF	3.3 k	660 kHz	± 10%
	5.0 k	484.1 kHz	± 14%
	10 k	267.63 kHz	± 15%
	160 k	29.44 kHz	± 19%

The frequencies are measured on DIP packages.

The percentage variation indicated here is part-to-part variation due to normal process distribution. The variation indicated is ±3 standard deviation from average value for VDD = 5 V.

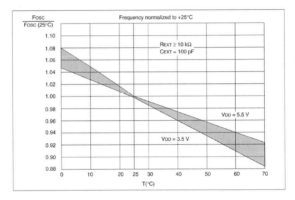

Figure 13.1

TABLE 13.2

Cext	Rext	Average Fosc @ 5V, 25°C	
20 pF	3.3k	4.68 MHz	± 27%
	5.1k	3.94 MHz	± 25%
	10k	2.34 MHz	± 29%
	100k	250.16 kHz	± 33%
100 pF	3.3k	1.49 MHz	± 25%
	5.1k	1.12 MHz	± 25%
	10k	620.31 kHz	± 30%
	100k	90.25 kHz	± 26%
300 pF	3.3k	524.24 kHz	± 28%
	5.1k	415.52 kHz	± 30%
	10k	270.33 kHz	± 26%
	100k	25.37 kHz	± 25%

*Measured in PDIP Packages. The percentage variation indicated here is part to part variation due to normal process distribution. The variation indicated is ±3 standard deviation from average value.

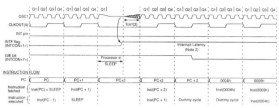

Figure 13.2

Note:
1: XT, HS or LP oscillator mode assumed.
2: TOST = 1024TOSC (drawing not to scale) This delay will not be there for RC osc mode.
3: GIE = '1' assumed. In this case after wake- up, the processor jumps to the interrupt routine. If GIE = '0', execution will continue in-line.
4: CLKOUT is not available in these osc modes, but shown here for timing reference.

LP Oscillator Operating Characteristics

The following table outlines different capacitor values for different "LP" oscillator executing frequencies using a crystal. Note that "LP" mode is active between 0 and 200 KHz only. The Idd (intrinsic) current requirements are quoted for 32.768 kHz and powered by 5 volts and are in micro-Amperes except where noted.

LP Operating Mode Part Specification and Idd Current Requirements			
Device	32.768 KHz	200 KHz	Idd Current
PIC12C5xx	15 pF	N/A	10 uA
PIC16C5x	15 pF	N/A	32 uA
PIC16C55x	68–100 pF	15–30 pF	32 uA
PIC16C6x	33–68 pF	15–47 pF	21 uA
PIC16C62x	68–100 pF	15–30 pF	32 uA
PIC16C7x	15–47 pF	15–33 pF	48 uA
PIC16F84	68–100 pF	15–33 pF	32 uA
PIC16F87x	33 pF	15 pF	48 uA
PIC18Cxx(x)	33 pF	15 pF	N/A

XT Oscillator Operating Characteristics

The following table outlines different capacitor values for different "XT" oscillator executing frequencies using a crystal or ceramic resonator. Note that "XT" mode is active between 0 and 4 MHz. The Idd (intrinsic) current requirements are quoted at 4 MHz and are in milli-Amperes except where noted.

XT Operating Mode Part Specification and Idd Current Requirements

```
Device        200 KHz     1 MHz       4 MHz      Idd Current
PIC12C5xx     47-68 pF    15 pF       15 pF      0.78 mA
PIC16C5x      15-30 pF    15-30 pF    15 pF      1.8 mA
PIC16C55x     N/A         15-68 pF    15-68 pF   3.3 mA
PIC16C6x      33-68 pF    15-68 pF    15-33 pF   1.6 mA
PIC16C62x     33-68 pF    15-58 pF    15-68 pF   3.3 mA
PIC16C7x      33-68pF     15-68 pF    15-33 pF   5 mA
PIC16F84      68-100 pF   15-33 pF    15-33 pF   4.5 mA
PIC16F87x     47-68 pF    15-68 pF    15 pF      2 mA
PIC17C4x      N/A         33-150 pF   15-68 pF   1.6 mA
PIC18Cxx(x)   47-68 pF    15 pF       15 pF      N/A
```

HS Oscillator Operating Characteristics

The following table outlines different capacitor values for different "HS" oscillator executing frequencies using a crystal or ceramic resonator. Note that "HS" mode is active for frequencies greater than 4 MHz. As a rule of thumb, the maximum speed for low-end and mid-range PICmicro® MCU EPROM program memory devices is 20 MHz. For Flash program memory parts, the maximum speed is usually 10 MHz, except where noted. For the PIC17Cxx, the maximum speed is 33 MHz and for the PIC18Cxx the maximum clock speed is 10 MHz. In the PIC18Cxx, the HS clock can be multiplied by four for an actual internal clock speed of 40 MHz.

Idd (intrinsic) current requirements are taken from the maximum speed and the PICmicro® MCU powered by 5 volts. Capacitor values are in pFs and Idd current is in milli-Amperes except where noted.

HS Operating Mode Part Specification and Idd Current Requirements					
Device	4 MHz	10 MHz	20 MHz	32 MHz	Idd Curr
PIC14C000	15-68 pF	10-47 pF	10-47 pF	N/A	4 mA
PIC16C5x	15 pF	15 pF	15 pF	N/A	5 mA
PIC16C55x	15-30 pF	15-30 pF	15-30 pF	N/A	20 mA
PIC16C6x	15-47 pF	15-47 pF	15-47 pF	N/A	35 mA
PIC16C62x	15-47 pF	15-30 pF	15-30 pF	N/A	20 mA
PIC16C7x	15-47 pF	15-47 pF	15-47 pF	N/A	30 mA
PIC16F84	15-33 pF	15-33 pF	N/A	N/A	10 mA
PIC16F07x	15 pF	15-33 pF	15-33 pF	N/A	20 mA
PIC17C4x	15-68 pF	15-47 pF	15-47 pF	0	40 mA
PIC18Cxx(x)	15 pF	15-33 pF	15-33 pF	N/A	45 mA

PICmicro® MCU Application Debugging Checklist

Debugging Checklist

Problem	Potential Causes	Check
PICmicro® MCU Application does Not Start	1 No/Bad Power	1a Make Sure Vdd is between 4.5 Volts and 5.5 Volts Relative to Vss
		1b Make Sure Vdd "ripple" is Less than 100 mV
	2 No/Bad Reset	2a Make Sure _MCLR" is pulled up to 4.5 Volts to 5.5 Volts
		2b Make Sure Disabled _MCLR Pin is not pulled below Ground
	3 Missing/Bad Decoupling Capacitor	3a Check for 0 01 uF to 0.1 uF Capacitor Close to PICmicro® MCU's Vdd Pin
	4 Part Orientation	4a Check that PICmicro® MCU Part is Installed Correctly
		4b Make Sure the PICmicro® MCU is NOT getting very hot
	5 Oscillator Not Running	5a Check both the OSC1 and OSC2 Pins With an Oscilloscope or Logic Probe
		5b If Internal Oscillator, Check Configuration Fuses For Correct Setting
		5c Check for Present and Correct Capacitors
	6 Device Programming Incorrect	6a Check/Verify Device Programming
		6b Look for I/O Pins being set high or low

7 Watchdog Timer Enabled

7a Check I/O pins for changing between input Output states

7b Check Actual Configuration Fuse Value

8 Uninitialized Variable/Value Incorrect

8a Check Variable initialization at application start

8b After resetting the simulated PICmicro® MCU, load file registers with a random value (such as 0x05A)

9 Interrupt Handler NOT allowing execution exit From Handler

9a Simulate Interrupt Handler and make sure that execution can return to Mainline before next Interrupt Request is Acknowledged

9b Make Sure that correct Interrupt Flag ("IF") is Reset in Handler

10 Variable Address Overlayed onto a Hardware I/O Register

10a Make Sure that the Variable "CBLOCK" statement is in the File Register area of the PICmicro® MCU and not in the Hardware I/O Area

11 Outputs switching too fast to See

11a Probe the Outputs using a Logic Probe or Oscilloscope

PICmicro® MCU Device Seems to Reset Itself Unexpectedly

1 Watchdog Timer Enabled

1a Check Configuration Fuse Values

1b Check for I/O pins changing state with Reset

Debugging Checklist (*Continued*)

Problem	Potential Causes		Check	
	2	High Internal Current and Inadequate Decoupling	2a	Check for Correlation to to Reset with changes in Load drawn by PICmicro® MCU
			2b	Check for Power Supply "sags" when the Load is drawn
	3	Check for a "noisy" Power Supply	3a	Check for greater than 100 mV "Ripple" from Power Supply
	4	Execution Jumps past Application End	4a	Check Code for Subroutine without "return" instruction or Table that is accessed past its End
	5	Uninitialized Variable/Value Incorrect	5a	Check Variable initialization for missed Variable
			5b	Set Variables to Random values (such as 0x05A) before starting simulation to find problem
Peripheral Hardware Not Active	1	Pin Programming not Correct	1a	Check Register Access Prerequisites
			1b	Check TRIS registers For Values which Prevent Peripheral Operation
	2	Incorrect Part Number	2a	Check to see if the Part Being Actually Used has hardware
			2b	Check to see that Part Being used Has Hardware Registers

...That Match Source

No Output Mode For I/O Pin

Possible Cause	Corrective Action
1 Pin NOT in Output Mode	1a Check Values Saved in TRIS Registers
	1b Check for inadvertently Execution
	1c "Float" Pin (disconnect from Circuit) to see if Pin is Actually in Output Mode with a Logic Probe
2 If Peripheral Hardware Built into Pin, Check for Activation	2a This problem may NOT be apparent in MPLAB Simulation because Peripherals are often not Modeled
	2b If Pin Shared with "TOCKI" in 12C5xx or 16C505, check for Correct State of "OPTION" Register
3 If Dual-Use Pin Check for Output Capability	3a If Pin Shared with "MCLR" in 12C5xx or 16C505, then no output Capabilities are built in
	3b If Pin Shared with "TOCKI" in 12C5xx or 16C505, check for Correct State of "OPTION" Register

Output Pin Not Changing State

Possible Cause	Corrective Action
1 Incorrect TRIS Specification	1a "Float" Pin (disconnect from Circuit) to see if Pin is Actually in Output Mode with a Logic Probe
	1b Check Causes for "No Output" Mode for I/O Pin
3 Hardware Switching too fast to observe	3a Check the Hardware using a Logic Probe or an Oscilloscope

Debugging Checklist (*Continued*)

Problem	Potential Causes	Check
	2 Pin being held by high Current Source/Sink	Check 2a "Float" Pin and see if State Changes are Possible with Pin Disconnected 2b Look for Shorts to Vcc/Gnd 2c Look for Missing/Incorrect Resistors or Components
	3 Output Changing State too Quickly to be Observed	3a Check output with a Logic Probe or an Oscilloscope
Pin Changes State Unexpectedly	1 Look for "bcf", "bsf" or "movf"/"movwf" Instruction Combinations that May Reset the Pin	1a Check Value written to I/O Port 1b Check Computed Values that are used to Modify Pin Values 1c Look for Saved Port Values that are Incorrect or Inappropriate
	2 Look for Hardware that "Backdrives" the Pin	2a "Float" Pin (disconnect from Circuit) to see if state is incorrect 2b Check Output Enable pins of Tri-State Drivers on the Pin's Net
	3 Variable Address Overlayed onto a Hardware I/O Register	3a Make Sure that the Variable "CBLOCK" s=atement is in the File Register area of the PICmicro® MCU and not in the Hardware I/O Area

Problem			Action
Output Timing Not as Expected	1	Delay Calculations Incorrect	1a Check to see if the Calculations match the Actual Output
			1b Use the Assembler Calculator to Calculate Delays and match to Developer Values
	2	Interrupt Handler Active during Timed Output	2a Check for Enabled Interrupts
			2b Put "bcf INTCON, GIE" Before Timed code and "bsf INTCON, GIE" after
	3	Check Instruction Timings	3a Note that "goto", "call", "return" and PCL Modifications require Two Instruction Cycles
Register Values Incorrect/Change Unexpectedly	1	Check for Interrupt Handler Active	1a Look for Instances in the Interrupt Handler when Register is Changed
			1b Mask Interrupt Handler During Critical Periods Of Register Operation
			1c Use another Register in the Interrupt Handler and update Mainline's version as appropriate
	2	Make sure Variables are not located in Hardware Register Space	2a Check actual Register Address from Listing File to Hardware Register Addresses
			2b Make sure Variables are in memory space "Above" the Hardware Registers for all PICmicro® MCU Family Devices the Application Runs on

Debugging Checklist (Continued)

Problem	Potential Causes		Check	
	3	Variable Registers in "shadowed" Memory Space	3a	Check File Register Addresses with "Shadowed" Registers
			3b	Mark unused "Shadow" Registers as "BADRAM"
LED Not Lighting	1	LED Polarity Incorrect	1a	Short PICmicro® MCU Pin to Ground to ensure LED Can Light
	2	Check PICmicro® MCU Pin for NOT Changing to Output	2a	Check PICmicro® MCU Pin as Specified Above
	3	PWM Active with setting that turns off LED	3a	Check PWM Output with an Oscilloscope or Logic Probe
			3b	Check for PWM Code Active
	4	PICmicro® MCU Not Working	4a	Check the PICmicro® MCU as Specified Above
	5	Output Changing too fast to Observe	5a	Check the output using a Logic Probe or an Oscilloscope
Button: No Response	1	Pin Pullup/Pull Down Incorrect	1a	Check the Wiring of the Button to the PICmicro® MCU, Vcc and Gnd
			1b	Check the operation of the PICmicro® MCU's Internal Pull Ups
	2	Pin in Output Mode	2a	Check to make sure the PICmicro® MCU I/O Pin is in "Input" Mode
			2b	Look for inadvertent Changes to the "TRIS" Register

Problem	Possible Cause		Action
Button: Strange Response	3	Output Changing to quickly to be Observed	3a Check Pin output using a Logic Probe or an Oscilloscope
	1	Poor Debounce	1a Check for Multiple Button Presses Recognized by Software
			1b Check Voltage Levels on Hardware to Ensure Button Press is within 0.2 Volts from Vcc or Gnd
	2	Interrupt Handler Response Incorrect	2a Check the Interrupt Handler's Operation with The Input Conditions
LCD: No Output	1	Check Wiring	1a Check Ground is on Pin 1
			1b Check Data Pins
			1c Make Sure "R/W" line is held low during Writes
	2	Check Contrast	2a Contrast Different for Different LCDs
	3	Check Timing	3a Make Sure that LCD "E" Strobes are a minimum of 450 nsecs in width
			3b Make Sure signals do not Change during "E" Strobes

PICmicro® MCU Application Software Development Tools

Microsoft Compatible Editor "Ctrl" Key Combinations

```
Keystrokes          Operation
Up Arrow            Move Cursor up one Line
Down Arrow          Move Cursor down one Line
Left Arrow          Move Cursor left one Character
Right Arrow         Move Cursor right on Arrow
Page Up             Move viewed Window Up
Page Down           Move viewed Window Down
Ctrl - Left Arrow   Jump to Start of Word
```

```
Ctrl - Right Arrow    Jump to Start of next Word
Ctrl - Page Up        Move Cursor to Top of viewed
                        Window
Ctrl - Page Down      Move Cursor to Bottom of viewed
                        Window
Home                  Move Cursor to Start of Line
End                   Move Cursor to End of Line
Ctrl - Home           Jump to Start of File
Ctrl - End            Jump to End of File
Shift - Left Arrow    Increase the Marked Block by one
                        character to the left
Shift - Right Arrow   Increase the Marked Block by one
                        character to the right
Shift - Up Arrow      Increase the Marked Block by one
                        line up
Shift - Down Arrow    Increase the Marked Block by one
                        line down
Ctrl Shift -          Increase the Marked Block by one
  Left Arrow            word to the left
Ctrl Shift -          Increase the Marked Block by one
  Right Arrow           word to the Right
```

MPSIM.INI

A typical MPSIM.INI File for an application is

```
;  MPSIM File for PROG2 - Turning on an LED
;
;  Myke Predko - 96.05.20
;
P 84               ;  Use a 16C84
SR X               ;  Hex Numbers in the Simulator
ZR                 ;  Zero the Registers
RE                 ;  Reset Elapsed Time and Step
                        Count
DW D               ;  Disable the WDT
V W,X,2            ;  Display: the "W" Register
AD F3,B,8          ;    Status Register
AD F4,X,2          ;    FSR Register
AD OPT,X,2         ;    Option Register
AD FB,B,8          ;    INTCON Register
AD F2,X,3          ;    PCL Register
AD FA,X,3          ;    PCLATH Register
AD F1,X,2          ;    TMR0 Register
```

```
AD IOA,X,2              ;   Port "A" Tris Register
AD F5,X,2               ;   Port "A" Register
AD IOB,X,2              ;   Port "B" Tris Register
AD F6,X,2               ;   Port "B" Register
AD FC,X,2               ;   "Test" Register
rs
sc 4                    ;   Set the Clock to 1MHz
lo prog2
di 0,0                  ;   Display the First Instruction
```

MPLAB

MPLAB is a complete "Integrated Development
Environment" ("IDE") for all the different PICmicro®
MCU architecture families that runs under Microsoft's
"Windows" version 3.1x or later operating systems.
MPLAB integrates the different operations of develop-
ing a PICmicro® MCU application. This is done from a
user configurable "desk top" (see Fig. 15.1) with differ-
ent capabilities built into the program.

MPLAB can integrate the following different functions:

- editor
- assemblers
- compilers
- linkers
- programmers
- emulators

The following files are accessed by MPLAB:

File Extension	Function
.asm	Application Source File
.$$$	Backup of the Application Source File
.cod	"Label Reference" for MPLAB Simulator/Emulator

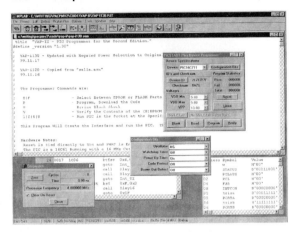

Figure 15.1 "MPLAB" IDE with PICStart Plus Interface

```
.err                Error Summary File
.lst                Listing File
.hex                Hex File to be loaded into the
                      PICmicro® MCU
.bkx                Backup of the Hex File
.pjt                Project file
```

MPLAB has the capability of displaying specific register and bit contents in the PICmicro® MCU. These windows, such as the one shown in Fig. 15.2 allow you to select the registers to monitor. To define a Watch Window or add more registers to it, the "Register Selection" Window is brought up for you to select the registers you would like to monitor. The "Properties" Window is selected from the "Register Selection" Window (as is shown in Fig. 15.3) to specify the characteristics of the register that is displayed.

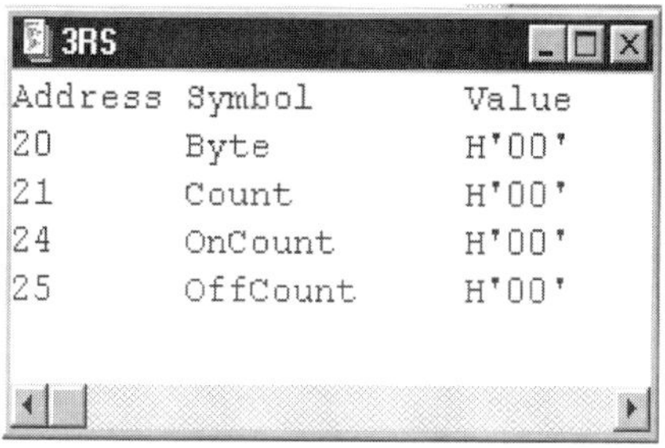

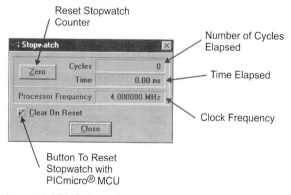

Figure 15.2 Sample "Watch Window"

The most basic input method is the "Asynchronous Stimulus" window (shown in Fig. 15.4) which consists of a set of buttons that can be programmed to drive any of the simulated PICmicro® MCU pins and can set the button to change the pin by:

Figure 15.3 MPLAB stopwatch

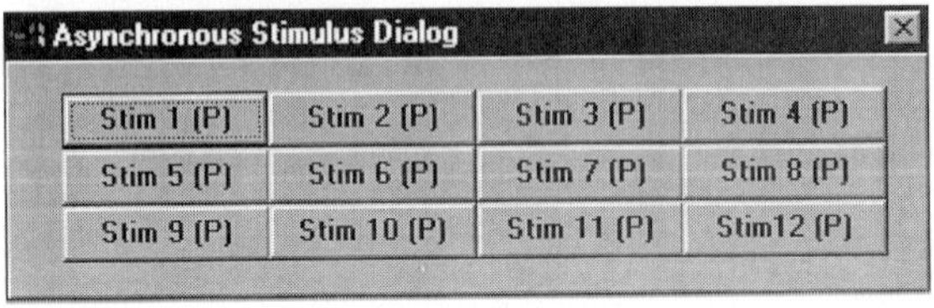

Figure 15.4 "Asynchronous Stimulus" Window

Pulse

Low

High

Toggle

The "Pulse" option, pulses the input pin to the complemented state and then back to the original state within one instruction cycle. This mode is useful for clocking TMR0 or requesting an external interrupt. Setting the pin "High" or "Low" will drive the set value onto the pin. To change the value of the pin between the two states, you can program two buttons in parallel with each other and each button changes the state. This can also be done with a single "Toggle" button, which changes the input state each time the button is pressed.

Clocks can be input into the simulated PICmicro® MCU by clicking on the "Debug" pull down, "Simulator Stimulus" and "Clock Stimulus . . ." selections. The clock stimulus dialog box (Fig. 15.5) can input regular clocks into a PICmicro® MCU by selecting the pin and then the "High" and "Low" time of the clock along with whether or not the clock is inverted (which means at reset, the clock will be low rather than high). The clock counts in Fig. 15.5 are in instruction cycles. Clock stim-

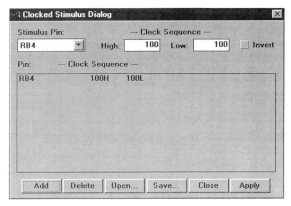

Figure 15.5 MPLAB "Clocked Stimulus"

ulus can be used for simple I/O tests, but it is really best suited for putting in repeating inputs that drive clocks or interrupts.

The "Register Stimulus" feature will store a two-digit hex value in a specified register every time a specific address is encountered in the simulated application execution. To load the operating parameters of the Register Stimulus method, the "debug" pull down is clicked, followed by "Simulator Stimulus" and then "Register Stimulus" is "Enabled". This brings up the small window shown in Fig. 15.6 on which you will select the address of the register to change as well as the address that this happens at. Once the addresses have been specified, the register stimulus file is selected by clicking on "Browse . . .". The register "Modify" Window (shown in Fig. 15.7) is available

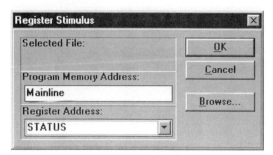

Figure 15.6 MPLAB "Register Stimulus" Specification

by clicking on the "Window" pull down and then selecting "Modify . . .". This window can access any register
in the simulated device, including "w", which cannot
be directly addressed in the low-end and mid-range
devices.

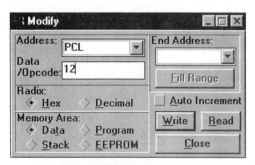

Figure 15.7 MPLAB "Register Modify" Window

Stimulus (.STI) files

Stimulus Files require a clock "Step" specification along with the pins to be driven. Comments in the file are preceded by a "!" character. The file below is a sample stimulus file (which, by convention, always ends in ".sti"):

```
!
!  Sample Stimulus File
!
Step    MCLR    RB4     !  Define the Bits to be
                              Controlled
   1       1      1     !  Initialize the Bit Values
!  Wait for the Program and Hardware to be
     Initialized
  500      0      1     !  Reset the PICmicro® MCU
 1000      1      1
 1500      1      0     ! Change the State of the Port
                              Bit
 2000      1      1     !  Restore it for rest of
                              program
```

Stimulus Files are the recommended method to simulate an application and understand what are the potential software problems.

MPLAB assembler directives

Directive	Usage Example	Comments
__BADRAM	__BADRAM Start, End	Flag a range of file registers which are unimplemented
BANKISEL	BANKISEL <label>	Update the "IRP" bit of the "STATUS" register before the "FSR" register is used to access a register indirectly. This directive is normally used with linked source files.
BANKSEL	BANKSEL Label	Update the "RPx" bits of the "STATUS" register before accessing a file register directly. This directive is not available for the low-end devices (for these devices, the FSR register should be used to access the specific address indirectly). This directive is also not available for the High-end PICmicro® MCUs which should use the "movlb" instruction.
CBLOCK	CBLOCK Address Var1, Var2 VarA:2 ENDC	Used to Define a starting address for variables or constants which require increasing values. To declare multiple byte variables or constants which increment by more than one, a colon (":") is placed after the label and before the number to increment by. This is shown for "VarA" in the usage example. The "ENDC" directive is required to "turn off" CBLOCK operation.
CODE	CODE [Address]	Used with an object file to define the start of application code in the source file. A "Label" can be

specified before the directive to give a specific label to the object file block of code. If no "Address" is specified, then MPLINK will calculate the appropriate address for the CODE statement and the instructions that follow it.

This directive is used to set the PICmicro® MCU's configuration bits to a specific value. "__CONFIG" automatically sets the correct address for the specific PICmicro® MCU. The "Value" is made up of constants declared in the PICmicro® MCU's ".inc" file.

Define a Constant Using one of the three formatting methods shown in usage example. The constant "Value" references to the "Label" and is evaluated when the "Label" is defined. For replacing a Label with a string, use "#DEFINE".

Set program memory words with the specified data values. If a "string" is defined, then each byte is put into its own word. The "DW" directive is recommended to be used instead of "DATA" or "DB" because its operation is less ambiguous when it comes to how the data is stored. Note that "DATA"/"DB"/"DW" do not store the data according as part of a "retlw" instruction. For the "retlw" instruction to be included with the data, the "DT" directive must be used. These directives are best suited for use in Serial EEPROM Source Files.

__CONFIG	__CONFIG Value
CONSTANT/ =/EQU	CONSTANT Label = Value or Value or Label = Value or Label EQU Value
DA/DATA/DB	DA Value|"string" or DATA Value|"string" DB Value|"string"

527

Directive	Usage Example	Comments
DE	ORG 0x02100 DE Value|'string"	This instruction is used to save initialization data for the PICmicro® MCU's built in Data EEPROM. Note that an "org 0x02100" statement has to precede the "de" directive to ensure that the PICmicro® MCU's program counter will be at the correct address for programming.
#DEFINE	#DEFINE Label [string]	Specify that any time "Label" is encountered, it is replaced by the string. Note that string is optional and the defined "Label" can be used for conditional assembly. If "Label" is to be replaced by a constant, then one of the "CONSTANT" declarations should be used. This directive is placed in the first column of the source file.
DT	DT Value [,Value...]|"string"	Place the "Value" in a "retlw" statement. If DT's parameter is part of a "string", then each byte of the string is given its own "retlw" statement. This directive is used for implementing read-only tables in the PICmicro® MCU.
DW	DW Value[,Value...]	Reserve program memory for the specified "Value". This value will be placed in a full program memory word.
ELSE		Used in conjunction with "IF", "IFDEF" or "IFNDEF" to provide an alternative path for conditional assembly. Look at these directives for examples of how "ELSE" is used.

528

END	END	End the program block. This directive is required at the end of all application source files.
ENDC		Used to end the "CBLOCK" Label constant value saving and updating. See "CBLOCK" for an example of how this directive is used.
ENDIF		Used to end an "if" statement conditional code block. See "IF", "IFDEF" or "IFNDEF" for an example of how this directive is used.
ENDM		Used to end the "MACRO" Definition. See "CBLOCK" for an example of how this directive is used.
ENDW		Used to end the block of code repeated by the "WHILE" Conditional Loop instruction. See "WHILE" for an example of how this directive is used.
ERROR	ERROR "string"	Force an "ERROR" into the code with the "string" message inserted into the Listing/Error Files.
ERRORLEVEL	ERRORLEVEL 0|1|2, +#|-#	Change the assembler's response to the specific "Error" ("2"), "Warning" ("1") or Message ("0") Number ("#"). Specifying "-" before the Number will cause any occurrences of the Error, Warning or Message to be ignored by the assembler and not reported. Specifying "+" before the Number will cause any occurrences of the Error Warning or Message to be output by the Assembler.
EXITM		For use within a MACRO to force the stopping of the MACRO expansion. Using this directive is not

Directive	Usage Example	Comments
		recommended except in the case where the MACRO's execution is in error and should rot continue until the error has been fixed. Using "EXITM" in the body of the MACRO could result in "Phase Errors" which can be very hard to find.
EXPAND	EXPAND	Enable printing MACRO Expansions in the listing file after they have been disabled by the "NOEXPAND" directive. Printing of MACRO Expansions is the default in MPLAB.
EXTERN	EXTERN Label	Make a program memory Label in an object file available to other object files.
FILL	FILL Value, Count	Put in "Value" for "Count" words. If "Value" is surrounded by parenthesis, then an instruction can be put in (ie "(goto 0)"). In earlier versions of MPLAB, "Fill" did not have a "Count" parameter and replaced any program memory address that does not have an instruction assigned to it or areas that is not reserved (using "RES") with the "Value".
GLOBAL	GLOBAL Label	Specify a Label within an object file that can be accessed by other object files. "GLOBAL" is different from "EXTERN" as it can only be put into the source after the label is defined.
IDATA	IDATA [Address]	Used to specify a data area within an object file. If no "Address" is specified, then the assembler

	calculates the address. A Label can be used with IDATA for referencing it.
__IDLOCS	__IDLOCS Value	Set the four ID Locations of the PICmicro® MCU with the four nybbles of "Value". This directive is not available for the 17Cxx devices.

IF	IF Parm1 COND Parm2
	; "True" Code
	ELSE
	; "False" Code
	ENDIF
		If "Parm1 COND Parm2" is "true", then insert and assemble the "True" code. Else, insert and assemble the "False" code. The "Else" directive and "False" codes are optional.

IFDEF	IFDEF Label
	; "True" Code
	ELSE
	; "False" Code
	ENDIF
		If the Label has been defined (using "#DEFINE"), then insert and assemble the "True" code. Else, insert and assemble the "False" code. The "Else" directive and "False" codes are optional.

IFNDEF	IFNDEF Label
	; "True" Code
	ELSE
	; "False" Code
	ENDIF
		If the Label has NOT been defined (using "#DEFINE"), then insert and assemble the "True" code. Else, insert and assemble the "False" code. The "Else" directive and "False" codes are optional.

INCLUDE	INCLUDE "FileName.Ext"	Load "FileName.Ext" at the current location within the source code.
LIST	LIST option[, ...]	Define the assembler options for the source file. The available options are:

Option   Default   Description
b=nnn   8   Set tab spaces.
c=nnn   132   Set column width.

531

Directive	Usage Example	Comments	Option	Default	Description
			f=format	INHX8M	Set the hex file output.
			free	FIXED	Use free-format parser.
			fixed	FIXED	Use fixed-format parser.
			mm=ON|OFF	ON	Print memory map in list file.
			n=nnn	60	Set lines per page.
			p=type	None	Set PICmicro® MCU type.
			r=radix	HEX	Set default radix (HEX, DEC, or OCT available).
			st=ON|OFF	ON	Print symbol table in list file.
			t=ON|OFF	OFF	Truncate lines of listing.
			w=0|1|2	0	Set the message level.
			x=ON|OFF	ON	Turn macro expansion on or off.
LOCAL	Fillup MACRO Size Local i i = 0	Define a Variable that is local to a MACRO and cannot be accessed outside of the MACRO.			

```
WHILE (i < Size)
DW 0x015AA
i = i + 1
ENDW
ENDM
```

Directive	Description

MACRO
```Label MACRO```
``` [Parm[, ...]]```
``` bsf Parm, 0```
``` ENDM```
	Define a block of code that will replace the "Label" every time it is encountered. The optional Parameters will replace the parameters in the Macro itself.

__MAXRAM ```__MAXRAM End```	Define the last File Register Address in a PICmicro® MCU that can be used.
MESSG ```MESSG "string"```	Cause "String" to be inserted into the source file at the "MESSG" statement. No errors or warnings are generated for this instruction.
NOEXPAND ```NOEXPAND```	Turn off Macro expansion in the Listing File.
NOLIST ```NOLIST```	Turn off Source Code Listing Output in Listing File.
ORG ```ORG Address```	Set the Starting Address for the following code to be placed at
PAGE ```PAGE```	Insert a Page Break before the "PAGE" directive.

PAGESEL
```PAGESEL Label```
```goto Label```
	Insert the Instruction Page of a Label before jumping to that Label or calling the subroutine at it.

Directive	Usage Example	Comments
PROCESSOR	PROCESSOR type	This directive is available for commonality with earlier Microchip PICmicro® MCU assemblers. The Processor option of the "LIST" directive should be used instead.
RADIX	RADIX Radix	This directive is available for commonality with earlier Microchip PICmicro® MCU assemblers. Available options are "HEX", "DEC" and "OCT". The default radix should be selected in the "LIST" directive instead.
RES	RES MemorySize	Reserve a block of program memory in an object file for use by another. A label may be placed before the RES directive to save what the value is.
SET	Label SET Value	"SET" is similar to the "CONSTANT", "EQU" and "=" directives, except that the "Label" can be changed later in the code with another "SET" directive statement.
SPACE	SPACE Value	Insert a Set number of blank lines into a listing file.
SUBTITLE	SUBTITLE "string"	Insert "string" on the line following the "TITLE" string on each page of a listing file.
TITLE	TITLE "string"	Insert "string" on the top line on each page of a listing file.

UDATA	UDATA [Address] Label1 RES 1 Label2 RES 2	Declare the beginning of an uninitialized data section. "RES" labels should follow to mark variables in the uninitialized data space. This command is designed for serial EEPROMs.
UDATA_ACS	UDATA_ACS [Address] Label1 RES 1 Label2 RES 2	Declare the beginning of an uninitialized data section in a 18Cxx P=Cmicro® MCU. "RES" labels should follow to mark variables in the uninitialized data space.
UDATA_OVR	UDATA_OVR [Address] Label1 RES 1 Label2 RES 2	Declare the beginning of an uninitialized data section that can be overwritten by other files (as an "Overlay"). "RES" labels should follow to mark variables in the uninitialized data space. This command is designed for serial EEPROMs.
UDATA_SHR	UDATA_SHR [Address] Label1 RES 1	Declare the beginning of data memory that is "shared" across all the register banks.
#UNDEFINE	#UNDEFINE Label	Delete a Label that was "#DEFINED".
VARIABLE	VARIABLE Label [= Value]	Declare an assembly-time variable that can be updated within the code using a simple assignment statement.
WHILE	WHILE Parm1 COND Parm2 ; while "True" ENDW	Execute code within the "WHILE"/"ENDW" directives while the "Parm1 COND Parm2" test is true. Note that in the listing file, the code will appear as if the code within the "WHILE"/"WEND" directives was repeated a number of times.

Standard Declaration and Include (".inc") Files

The Microchip developed standard "include" files are recommended to be included into source code rather than using developer supplied register definitions. There is an ".inc" file for every PICmicro® MCU part number in the format:

```
p<I>PICmicro® MCU</I>.inc
```

where "*PICmicro® MCU*" is the PICmicro® MCU part number.

Linking–Linked Applications

Before creating the linked application object, the source files to be linked together have to be created. Once the files are created, "links" are created to addresses that have to be accessed between the different files. It is important to remember that variables as well as instruction addresses have to be common.

The following two example source code files show how addresses are linked together. The first is the "mainline":

```
 TITLE - Test3 - Jump to Test3A
;
;  Example Application using the MPLAB Linker
;
;  Myke Predko
;  2000.02.02
;
;  Hardware Notes:
;  PIC16F84 running in a Simulator
;
```

```
    LIST       R=DEC                    ; list directive to
                                          define processor
    #include "p16F84.inc"               ; processor specific
                                          variable
definitions

    __CONFIG _CP_OFF & _WDT_ON & _PWRTE_ON & _XT_OSC

    EXTERN TEST3A                       ; Specify Mainline
                                          Location
    GLOBAL TEST3AStart
    GLOBAL flag                         ; Variable passed to
                                          Linked File

;***** VARIABLE DEFINITIONS (examples)

; example of using Uninitialized Data Section
INT_VAR         UDATA      0x0C
w_temp          RES        1            ; variable used for
                                          context saving
status_temp     RES        1            ; variable used for
                                          context saving
flag            RES        2            ; temporary variable
                                          (shared locations
                                          - G_DATA)
;************************************************************
RESET_VECTOR CODE      0x000            ; processor reset
                                          vector
                 goto      start        ; go to beginning of
                                          program

INT_VECTOR    CODE     0x004            ; interrupt vector
                                          location
                 movwf w_temp
PROG CODE 0x005
                 movf      STATUS, w
                 movwf     status_temp

; isr code can go here or be located as a call
  subroutine elsewhere

                 movf      status_temp, w  ; Restore
                                             Context
                                             Registers

                 movwf     STATUS
                 swapf     w_temp, f
```

```
                swapf    w_temp, w
                retfie

start
      goto   TEST3A

TEST3AStart

              END                   , directive 'end
                                      of program'
```

The second is the file that is linked to the "mainline":

```
 TITLE "Test3A - Actually Execute the Code"
 ;
 ;  Example Application using the MPLAB Linker
 ;
 ;  Myke Predko
 ;  2000.02.02
 ;
 ;  Hardware Notes:
 ;  PIC16F84 running in a Simulator
 ;

 EXTERN flag                    ;  External Values
                                   Linked into Code
 EXTERN TEST3AStart

 list r=dec
 #include "p16f84.inc"

TEST3ACODE CODE
TEST3A
 GLOBAL TEST3A                  ;  Address to Pass to
                                   Linked File
        banksel flag            ; example
        clrf flag               ; example
; remaining code goes here

 movlw 77
 movwf flag
 movlw 0x001
 subwf flag, f
 btfss STATUS, Z
  goto $ - 2
```

```
    goto $                          ; Loop Here Forever

                END                     ; directive
                                          'end of
                                          program'
```

Application Code Template

The following file should be used as a mid-range
PICmicro® MCU application source code template.

```
   title "FileName - One Line Description"
#define _version "x.xx"
 ;
 ; Update History:
 ;
 ; Application Description/Comments
 ;
 ; Author
 ,
 ; Hardware Notes:
 ;
  LIST R=DEC                ; Device Specification
   INCLUDE "p16cxx.inc"     ; Include Files/Registers

 ; Variable Register Declarations

 ; Macros

__CONFIG _CP_OFF & _XT_OSC & _PWRTE_ON & _WDT_OFF &
   _BODEN_OFF

 org     0
Mainline

 goto    Mainline_Code

 org     4                  ; Interrupt Handler at
                               Address 4
Int

MainLine_Code

 ;   Subroutines

  end
```

The BASIC Language

BASIC variables do not have to be declared except in specialized cases. The variable name itself follows normal conventions of a letter or "_" character as the first character, followed by alphanumeric characters and "_" for variable names. Variable (and Address "Label") names may be case sensitive, depending on the version.

To Specify Data Types, a "Suffix" character is added to the end of the Variable name:

```
Suffix              Function
$                   String Data
%                   Integer
&                   Long Integer (32 Bits) - Microsoft
                       BASIC Extension
!                   Single Precision (32 Bits) -
                       Microsoft BASIC Extension
#                   Double Precision (64 Bits) -
                       Microsoft BASIC Extension
```

The following table lists the different BASIC functions:

```
Statement           Function
BASE                Starting Array Element
DATA                Data Block Header
DIM                 Dimension Array Declaration
OPTION              Starting Array Element
LET                 Assignment Statement (Not
                       Mandatory)
RANDOMIZE           Reset Random Number "Seed"
INPUT [Prompt ,] Variables
                    Get Terminal Input
PRINT               Output to a Terminal
?                   Output to a Terminal
READ                Get "Data" Information
GOTO                Jump to Line Number/Label
GOSUB               Call Subroutine at Line
                       Number/Label
RETURN              Return to Caller from Subroutine
```

```
IF Condition [THEN] Statement
                    Conditionally Execute the
                    "Statement"
FOR Variable = Init TO Last [STEP Inc] ... NEXT
                    [Variable] Loop Specified Number
                    of Times
ON Event GOTO       On an Event, Jump to Line
                    Number/Label
RESTORE             Restore the "DATA" Pointer
STOP                Stop Program Execution
END                 End Program Execution
`                   Comment - Everything to the Right
                    is Ignored
REM                 Comment - Everything to the Right
                    is Ignored
ABS                 Get Absolute Value of a Number
SGN                 Return the Sign of a Number
COS                 Return Cosine of an Angle (input
                    usually in Radians)
SIN                 Return Sine of an Angle (input
                    usually in Radians)
CIN                 Return Tangent of an Angle (input
                    usually in Radians)
ATN                 Return the Arc Tangent of a Ratio
INT                 Convert Real Number to Integer
SQR                 Return the Square Root of a Number
EXP                 Return the Power of e for the
                    input
LOG                 Return the Natural Logarithm for
                    the Input
RND                 Return a Random Number
TAB                 Set Tab Columns on Printer
```

For assignment and "if" statements, the following operators are available in BASIC:

```
Operator            Operation
+                   Addition
-                   Subtraction
*                   Multiplication
/                   Division
^                   Exponentiation
"                   Start/End of Text String
,                   Separator
```

```
;                    Print Concatenation
$                    String Variable Identifier
=                    Assignment/Equals To Test
<                    Less than
<=                   Less than or Equals To
>                    Greater than
>=                   Greater than or Equals To
<>                   Not Equals
```

BASIC's Order of Operations is quite standard for programming languages:

```
Operators            Priority   Type
Functions                       Expression Evaluation
= <> < <= > >=       Highest    Conditional Tests
^                               Exponentiation
* /                             Multiplication/Division
+ -                  Lowest     Addition/Subtraction
```

Microsoft BASIC Enhancements

The following functions are available in Microsoft versions of BASIC for the PC as well as some BASICs for the
PICmicro® MCU:

```
Function             Operation
AND                  AND Logical Results
OR                   OR Logical Results
XOR                  XOR Logical Results
EQV                  Test Equivalence of Logical
                       Results
IMP                  Test Implication of Logical
                       Results
MOD                  Get the Modulus (remainder)
                       of an Integer Division
FIX                  Convert a Floating Point
                       Number to Integer
DEFSTR Variable      Define the Variable as a
                       String (instead of the
                       "DIM" Statement)
DEFINT Variable      Define the Variable as an
                       Integer (instead of the
                       "DIM" Statement)
```

```
DEFLNG Variable              Define the Variable as a
                             "long" Integer (instead of
                             the "DIM" Statement)
DEFSNG Variable              Define the Variable as a
                             Single Precision Floating
                             Point Number (instead of
                             the "DIM" Statement)
DEFDBL Variable              Define the Variable as a
                             Double Precision Floating
                             Point Number (without using
                             the "DIM" Statement)
REDIM Variable([low TO] High[, [low TO] High...])
                             [AS Type]
                             Redefine a Variable
ERASE                        Erase an Array Variable from
                             Memory
LBOUND                       Return the First Index of an
                             Array Variable
UBOUND                       Return the Last Index of an
                             Array Variable
CONST Variable = Value   Define a Constant Value
DECLARE Function | Subroutine
                             Declare a Subroutine/
                             Function Prototype at
                             Program Start
DEF FNFunction( Arg[, Arg...])
                             Define a Function
                             ("FNFunction") that returns
                             a Value. If a Single Line,
                             then "END DEF" is not
                             required
END DEF                      End the Function Definition
FUNCTION Function( Arg[, Arg...])
                             Define a Function. Same
                             Operation, Different Syntax
                             as "DEF FNFunction"
END FUNCTION                 End a Function Declaration
SUB Subroutine( Arg[, Arg...])
                             Define a "Subroutine" which
                             does not return a Value. If
                             a Single Line, then "END
                             DEF" is not required
END SUB                      End the Subroutine
                             Definition
DATA Value[, Value...]   Specify File Data
```

```
READ Variable[, Variable...]
                        Read from the "Data" File
                        Data
IF Condition THEN Statements ELSE Statements END IF
                        Perform a Structured
                        If/Else/Endif
ELSEIF                  Perform a Condition
                        Test/Structured
                        If/Else/Endif instead of
                        simply "Else"
ON ERROR GOTO Label     On Error Condition, Jump to
                        Handler
RESUME [Label]          Executed at the End of an
                        Error Handler. Can either
                        return to current location,
                        0 (Start of Application) or
                        a specific label
ERR                     Return the Current Error
                        Number
ERL                     Return the Line the Error
                        Occurred at
ERROR #                 Execute an Application-
                        Specific Error (Number "#")
DO WHILE Condition Statements LOOP
                        Execute "Statements" while
                        "Condition" is True
DO Statements LOOP WHILE Condition
                        Execute "Statements" while
                        "Condition" is True
DO Statements LOOP UNTIL Condition
                        Execute "Statements" until
                        "Condition" is True
EXIT                    Exit Executing "FOR",
                        "WHILE" and "UNTIL" Loops
                        without executing Check
SELECT Variable         Execute based on "Value"
                        "CASE" Statements used to
                        Test the Value and Execute
                        Conditionally
CASE Value              Execute within a "SELECT"
                        Statement if the "Variable"
                        Equals "Value". "CASE ELSE"
                        is the Default Case
END SELECT              End the "SELECT" Statement
LINE INPUT              Get Formatted Input from the
                        User
```

INPUT$(#)	Get the Specified Number ("#") of Characters from the User
INKEY$	Check Keyboard and Return Pending Characters or Zero
ASC	Convert the Character into an Integer ASCII Code
CHR$	Convert the Integer ASCII Code into a Character
VAR	Convert the String into an Integer Number
STR$	Convert the Integer Number into a String
LEFT$(String, #)	Return the Specified Number ("#") of Left Most Characters in "String"
RIGHT$(String, #)	Return the Specified Number ("#") of Right Most Characters in "String"
MID$(String, Start, #)	Return/Overwrite the Specified Number ("#") of Characters at Position "Start" in "String"
SPACE$(#)	Returns a String of the Specified Number ("#") of ASCII Blanks
LTRIM$	Remove the Leading Blanks from a String
RTRIM$	Remove the Trailing Blanks from a String
INSTR(String, SubString)	Return the Position of "SubString" in "String"
UCASE$	Convert all the Lower Case Characters in a String to Upper Case
LCASE$	Convert all the Upper Case Characters in a String to Upper Case
LEN	Return the Length of a String
CLS	Clear the Screen
CSRLIN	Return the Current Line that the Cursor is On

```
POS                          Return the Current Column
                             that the Cursor is On
LOCATE X, Y                  Specify the Row/Column of
                             the Cursor (Top Left is
                             1,1)
SPC                          Move the Display the
                             Specified Number of Spaces
PRINT USING "Format"         Print the Value in the
                             Specified Format. "+", "#",
                             ".", "^" Characters are
                             used for number formats
SCREEN mode[, [Color] [,[Page] [,Visual]
                             Set the Screen Mode. "Color"
                             is 0 to display on a
                             "Color" display, 1 to
                             display on a
                             "Monochrome". "Page" is the Page that
                             receives I/O and "Visual"
                             is the Page that is
                             currently active.
COLOR [foreground] [,[background] [,border]]
                             Specify the Currently Active
                             Colors
PALETTE [attribute, color]
                             Change Color Assignments.
VIEW [[SCREEN] (x1,y1) - (x2,y2)[,[color]]
  [,border]]]
                             Create a small Graphics
                             Window known as a
                             "Viewport"
WINDOW [[SCREEN] (x1,y1) - (x2,y2)]
                             Specify the Viewport's
                             logical location on the
                             Display
PSET (x,y)[,color]           Put a Point on the Display
PRESET (x,y)                 Return the Point to the
                             Background Color
LINE (x1,y1)-(x2,y2)[,[Color][,[B|BF][,style]]]
                             Draw a Line between the two
                             specified points. If "B" or
                             "BF" specified, Draw a Box
                             ("BF" is "Filled")
CIRCLE (x,y), radius[,[color] [,[start] [,end]
  [,aspect]]]
                             Draw the Circle at center
                             location and with the
```

	specified "radius". "start" and "end" are starting and ending angles (in radians). "aspect" is the circle's aspect for drawing ellipses
DRAW CommandString	Draw an arbitrary Graphics Figure. There should be spaces between the commands Commands:
	U# - Moves Cursor up # Pixels
	D# - Moves Cursor down # Pixels
	E# - Moves Cursor up and to the right # Pixels
	F# - Moves Cursor down and to the right # Pixels
	G# - Moves Cursor down and to the Left # Pixels
	H# - Moves Cursor up and to the left # Pixels
	L# - Moves Cursor Left # Pixels
	R# - Moves Cursor Right # Pixels
	Mxy - Move the Cursor to the Specified x,y Position
	B - Turn Off Pixel Drawing
	N - Turns On Cursor and Move to Original Position
Position	A# - Rotate Shape in 90 Degree Increments
	C# - Set the Drawing Color
	P#Color#Border - Set the Shape Fill and Border Colors
	S# - Set the Drawing Scale
	T# - Rotates # Degrees
LPRINT	Send Output to the Printer
BEEP	"Beep" the Speaker
SOUND Frequency, Duration	
	Make the Specified Sound on the PC's Speaker
PLAY NoteString	Output the Specified String of "Notes" to the PC's Speaker

```
DATE$                    Return the Current Date
TIME$                    Return the Current Time
TIMER                    Return the Number of Seconds
                         since Midnight
NAME FileName AS NewFileName
                         Change the Name of a File
KILL FileName            Delete the File
FILES [FileName.Ext]     List the File (MS-DOS
                         "dir"). "FileName.Ext"
                         can contain "Wild
                         Cards"
OPEN FileName [FOR Access] AS #Handle
                         Open the File as the
                         Specified Handle (Starting
                         with the "#" Character).
                         Access:
                         I - Open for Text Input
                         O - Open for Text Output
                         A - Open to Append Text
                         B - File is Opened to Access
                         Single Bytes
                         R - Open to Read and Write
                         Structured Variables
CLOSE #Handle            Close the Specified File
RESET                    Close all Open Files
EOF                      Returns "True" if at the End
                         of a File
READ #Handle, Variable   Read Data from the File
GET #Handle, Variable    Read a Variable from the
                         File
INPUT #Handle, Variable  Read Formatted Data from the
                         File using "INPUT","INPUT
                         USING" and "INPUT$" Formats
WRITE #Handle, Variable  Write Data to the File
PUT #Handle, Variable    Write a Variable to a File
PRINT #Handle, Output    Write Data to the File using
                         the "PRINT" and "PRINT
                         USING" Formats
SEEK #Handle, Offset     Move the File Pointer to the
                         Specified Offset within the
                         File
```

PicBasic

microEngineering Labs, Inc.'s ("meLab's"), "PicBasic" is an excellent tool for learning about the PICmicro® MCU, before taking the big plunge into assembly language programming. The source code required by the compiler is similar to the Parallax Basic Stamp BS2's "PBASIC" with many improvements and changes to make the language easier to work with and support different PICmicro® MCUs.

PicBasic does not currently have the ability to link together multiple source files which means that multiple source files must be "included" in the overall source. Assembly language statements are inserted in line to the application. PicBasic produces either assembler source files or completed ".hex" files. It does not create object files for linking modules together.

For additional information and the latest device libraries, look at the microEngineering Labs, Inc., Web page at:

http://www.melabs.com/mel/home.htm

PicBasic Pro is an MS-DOS command line application that is invoked using the statement:

```
PBP[W] [options...] source
```

"Options" are compiler execution options and are listed in the table below:

```
Option         Function
-h/-?          Display the help screen. The help
                 screen is also displayed if no options
                 or source file name is specified
```

```
-ampasm        Use the MPASM Assembler and not the
               PicBasic Assembler
-c             Insert Comments into PicBasic Compiler
               produced Assembler Source File. Using
               this option is recommended if you are
               going to produce MPASM Assembler
               Source from PicBasic
-iPath         Specify a new directory "Path" to use
               for include files in PicBasic
-lLibrary      Specify a different library to use
               when compiling. Device specific
               libraries are provided by PicBasic
               when the processor is specified
-od            Generate a listing, symbol table and
               map files
-ol            Generate a listing file
-pPICmicro®    Specify the "PICmicro® MCU" that the
 MCU           source is to be compiled into. If
               this parameter is not specified, then
               a PIC16F84 is used as the processor.
               "PICmicro® MCU" is in the format:
               16F84, where the "PIC" at the start
               of the Microchip part number is not
               specified.
-s             Do not assemble the compiled code
-v             Turn on "Verbose Mode" which provides
               additional information when the
               application is compiled
```

PicBasic does assume a constant set of configuration values. For most PICmicro® MCUs the configuration fuses are set as listed in the table below:

```
Feature        PicBasic Setting
Code Protect   Off
Oscillator     XT - or Internal RC if 12Cxxx
WDT            On
PWRTE          Off
```

Each byte takes place in one of the words; for example, "b4" is the least significant byte of "w2". The 16-bit variables are defined as being a part of the 16-bits taken up by "w0" ("b0" and "b1"). This method works well, but care has to be taken to make sure that the overlapping variables are kept track of and not used incorrectly. The most common problem for new Basic Stamp developers is defining a variable on b0 and w0 and having problems when a write to one variable overwrites the other.

To provide these variables to the PicBasic application, the Basic Stamp variable declaration files are defined in the following two "include" files that are shown within "include" statements below. Only one of these statements can be used in an application.

```
include "bs1defs.bas"
include "bs2defs.bas"
```

A much better way of declaring variables is to use the "var" directive to select the different variables at the start of the application and let the PicBasic compiler determine where the variables belong and how they are accessed (i.e., put in different variable pages). Along with the "var" directive, the "word", "byte", and "bit" directives are used to specify the size of the variable. Some example variable declarations are

```
WordVariable var word    ' Declare a 16 Bit Variable
ByteVariable var byte    ' Declare an 8 Bit Variable
BitVariable var bit      ' Declare a single byte
                           Variable
```

Initial values for the variables cannot be made at the variable declarations.

Along with defining variables, the "var" directive can be used to define variable labels built out of previously defined variables to specify specific data. Using the variables above, I can break "WordVariable" up into a top half and bottom half and "ByteVariable" into specific bytes with the statements:

```
WordVariableTop        var WordVariable.byte1
WordVariableBottom     var WordVariable.byte0

BitVariableMSB         var BitVariable.bit7
BitVariableLSB         var BitVariable.0
```

Variables can also be defined over registers. When the PicBasic libraries are merged with the source code, the standard PICmicro® MCU register names are available within the application. Using this capability, labels within the application can be made to help make the application easier to work with. For example, to define the bits needed for an LCD, the declarations below could be used:

```
LCDData var PORTB      ' PORTB as the 8 Bits of Data
LCDE var PORTA.0       ' RA0 is "E" Strobe
LCDRS var PORTA.1      ' RA1 is Data/Instruction
                        Select
LCDRW var PORTA.2      ' RA2 is the Read/Write
                        Select Bit
```

When variables are defined using the "var" and "system" directives, specific addresses can be made in the application. For example, the statement:

```
int_w var byte $0C system
```

will define the variable "_w" at address 0x00C in the system. This reserves address 0x00C and does not allow its use by any other variables. The bank of a variable can be specified using the "system" directive like:

```
int_status var byte bank0 system
```

These two options to the "var" directive are useful when defining variables for interrupt handlers as discussed below.

Along with redefining variables with the "var" statement, PicBasic also has the "symbol" directive. The symbol directive provides the same capabilities as the "var" statement and it is provided simply for compatibility with the BS1. If you were only developing PicBasic applications, I would recommend only using "var" and avoiding the "symbol" directive. Single dimensional arrays can be defined within PicBasic for each of the three data types when the variable is declared.

```
WordArray var word[10]   ' Ten Word Array
ByteArray var byte[11]   ' Eleven Byte Array
BitArray var bit[12]     ' Twelve Bit Array
```

Note that bits can be handled as an array element. Depending on the PICmicro® MCU part number, the maximum array sizes are

Variable Type	Maximum Number of Elements
Word	32
Byte	64
Bit	128

As part of the "bit" definition, I/O port pins are predefined within PicBasic. Up to 16 pins (addressed using the "Pin#" format, where "#" is the pin number) can be accessed, although how they are accessed changes according to the PICmicro® MCU part number the application is designed for. The pins for different parts are defined as:

Number of Pins	Pins 0 - 7	Pins 8 - 15
8	GPIO	Mapped onto Pins 0 - 5
18	PORTB	PORTA

```
28 - Not PIC14C000   PORTB        PORTC
14C000               PORTC        PORTD
40 Pin               PORTB        PORTC
```

Note that not all the ports that have 8 pins are specified. For example, accessing "RA6" in an 18-pin device (which does not have an "RA6" bit) will not do anything.

Constants are declared in a similar manner to variables, but by using the "con" directive with a constant parameter:

```
SampleConstant con 3 + 7   ' Define a Sample Constant
```

Constant values can be in four different formats. The table below lists the different formats and the modifiers to indicate to the PicBasic compiler which data type is being specified:

```
Data Type   Modifier    Comments
Decimal     None        PicBasic Default is Decimal
Hex         $           "$" is placed before the
                        Number
Binary      %           "%" is placed before the
                        Number
ASCII       "           Double Quotes placed around
                        a Single Character
```

In the table above, note that only an ASCII byte can be passed within double quotes. Some instructions (described below) can be defined with strings of characters that are enclosed within double quotes.

The "define" statement is used to change constants given defaults within the PICmicro® MCU when a PicBasic compiled application is running. The format is

```
DEFINE Label NewValue
```

The labels, their default values, and their values are listed in the table below:

Define	Default	Optional Values	Comments
BUTTON_PAUSE	10	Any Positive Int	Button Debounce Delay in msecs
CHAR_PACING	1000	Any Positive Int	Time between SerOut Characters
DEBUG_BAUD	2400	Any	Specified Data Rate of Debug information
DEBUG_BIT	0	0 - 7	Output Pin for Debug Serial Output
DEBUG_MODE	1	0, 1	Polarity of Debug NRZ Output data. "0" Positive, "1" Inverted
DEBUG_PACING	1000	Any Positive Int	Time between Output Characters for DEBUG Statements
DEBUG_REG	PORTB	Any PORT	Port Debug Output Pin is Connected to
DEBUGIN_BIT	0	0 - 7	Input Pin for Debug Serial Output
DEBUGIN_MODE	1	0, 1	Polarity of Debug NRZ Input data. "0" Positive, "1" Inverted
DEBUGIN_REG	PORTB	Any PORT	Port Debug Input Pin is Connected to
HSER_BAUD	2400	Any	Hardware Serial Port's Data Rate
HSER_SPBRG	25	0 - 0xFF	Hardware Serial Port's SPBRG Register Value
HSER_RCSTA	0x090	0 - 0xFF	Hardware Serial Port's Initialization value for "RCSTA" register. Default set for Asynchronous Communications
HSER_TXSTA	0x020	0 - 0xFF	Hardware Serial Port's Initialization value for "TXSTA" register. Default set for Asynchronous Communications
HSER_EVEN	1	0, 1	Hardware Serial Port's Parity Select Values. Only used if Parity checking is desired
HSER_ODD	1	0, 1	Hardware Serial Port's Parity Select Values. Only used if Parity checking is desired

Define	Default	Optional Values	Comments
I2C_HOLD	1	0, 1	Stop I2C transmission while the SCL line is held low
I2C_INTERNAL	1	0, 1	Set to use Internal EEPROM in the 12Cexx PICmicro® MCUs
I2C_SCLOUT	1	0, 1	Use a Bipolar Driver instead of an Open-Drain I2C Clock Driver
I2C_SLOW	1	0, 1	Run the I2C at no more than 100 kbps data rate
LCD_BITS	4	4, 8	Number of Bits for LCD Interface
LCD_DBIT	0	0, 4	Specify the Data bit for LCD Data
LCD_DREG	PORTA	Any PORT	Select the LCD Data Port
LCD_EBIT	3	0 - 7	Specify the Bit for the LCD Clock
LCD_EREG	PORTB	Any PORT	Specify the Port for the LCD "E" Clock Bit
LCD_LINES	2	1, 2	Specify the Number of Lines on the LCD. Check Information on LCDs for how the multiple line specification is used in some single line LCDs.
LCD_RSBIT	4	Any PORT	LCD RS Bit Selection
LCD_RSREG	PORTA	Any PORT	LCD RS Bit Select Register
OSC	4	3, 4, 8, 10, 12, 16, 20	Specify PICmicro® MCU Operating Speed in MHz. Note "3" is actually 3.58
OSCCAL_1K	1	0, 1	Set OSCCAL for PIC12C672
OSCCAL_2K	1	0, 1	Set OSCCAL for PIC12C672
SER2_BITS	8	4 - 8	Specify Number of bits sent with "SERIN2" and "SEROUT2" instructions

The "OSC" define should be specified if serial I/O is going to be implemented in the PICmicro® MCU. This value is used by the compiler to calculate the time delays necessary for each bit.

Assembly language can be inserted at any point within a PicBasic application. Single instructions can be inserted by simply starting the line with a "@" character:

```
@bcf INTCON, T0IF       ; Reset T0IF Flag
```

Multiple lines of assembly language are prefaced by the "asm" statement and finished with the "endasm". An example of this is shown below:

```
asm
  movlw 8              ; Loop 8x
Loop
  bsf PORTA, 0         ; Pulse the Bit
  bcf PORTA, 0
  addlw $0FF           ; Subtract 1 from "w"
  btfss STATUS, Z      ; Do 8x
   goto Loop
endasm
```

Note that labels inside the assembler statements do not have a colon at the end of the string and that the traditional assembly language comment indicator (the semicolon [";"]) is used.

Implementing interrupt handlers in PicBasic can be done in one of two ways. The simplest way of implementing it is using the "ON INTERRUPT GOTO Label" statement. Using this statement, any time an interrupt request is received, the "Label" specified in the ON INTERRUPT statement will be executed until there is a "resume" instruction, which returns from an interrupt. Using this type of interrupt handler, straight PicBasic statements can be used and assembly language statements avoided.

The basic operation looks like:

```
    :

ON INTERRUPT GOTO IntHandler

    :

IntHandler:
    disable        ' Turn off interrupt and debug
                     requests
       :           ' Process Interrupt
    enable         ' Enable other Interrupts and debug
                   ' requests
    resume         ' Return to the executing code
```

The problem with this method is the interrupt handler is executed once the current instruction has completed. If a very long statement is being executed (say a string serial send), then the interrupt will not be serviced in a timely manner.

The best way of handling an interrupt is to add the interrupt handler as an assembly language routine. To reference the interrupt handler, the "define INTHAND Label" instruction is used to identify the label where the assembly language code is listed. The interrupt handler will be moved to start at address 0x004 in the mid-range devices.

A code template for generic mid-range PICmicro® MCU interrupt handlers is

```
    int_w var byte 0x020 system      ' Define the
                                         Context Save
Variables
    int_status var byte bank0 system
    int_fsr var byte bank0 system
    int_pclath byte bank0 system
    :
```

```
      define INTHAND IntHandler          '  Specify what the
                                             Interrupt
                                       '   Handler is
       :

'  Interrupt Handler - to be relocated to 0x00004
asm
IntHandler
   movwf    int_w                    ;  Save the Context
                                         Registers
   movf     STATUS, w
   bcf      STATUS, RP0              ;  Move to bank 0
   bcf      STATUS, RP1
   movwf    int_status
   movf     FSR, w
   movwf    int_fsr
   movf     PCLATH, w
   movwf    int_pclath
   clrf     PCLATH

;  #### - Execute Interrupt Handler Code Here

   movf int_pclath, w               ;  Finished,
                                        restore the
                                        Context
   movwf    PCLATH                   ;  Registers
   movf     int_fsr, w
   movwf    FSR
   movf     int_status, w
   movwf    STATUS
   swapf    int_w, f
   swapf    int_w, w
   retfie
endasm
```

In the interrupt template, note that the "worst case" condition context register save is presented.

Mathematical operators used in assignment statements and PicBasic instructions are very straightforward in PicBasic and work conventionally. In Basic Stamp PBASIC, you must remember that the operations execute from left to right. This means that the statement:

$$A = B + C * D$$

which would be expected to operate as:

1. Multiply "C" and "D"
2. Add the results from one to "B"

in Parallax PBASIC, returns the result of:

1. Get the Sum of "B" and "C"
2. Multiply the results from one with "D"

PicBasic does not follow the PBASIC evaluation convention and returns the "expected" result from complex statements like the one above. This means that in PicBasic, you do not have to break complex statements up into single operations, like you do in PBASIC, to avoid unexpected expression evaluation. If you are using a Basic Stamp to "prototype" PicBasic applications, then break up the complex statements and use the temporary values.

The mathematical operators used are listed in the table below along with their execution priority and parameters. All mathematical operators work with 16-bit values.

Priority	Operator	Operation
Lowest	Parm1 + Parm2	Return the Sum of "Parm1" and "Parm2"
	Parm1 - Parm2	Return the Result of "Parm2" Subtracted from "Parm1"
	Parm1 * Parm2	Return the least-significant sixteen bits of the product of "Parm1" and "Parm2". This is often referred to as Bytes 0 and 1 of the result
	Parm1 */ Parm2	Return the middle sixteen bits of the product of "Parm1" and "Parm2". This is often referred to as Bytes 1 and 2 of the result
	Parm1 ** Parm2	Return the most significant sixteen bits of the product of "Parm1" and "Parm2". This is often referred to as Bytes 2 and 3 of the result
	Parm1 / Parm2	Return the number of times Parm2 can be divided into Parm1 evenly
	Parm1 // Parm2	Return the remainder from dividing Parm2 into Parm1. This is known as the "Modulus".
	Parm1 & Parm2	Return the bitwise value of "Parm1" AND "Parm2"
	Parm1 | Parm2	Return the bitwise value of "PARM1" OR "Parm2"
	Parm1 ^ Parm2	Return the bitwise value of "PARM1" XOR "Parm2"
	~ Parm1	Return the inverted bitwise value of "PARM1"
	Parm1 &/ Parm2	Return the inverted bitwise value of "Parm1" AND "Parm2"
	Parm1 |/ Parm2	Return the inverted bitwise value of "Parm1" OR "Parm2"
	Parm1 ^/ Parm2	Return the inverted bitwise value of "PARM1" XOR "Parm2"
	Parm1 << Parm2	Shift "Parm1" to the left "Parm2" bits. The new least significant bits will all be zero
	Parm1 >> Parm2	Shift "Parm1" to the right "Parm2" bits. The new most significant bits will all be zero
	ABS Parm1	Return the magnitude of a number. ("ABS -4" is equal to "ABS 4" and returns "4")

Priority	Operator	Operation
	Parm1 MAX Parm2	Return the higher Parameter
	Parm1 MIN Parm2	Return the lower Parameter
	Parm1 DIG Parm2	Return Digit Number "Parm2" (Zero Based) of "Parm1". ("123 DIG 1" returns "2")
	DCD Parm1	Return a value with only the "Parm1" bit Set. ("DCD 4" returns "%00010000")
	NCD Parm1	Return the Bit number of the highest set bit in "Parm1"
	Parm1 REV Parm2	Reverse the Bits in "Parm1" from zero to "Parm2". ("%10101100 REV 4" will return "%10100011")
	SQR Parm1	Return the Integer Square Root of "Parm1"
	SIN Parm1	Return the Trigonometric "Sine" of "Parm1". The returned value will be based on a circle of radius 127 and 256 degrees (not the traditional 360)
Highest	COS Parm1	Return the Trigonometric "Cosine" of "Parm1". The returned value will be based on a circle of radius 127 and 256 degrees (not the traditional 360)

Along with the mathematical operators, the "if" statement provides the following "Test Conditions". This is listed in the following table. Note that both the "BASIC" standard labels as well as the "C" standard labels are used. "Parm1" and "Parm2" are constants, variables, or statements made up of expression statements along with the different mathematical operators and test conditions.

When a test condition is true, a nonzero is returned, if it is false, then a zero is returned. Using this convention, single variable parameters can be tested in "if" statements rather than performing comparisons of them to zero.

Test Condition	Description
Parm1 = Parm2	Return a Non-Zero if "Parm1" equals "Parm2"
Parm1 == Parm2	Return a Non-Zero if "Parm1" equals "Parm2"
Parm1 <> Parm2	Return a Non-Zero if "Parm1" does not equal "Parm2"
Parm1 != Parm2	Return a Non-Zero if "Parm1" does not equal "Parm2"
Parm1 < Parm2	Return a Non-Zero if "Parm1" is less than "Parm2"
Parm1 <= Parm2	Return a Non-Zero if "Parm1" is less than or equal to "Parm2"
Parm1 > Parm2	Return a Non-Zero if "Parm1" is greater than "Parm2"
Parm1 >= Parm2	Return a Non-Zero if "Parm1" is greater than or equal to "Parm2"
Parm1 AND Parm2	Return a Non-Zero if "Parm1" is Non-Zero and "Parm2" is Non-Zero
Parm1 && Parm2	Return a Non-Zero if "Parm1" is Non-Zero and "Parm2" is Non-Zero
Parm1 OR Parm2	Return a Non-Zero if "Parm1" is Non-Zero or "Parm2" is Non-Zero

```
Parm1 || Parm2          Return a Non-Zero if "Parm1"
                        is Non-Zero or "Parm2" is
                        Non-Zero
Parm1 XOR Parm2         Return a Non-Zero if "Parm1"
                        and "Parm2" are different
                        logical values.
Parm1 ^^ Parm2          Return a Non-Zero if "Parm1"
                        and "Parm2" are different
                        logical values.
Parm1 NOT AND Parm2     Return Zero if "Parm1" is
                        Non-Zero and "Parm2" is
                        Non-Zero
Parm1 NOT OR Parm2      Return Zero if "Parm1" is
                        Non-Zero or "Parm2" is
                        Non-Zero
Parm1 NOT XOR Parm2     Return a Non-Zero if "Parm1"
                        and "Parm2" are in the same
                        logical state.
```

The PicBasic instructions are based on the Parallax Basic Stamp "PBASIC" language and while there are a lot of similarities, they are really two different languages. In the following table, all the PicBasic instructions are listed with indications of any special considerations that should be made for them with respect to being compiled in a PICmicro® MCU.

These "instructions" are really "library routines" that are called by the mainline of the application. I am mentioning this because you will notice that the size of the application changes based on the number of instructions that are used in the application. Program memory size can be drastically reduced by looking at the different instructions that are used and changing the statements to assembler or explicit PicBasic statements.

When the various instructions are specified, note that the square brackets ("[" and "]") are used to specify data tables in some instructions. For this reason, optional values use braces ("{" and "}"), which breaks with the conventions used in the rest of the book.

Instruction	Description
BRANCH Index,[Label (,Label. . .)]	Jump to the "Label" specified by the value in "Index". "Index" is zero based, so an Index of zero will cause execution jump to the first "Label", an "Index" of one will cause execution to jump to the second "Label" and so on. This instruction only jumps within the current page; if a PICmicro® MCU with more than one page of program memory is used, then the "BRANCHL" instruction is recommend
BRANCHL Index,[Label (,Label. . .)]	Jump to the "Label" specified by the value in "Index". "Index" is zero based, so an Index of zero will cause execution jump =o the first "Label", an "Index" of one will cause execution to jump to the second "Label" and so on. This instruction can jump anywhere in PICmicro® MCU program memory
BUTTON Pin, Down, Delay, Rate, Bvar, Action, Label	Jump to "Label" when the Button has been pressed for the specified number of milliseconds. "Rate" is how many invocations after the first "BUTTON" jump is true that an "autorepeat" happens. "Bvar" is a byte sized variable only used in this function. "Action" is whether or not you want the jump =o take place when the key is pressed ("1") or released ("0")
CALL Label	Execute the assembly language "call" instructions
CLEAR	Load all the Variables with Zero
COUNT Pin, Period, Variable	Count the number of pulses on "Pin" that occur in "Period" msecs
DATA @Location, Constant (,Constant. . .)	Store Constants in Data EEPROM starting at "Location" when the PICmicro® MCU is programmed. For data at different addresses, use multiple "DATA" statements.

Instruction	Description
DEBUG Value (,Value. . .}	Define the "DEBUG" pin as output with the serial output parameters used in the "DEBUG" defines at reset. When this instruction is executed, pass the parameter data. If an ASCII "#" (0x023) is sent before a "Value", the decimal numeric is sent, rather than the ASCII byte. This instruction (and "DEBUGIN") can be used for serial I/O as they take up less space than the "SERIN" and "SEROUT" instructions
DEBUGIN {TimeOut, Label}[Variable (,Variable. . .}]	Define the "DEBUGIN" pin as an input with the serial input parameters used in the "DEBUGIN" defines at reset. When this instruction is executed, wait for a data byte to come in or jump to the label if the "TimeOut" value (which is specified in msecs) is reached
DISABLE	DISABLE Interrupts and Debug Operations. Interrupts will still be acknowledged by "ON INTERRUPT GOTO" Will not execute
DISABLE INTERRUPT	DISABLE Interrupts and Debug Operations. Interrupts will still be acknowledged by "ON INTERRUPT GOTO" Will not execute
DTMFOUT Pin,(On,Off.) [Tone(,Tone. . .}]	Output the Touch tone sequence on the specified pin. Tones "0" through "9" are the same as on the telephone keypad. Tone 10 is the "*" key and tone 11 is the "#" key. Tones 12 through 15 correspond to the extended key standards for "A" to "D". Filtering is required on the pin output to "smooth" out the signal output
EEPROM Location, [Constant (,Constant. . .}]	Store new values in EEPROM when the PICmicro® MCU is programmed. This instruction is the same as "DATA"
ENABLE	Enable debug and interrupt processing that was stopped by "DISABLE"

Instruction	Description
`ENABLE DEBUG`	Enable debug operations that were stopped by "DISABLE"
`ENABLE INTERRUPT`	Enable Interrupt operations that were stopped by the "DISABLE" and "DISABLE INTERRUPT" instructions
`END`	Stop processing the application and put the PICmicro® MCU in a low power "Sleep" mode
`FOR Variable = Start TO Stop (STEP Value) : NEXT (Variable)`	Execute a Loop, first initializing "Variable" to the "Start" value until it reaches the "Stop" value. The increment value defaults to one if no "STEP" value is specified. When "NEXT" is encountered "Variable" is incremented and tested against the "Stop" Value
`FREQOUT Pin, On, Frequency (,Frequency)`	Output the specified "Frequency" on the "Pin" for "On" msecs. If a second "Frequency" is specified, output this at the same time. Filtering is required on the pin output to "smooth" out the signal output
`GOSUB Label`	Call the subroutine that starts at address "Label". The existence of "Label" is checked at compile time
`GOTO Label`	Jump to the code that starts at address "Label".
`HIGH Pin`	Make "Pin" an Output and drive it High
`HSERIN (ParityLabel,) (TimeOut,Label,) [Variable ((,Variable. . .)]`	Receive one or more bytes from the built in USART (if present). The `ParityLabel` will be jumped to if the parity of the incoming data is incorrect. To use "ParityLabel", make sure the "HSER_EVEN" or "HSER_ODD" defines have been specified
`HSEROUT [Value (,Value. . .)]`	Transmit one or more bytes from the built in USART (if present)
`I2CREAD DataPin, ClockPin, ControlByte, (Address,)`	Read a Byte string from an I2C device. The "ControlByte" is used to access the device with block or device select bits. This instruction can be used to access internal EEPROM in the

Instruction	Description
	PIC12CExxx devices by entering the "define I2C_INTERNAL 1" statement at the start of the application code.
I2CWRITE DataPin, ClockPin, Control, (Address,) [Value(,Value. . .)] (,NoAckLabel)	Send a byte string to an I2C device. The "ControlByte" is used to access the device with block or device select bits. This instruction can be used to access internal EEPROM in the PIC12CExxx devices by entering the "define I2C_Internal 1" statement at the start of the application code.
IF Comp THEN Label Label	Evaluate the "Comp" Comparison Expression and if it is not equal to zero then jump to "Label"
IF Comp THEN Statement : {ELSE Statement :} ENDIF	Evaluate the "Comp" Comparison Expression and if it is not equal to zero then execute the "Statements" below until either an "ELSE" or an "ENDIF" statement is encountered. If an "ELSE" statement is encountered, then the code after it, to the "ENDIF" instruction is ignored. If "Comp" evaluates to zero, then skip over the "statements" after the "IF" statement are ignored to the "ELSE" or "ENDIF" Statements, after which any Statements are executed
INCLUDE "file"	Load in "file.bas" in the current directory and insert it at the current location in the source file
INPUT Pin	Put the specified pin into "Input Mode"
(LET) Assignment	Optional instruction value for an Assignment statement
LCDOUT Value(,Value...)	Send the specified Bytes to the LCD connected to the PICmicro® MCU. The LCD's operating parameters are set with the "LCD" defines. To send an instruction byte to the LCD, a $0FE byte is sent first
LOOKDOWN offset, [Constant	Go through a list of Constants with an "offset" and store the constant value at the offset in the second "Variable".

(,Constant. . .)],
Variable

LOOKDOWN2 offset,
[Test][Constant
(,Constant. . .)],
Variable

LOOKUP Variable,
[Constant
(,Constant. . .)],
Variable

LOOKUP2 Variable,
[Value
(,Value. . .)],
Variable

LOW Pin

NAP Period

If the "offset" is greater than the number of constants, then Zero is returned in "Variable". "Offset" is Zero based, the first constant is returned if "offset" is equal to zero.

Search the list and find the Constant value that meets the condition "Test". If "Test" is omitted, then the "LOOKDOWN2" instruction behaves like the "LOOKDOWN" instruction with the "Test" is assumed to be and equals sign ("=").

Compare the first "Variable" value with a constant string and return the offset into the constant string in the second "Variable". If there is no match, then the second "Variable" is not changed

Compare the first "Variable" value with a "Value" string and return the offset into the "Value" string in the second "Variable". If there is no match, then the second "Variable" is not changed. LOOKUP2 differs from LOOKUP as the "Values" can be sixteen bit variable values

Make "Pin" an Output pin and drive it with a "High" Voltage

Put the PICmicro® MCU to "sleep" for the period value which is given in the table below:

Period	Delay
0	18 msecs
1	36 msecs
2	73 msecs
3	144 msecs
4	288 msecs
5	576 msecs
6	1,152 msecs
7	2,304 msecs

Instruction	Description
ON DEBUG GOTO Label	When invoked, every time an instruction is about to be invoked, the Debug monitor program at "Label" is executed. Two Variables, the "DEBUG ADDRESS" word and "DEBUG_STACK" byte must be defined as "bank0 system" bytes. To return from the debug monitor, a "RESUME" instruction is used
ON INTERRUPT GOTO Label	Jump to the Interrupt Handler starting at "Label". When the interrupt handler is complete, execute a "RESUME" instruction
OUTPUT Pin	Put "Pin" into Output Mode
PAUSE Period	Stop the PICmicro® MCU from executing the next instruction for "Period" milliseconds. "PAUSE" does not put the PICmicro® MCU to "sleep" like "NAP" does
PAUSEUS Period	Stop the PICmicro® MCU from executing the next instruction for "Period" microseconds
PEEK Address, Variable	Return the Value at the register "Address" in "Variable"
POKE Address, Value	Write the register "Address" with the "Value"
POT Pin, Scale, Variable	Read a Potentiometer's wiper when one of its pins is connected to a capacitor. "Scale" is a value which will change the returned value until it is in the range of 0 to 0x0FF (255)
PULSIN Pin, State, Variable	Measure an incoming pulse width of "Pin". "State" indicates the state of the expected Pulse. If a 4 MHz clock is used with the PICmicro® MCU, the time intervals have a granularity of 10 usecs
PULSOUT Pin, Period	Pulse the "Pin" for the "Period". If the PICmicro® MCU is run with a 4 MHz clock, then the pulse "Period" will have a granularity of 10 usecs
PWM Pin, Duty, Cycle	Output a Pulse Width modulated signal on "Pin". Each cycle is 5 msecs long for a PICmicro® MCU running at 4 MHz. "Duty"

selects the fraction of the cycles (zero to 255) that the PWM is active. "Cycle" specifies the number of cycles that is output.

RANDOM Variable	Load "Variable" with a pseudo-random Variable
RCTIME Pin, State, Variable	Measure the absolute time required for a signal to be delayed in a RC Network. If a 4 MHz oscillator is used with the PICmicro® MCU, then the value returned will be in 10 usec increments
READ Address,Variable	Read the Byte in the built in Data EEPROM at "Address" and return its value into "Variable". This instruction does not work with the built in EEPROM of PIC12CExx parts
RESUME {Label}	Restore execution at the instruction after the "ON DEBUG" or "ON INTERRUP" instruction handler was executed. If a "Label" is specified then the hardware is returned to its original state and execute jumps to the code after "Label"
RETURN	Return to the instruction after the "GOSUB" instruction
REVERSE Pin	Reverse the function of the specified "Pin". For example, if it were in "output mode", it is changed to "input mode"
SERIN Pin, Mode,{Timeout,Label,} {Qual,...} [Variable [,Variable...]]	Receive one or more asynchronous data bytes on "Pin". The "Pin" can be defined at run time. The "Qual" bytes are test qualifiers that only pass following bytes when the first byte of the incoming string match them. The "Timeout" value is in msecs and execution jumps to "Label" when the "Timeout" interval passes without any data being received. "Mode" is used to specify the operation of the Pin and is defined in the table below:

Mode	Baud Rate	State
T300	300	Positive
T1200	1230	Positive

Instruction	Description		
	T2400	2400	Positive
	T9600	9600	Positive
	N300	300	Negative
	N1200	1200	Negative
	N2400	2400	Negative
	N9600	9600	Negative

SERIN2 Pin{\FlowPin},
Mode,(ParityLabel,}
(Timeout,Label,}
[Specification]

Receive one or more asynchronous data bytes on "Pin".
"FlowPin" is used to control the input of data to the
PICmicro® MCU to make sure there is no overrun. If Even Parity
is selected in the "Mode" Parameter, then any time an
invalid byte is received, execution will jump to the
"ParityLabel". Input Timeouts can be specified in 1 msec
intervals with no data received in the specified period
causing execution to jump to "Label". "Mode" selection is
made by passing a sixteen bit variable to the SERIN2
instruction. The bits are defined as:

Bit	Function
15	Unused
14	Set if Input Data is Negative
13	Set if Even Parity is to be used with the Data
12-0	Data Rate Specification, found by the formula:
	Rate = (1,000,000/Baud) - 20

The "Specification" is a string of data qualifiers/modifiers
and destination variables that are used to filter and
process the incoming data. The qualifiers/modifiers are
listed in the table below:

Modifier	Operation
Bin{1...16) Var	Receive Up to 16 Binary Digits and store in "Var"

DEC(1. . 5) Var	Receive Up to 5 Decimal Digits and store in "Var"
HEX(1. . 4) Var	Receive Up to 4 Hexadecimal Digits and store in "Var"
SKIP #	Skip "#" Received Characters
STR Array\n\c	Receive a string of "n" characters and store in "Array". Optionally ended by character "c"
WAIT("String")	Wait for the Specified "String" of Characters
WAITSTR Array\n	Wait for a Character String "n" characters long

SEROUT Pin,Mode,
[Value(,Value. . .)]

Send one or more asynchronous data bytes on "Pin". The "Pin" can be defined at run time. "Mode" is used to specify the operation of the Pin and the output driver and is defined in the table below

Mode	Baud Rate	State	Driver
T300	300	Positive	CMOS
T1200	1200	Positive	CMOS
T2400	2400	Positive	CMOS
T9600	9600	Positive	CMOS
N300	300	Negative	CMOS
N1200	1200	Negative	CMOS
N2400	2400	Negative	CMOS
N9600	9600	Negative	CMOS
OT300	300	Positive	Open-Drain
OT1200	1200	Positive	Open-Drain
OT2400	2400	Positive	Open-Drain
OT9600	9600	Positive	Open-Drain

Instruction	Description
	Mode Baud Rate State Driver
	ON300 300 Negative Open-Drain
	ON1200 1200 Negative Open-Drain
	ON2400 2400 Negative Open-Drain
	ON9600 9600 Negative Open-Drain
SEROUT2 Pin{FlowPin), Mode,{Pace,} {Timeout,Label,} [Specification]	Send one or more asynchronous data bytes on "Pin". "FlowPin" is used to control the output of data to the PICmicro® MCU to make sure there is no overrun. Timeouts can be specified in 1 msec intervals with no "Flow" control on the receiver the specified period causing execution to jump to "Label". The optional "Pace" parameter is used to specify the length of time (measured in usecs) that the PICmicro® MCU delays before sending out the next character. "Mode" selection is made by passing a sixteen bit variable to the SERIN2 instruction. The bits are defined as:
	Bit Function
	15 CMOS/Open Drain Driver
	Specification. If Set, Open Drain Output
	14 Set if Input Data is Negative
	13 Set if Even Parity is to be used with the Data
	12-0 Data Rate Specification, found by the formula:
	Rate = (1,000,000/Baud) - 20
	The "Specification" is a string of data qualifiers/modifiers and source values that are used to format the outgoing data. The output format data can be specified with an "I" prefix to indicate that the data type is to be sent before the data and the "S" prefix indicates that a sign ("-") indicator is sent for negative values. The qualifiers/modifiers are listed in the table below:

Modifier	Operation
Bin(1. . .16} Var	Receive Up to 16 Binary Digits and store in "Var"
DEC(1. . .5) Var	Receive Up to 5 Decimal Digits and store in "Var"
HEX(1. . .4) Var—	Receive Up to 4 Hexadecimal Digits and store in "Var"
SKIP #	Skip "#" Received Characters
STR Array\n\c	Receive a string of "n" characters and store in "Array". Optionally ended by character "c"
WAIT("String")	Wait for the Specified "String" of Characters
WAITSTR Array\r	Wait for a Character String "n" characters long

```
SHIFTIN DataPin,
ClockPin,Mode,
[Variable(\Bits}
{,Variable. . .}]
```

Synchronously shift data into the PICmicro® MCU. The "Bits" Parameter is used to specify the number of bits that are actually shifted in (if "Bits" is not specified, the default is 8). The "Mode" Parameter is used to indicate how the data is to be transferred and the values are listed in the table below:

Mode	Function
MSBPRE	Most Significant Bit First, Read Data before pulsing Clock
LSBPRE	Least Significant Bit First, Read Data before pulsing Clock
MSBPOST	Most Significant Bit First, Read Data after pulsing Clock

Instruction	Description	Function
	Mode	Function
	LSBPOST	Least Significant Bit First, Read Data after pulsing Clock
SHIFTOUT DataPin, ClockPin,Mode, [Variable\(Bits\) . .}]	Synchronously shift data out of the PICmicro® MCU. The "Bits" Parameter is used to specify how many bits are to be shifted out in each word (if not specified, the default is 8). The "Mode" parameter is used to specify how the data is to be shifted out and the values are listed in the table below:	
	Mode	Function
	LSBFIRST	Least Significant Bit First
	MSBFIRST	Most Significant Bit First
SLEEP Period	Put the PICmicro® MCU into "Sleep" mode for "Period" seconds	
SOUND Pin, [Note,Duration {,Note,Duration. . .}]	Output a string of Tones and Durations (which can be used to create a simple tune) on the "Pin". Note "0" is silence and Notes "128" to "255" are "white noise". Note "1" (78.5 Hz for a 4 MHz PICmicro® MCU) is the lowest valid tone and note "127" is the highest (10 kHz in a 4 MHz PICmicro® MCU). Duration is specified in 12 msec increments	
STOP	Place the PICmicro® MCU into an endless loop. The PICmicro® MCU is not put into "sleep" mode	
SWAP Variable, Variable	Exchange the values in the two Variables	
TOGGLE Pin	Toggle the Output Value of the Specified Pin	
WHILE Cond : WEND	Execute the code between the "WHILE" and the "WEND" statements while the "Cond" condition returns a non-zero value. Execution exits the loop when "Cond" is evaluated to Zero	
WRITE Address, Value	Write the Byte "Value" into the built in Data EEPROM. This instruction will not work with the built in EEPROM in the PIC12CExxx devices	

```
XIN DataPin,ZeroPin,        Receive data from X-10 devices. "ZeroPin" is
[Timeout,Label,}            used to detect the "Zero Crossing" of the
[Variable                   input AC Signal. Both "DataPin" and "ZeroPin"
{,Variable...}]             should be pulled up with 4.7 K resistors. The optional
                            Timeout (specified in 8.33 msec intervals) will cause
                            execution to jump to "Label" if no data is received by the
                            specified interval. If the first Variable data destination
                            is sixteen bits, then both the "House Code" and the "Key
                            Code" will be saved. If the first Variable is eight bits in
                            size, then only the "Key Code" will be saved.

XOUT DataPin,ZeroPin,       Send X-10 data to other devices. The "ZeroPin" is an input and
[HouseCode\KeyCode          should be pulled up with a 4.7K resistor. "HouseCode" is a
{\Repeat}{,Value...}]       number between 0 and 15 and corresponds to the "House Code"
                            set on the X-10 Modules A through P. The "KeyCode" can either
                            be the number of a specific X-10 receiver or the function to
                            be performed by the module.
```

Visual Basic

Microsoft's "Visual Basic" is probably the fastest way to get into Microsoft "Windows" application programming. The ease of using the language and development system also makes it great as a "what if" tool and allows you to write an application quickly to try out new ideas.

To create an application, the Primary dialog box (which is known as a "form" and is shown in Fig. 15.8) is created first, with different features (I/O boxes, buttons, etc.). These features are known as "controls" within Visual Basic. With the Window defined, by simply clicking on the different controls, subroutine prototypes to handle "events" (such as mouse clicks over these features) are automatically created. Additional features in Visual Basic's source code editor allow

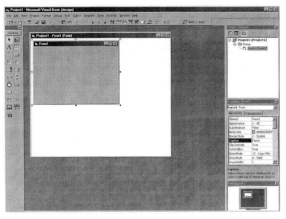

Figure 15.8 "Visual Basic" Development System

you to specify the control parameters (known as "properties"). Visual Basic applications are built around "The Dialog Box Editor" desktop. When application development is started, Visual Basic provides you with the initial "Dialog" box of the application that can be seen in Fig. 15.9. From here, "Dialog Resources" are selected from the "ToolBox" and placed upon the dialog.

Control attributes (also known as "Properties") can be set globally from the Integrated Development Environment or from within the "Event Handlers". The event handler's code is written in pretty standard Microsoft BASIC. Once the handler prototypes are created by Visual Basic, it is up to the application developer to add the response code for the application. Visual Basic provides a large number of built-in functions, including trigonometric functions, logarithm

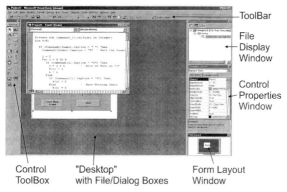

Control   "Desktop"   Form Layout
ToolBox   with File/Dialog Boxes   Window

Figure 15.9 Visual Basic Desktop

functions, and the ability to interface with the file system and dialog controls.

Variables in Visual Basic are typically "Integer", which is to say they are sixteen bit values in the ranges -32768 to $+32765$. Thirty-two bit integer variables can be specified by putting a "%" character at the end of the variable label. One important thing to note about variables is that they are local to the event routine they are used in unless they are declared globally in the "General Module", which executes at the beginning of the application and is not specific to any controls.

There are a number of controls that are basic to Visual Basic with others being available for download off the Internet or bought which can make your Visual Basic applications more impressive and lend "pizzazz" to Windows applications.

```
Control          Description
Pull Downs       Selected from the "Menu Editor"
                 icon on the "ToolBar"
PictureBox       Display Bitmaps and other graphic
                 files on the Dialog Box
Label            Put Text in the Dialog Box
TextBox          Input/Output Text Box
Frame            Put a Frame around Resources
CommandButton    Button for Code Operation
CheckBox         For Checking Multiple Selections
OptionButton     Also known as the "Radio Button".
                 For Checking one selection for a
                 list of Multiple options
ComboBox         Select or Enter Test in a Box/List
ListBox          List Data (with User controlled
                 Scrolling)
HScrollBar       Provide Horizontal Scrolling in a
                 Text or Graphic Output Control
VscrollBar       Provide Vertical Scrolling in a
                 Text or Graphic Output Control
Timer            Cause a periodic interrupt
```

```
DriveListBox      Select a Drive for File I/O
DirListBox        Select a Subdirectory for File I/O
                    on a Specific Drive
FileListBox       Display Files for a Specific
                    Subdirectory on a Specific Drive
Shape             Put a Graphics Shape on the Dialog
                    Box
Line              Draw a Line on the Dialog Box
Image             Display an Image File on the Dialog
                    Box
OLE               Insert OLE Objects to the Dialog
```

There are a number of controls that cannot be activated with a left button click and cannot be "seen" on the application's form. The one that is used the most is the "Timer". This control causes an event after a set period of microseconds. This control can be set within the dialog editor or modified within the application itself. The Timer can provide many different advanced functions without requiring any interrupt interfaces.

MSComm control

The MSComm control recommended initialization sequence is

1. Specify the Hardware Serial Port to be used.

2. Set the speed and data format to be used.

3. Define the buffer size.

4. Open the port and begin to use it.

The instructions used to perform these functions are placed in the "Form_Load" subroutine, which means the port is enabled before the primary dialog box is executing. The following code is an example of an MSComm object initialization:

```
Private Sub Form_Load()
'  On Form Load, Setup Serial Port 3 for YAP
'  Programmer

   MSComm3.CommPort = 3

   MSComm3.Settings = "1200,N,8,1"

   MSComm3.InputLen = 0

   MSComm3.PortOpen = True

   Text1.Text = "Turn on YAP Programmer"

End Sub
```

A 50-msec timer can be used to continually poll the serial port and display data in the "Text" Box when it is received:

```
Private Sub Timer1_Timer()
'  Interrupt every 50 msecs and Read in the Buffer
Dim InputString

   InputString = MSComm3.Input

   If (InputString <> "") Then
     If (Text1.Text = "Turn on YAP Programmer") Then
       Text1.Text = "" ' Clear the Display Buffer
     End If
     Text1.Text = Text1.Text + InputString
   End If

End Sub
```

Once the "MSComm" control is placed on the display, the following properties are used to control it:

Property	Setting	Description
Break	True/False	When set to "True", Break Sends a "0" break signal until the property is changed to "False".

CDHolding	True/False	Read only property that indicates if the "Carrier Detect" line is active. This is an important line to poll in applications which use modems.
CommEvent	Integer	Read only property that is only available while the application is running. If the application is running without any problems, this property returns zero. This property is read by the "OnComm" event handler code to process the reason why the "event" was caused.
CommID	Object	Read only property that returns an identifier for the serial port assigned to the MSComm control.
CommPort	Integer	Specify the "COMx" (1–9) serial port that is used by the MSComm control.
CTSHolding	True/False	Read only property that returns the current state of the serial port's "Clear To Send" line.
DSRHolding	True/False	Read only property that returns the current state of the serial port's "Data Set Ready" line.
DTREnable	True/False	Property used to specify the state of the "Data Terminal Ready" line.
EOFEnable	True/False	Specify whether or not an "OnComm" event will be generated if an "End Of File" character (0x01A) is encountered.

Handshaking	0, 1, 2 or 3	Sets the current handshaking protocol for the serial port: 0 - No handshaking (default) 1 - XON/XOFF Handshaking 2 - RTS/CTS (Hardware) Handshaking 3 - Both XON/XOFF and RTS/CTS Handshaking
InBufferCount	Integer	Read only property indicating how many characters have been received by the serial port.
InBufferSize	Integer	Property used to specify the number of bytes available for the Input Data Buffer. The default size is 1024 bytes.
Input	String	Return a String of Characters from the Input Buffer.
InputMode	Integer	Specify how data is to be retrieved using the "Input" property. Zero specifies data will be received as Text (Default). One will specify that data will be passed without editing ("Binary" format).
InputLen	Integer	Sets the Maximum Number of characters that will be returned when the "Input" property is accessed. Setting this value to zero will return the entire buffer.

NullDiscard	True/False	Specify whether or not Null Characters are transferred from the port to the receiver buffer.
OutBufferCount	Integer	Read only property that returns the Number of Characters waiting in the Output Buffer.
OutBufferSize	Integer	Specify the size of the Output Buffer. The default is 512 Bytes.
Output	Integer	Output a string of characters through the serial port.
ParityReplace	Integer	Specify the character that will replace characters which have a "Parity" Error. The default character is "?" and the ASCII code for the replacement character must be specified.
PortOpen	True/False	Specify whether or not the data port is to be transmitting and receiving data. Normally a port is closed ("False").
Rthreshold	Integer	Specify the number of characters before there is an "OnComm" event. The default value of zero disables event generation. Setting the "Rthreshold" to one will cause an "OnComm" event each time a character is received.

RTSEnable	True/False	Specify the value output on the "Request To Send" line.
Settings	String	Send a String to the Serial Port to specify its operating characteristics. The String is in the format "Speed, Parity, Length, Stop" with the following valid parameter values: Speed: Data Rate of the Communication

    110
    300
    600
   1200
   2400
   9600 (Default)
 14400
 19200
 28800
 38400
 56000
128000
256000

Parity: The type of error checking sent with the byte
E - Even Parity
M - Mark Parity
N - No Parity (Default)
O - Odd Parity
S - Space Parity

Length: The number of bits transmitted at a time
4 - 4 Bits
5 - 5 Bits
6 - 6 Bits
7 - 7 Bits
8 - 8 Bits (Default)

Stop: The number of stop bits transmitted with the byte

```
                                    1 - 1 Stop Bit (Default)
                                    1.5 - 1.5 Stop Bits
                                    2 - 2 Stop Bits

Sthreshold   Integer               Specify the number of
                                    bytes to be transmitted
                                    before an "OnComm" event
                                    is generated. The
                                    default is zero (which
                                    means no "OnComm" event
                                    is generated for
                                    transmission). Setting
                                    this value to one will
                                    cause an "OnComm" event
                                    after each character is
                                    transmitted.
```

Further enhancing the usefulness of the "MSComm" control is the "OnComm" event. This routine is similar to an interrupt, as it is requested after specified events in the serial port. The "CommEvent" property contains the reason code for the event. These codes include:

```
CommEvent Identifier   CommEvent Code   Description
comEvSend                    1          Specified Number
                                          of Characters
                                          Sent
comEvReceive                 2          Specified Number
                                          of Characters
                                          Received
comEvCTS                     3          Change in the
                                          "Clear To Send"
                                          line
comEvDSR                     4          Change in the
                                          "Data Set Ready"
                                          line
comEvCD                      5          Change in the
                                          "Carrier Detect"
                                          line
comEvRing                    6          Ring Detect is
                                          Active
comEvEOF                     7          "End Of File"
                                          Character
                                          Detected
```

comEventBreak	1001	Break Signal Received
comEventFrame	1004	Framing Error in incoming data
comEventOverrun	1006	Receive Port Overrun
comEventRxOver	1008	Receive Buffer Overflow
comEventRxParity	1009	Parity Error in Received Data
comEventTxFull	1010	Transmit Buffer Full
comEventDCB	1011	Unexpected Device Control Block Error

The "C" Language

Declarations

Constant declaration:

```
const int Label = Value;
```

Variable declaration:

```
type Label [= Value];
```

"Value" is an optional Initialization Constant, where "type" can be:

```
char
int
unsigned int
float
```

Note that "int" is defined as the "word size" of the processor/operating system. For PCs, an "int" can be a Word (16 bits) or a Double Word (32 bits). For the PICmicro® MCU, an "int" is normally 8 bits (a byte).

There may also be other basic types defined in the language implementation. Single dimensional arrays are declared using the form:

```
type Label[ Size ] [= { Initialization Values..}];
```

Note that the array "Size" is enclosed within square brackets ("[" and "]") and should not be confused with the optional "Initialization Values".

Strings are defined as single dimensional ASCIIZ arrays:

```
char String[ 17 ] = "This is a String";
```

where the last character is an ASCII "NUL."

Strings can also be defined as pointers to characters:

```
char *String = "This is a String";
```

although this implementation requires the text "This is a String" to be stored in two locations (in code and data space). For the PICmicro® MCU and other Harvard architected processors, the text data could be written into data space when the application first starts up as part of the language's initialization.

Multidimensional Arrays are defined with each dimension separately identified within Square Brackets ("[" and "]"):

```
int ThreeDSpace[ 32 ][ 32 ][ 32 ];
```

Array Dimensions must be specified unless the Variable is a pointer to a Single Dimensional Array.

Pointers are declared with the "*" character after the "type"

```
char * String = "This is a String";
```

Accessing the address of the Pointer in Memory is accomplished using the "&" character:

```
StringAddr = &String;
```

Accessing the address of a specific element in a String is accomplished using the "&" character and a String Array Element:

```
StringStart = &String[ n ];
```

In the PC running MS-DOS, it is recommended that "far" (32 bit) pointers be always used with absolute offset:segment addresses within the PC memory space to avoid problems with varying segments. In the PICmicro® MCU, all addresses can be specified with two bytes.

The Variable's "Type" can be "overridden" by placing the new type in front of the variable in single brackets:

```
(long) StringAddr = 0x0123450000;
```

Statements

Application "Mainline".

```
main(envp)
  char *envp;
{  //  Application Code

   :                         // Application Code

}  //  End Application
```

Function format.

```
Return_Type Function( Type Parameter [, Type
Parameter..])
{  //  Function Start

    :                           // Function Code

    return value;

} //  End Function
```

Function prototype.

```
Return_Type Function( Type Parameter [, Type
Parameter..]);
```

Expression.

```
[(..] Variable | Constant [Operator [(..] Variable |
Constant ][)..]]
```

Assignment statement.

```
Variable = Expression;
```

"C" conditional statements (consisting of "if", "?", "while", "do", "for" and "switch").

The "if" statement is defined as

```
   if ( Statement )
;  | { Assignment Statement | Conditional
  Statement.. } | Assignment Statement | Conditional
  Statement
   [else ;| { Assignment Statement | Conditional
  Statement..} | Assignment Statement | Conditional
  Statement ]
```

The "? :" statement evaluates the statement (normally a comparison) and if not equal to zero, executes the first statement, else executes the statement after the ":".

```
Statement ? Assignment Statement | Conditional
Statement : Assignment Statement | Conditional
Statement
```

The "while" statement is added to the application following the definition below:

```
while ( Statement ) ; | { Assignment Statement |
Conditional Statement.. } | Assignment Statement |
Conditional Statement
```

The "for" statement is defined as:

```
for ( initialization (Assignment) Statement;
Conditional Statement; Loop Expression (Increment)
Statement )
; | { Assignment Statement | Conditional Statement..
} | Assignment Statement | Conditional Statement
```

To jump out of a currently executing loop, "break" statement

```
break;
```

is used.

The "continue" statement skips over remaining code in a loop and jumps directly to the loop condition (for use with "while", "for" and "do/while" Loops). The format of the statement is

```
continue;
```

For looping until a condition is true, the "do/while" statement is used:

```
do

   Assignment Statement | Conditional Statement..

while ( Expression );
```

To conditionally execute according to a value, the "switch" statement is used:

```
switch( Expression ) {
  case Value:                    // Execute if "Statement"
                                 //    == "Value"
    [ Assignment Statement | Conditional Statement.. ]
    [break;]
  default:                       // If no "case" Statements
                                 //    are True
    [ Assignment Statement | Conditional Statement.. ]
} // End switch
```

Finally, the "goto Label" statement is used to jump to a specific address:

```
  goto Label;

Label:
```

To return a value from a function, the "return" statement is used:

```
return Statement;
```

Operators

Statement operators.

```
Operator         Operation
!                Logical Negation
!                Bitwise Negation
&&               Logical AND
&                Bitwise AND, Address
||               Logical OR
|                Bitwise OR
^                Bitwise XOR
+                Addition
++               Increment
```

-	Subtraction, Negation
--	Decrement
*	Multiplication, Indirection
/	Division
%	Modulus
==	Equals
!=	Not Equals
<	Less Than
<=	Less Than or Equals To
<<	Shift Left
>	Greater Than
>=	Greater Than or Equals To
>>	Shift Right

Compound assignment operators.

Operator	Operation
&=	AND with the Variable and Store Result in the Variable
|=	OR with the Variable and Store Result in the Variable
^=	XOR with the Variable and Store Result in the Variable
+=	Add to the Variable
-=	Subtract from the Variable
*=	Multiply to the Variable
/=	Divide from the Variable
%=	Get the Modulus and Store in the Variable
<<=	Shift Left and Store in the Variable
>>=	Shift Right and Store in the Variable

Order of operations.

Operators	Priority	Type
() [] . ->	Highest	Expression Evaluation
- ~ ! & * ++ --		Unary Operators
* / %		Multiplicative
+ -		Additive
<< >>		Shifting
< <= >= >		Comparison

```
==  !=                          Comparison
&                               Bitwise AND
^                               Bitwise XOR
|                               Bitwise OR
&&                              Logical AND
||                              Logical OR
?:                              Conditional
                                  Execution
= &= |= ^= += -= *= /= %= >>= <<=  Assignments
,                   Lowest      Sequential
                                  Evaluation
```

Directives

All Directives start with "#" and are executed before the code is compiled.

```
Directive                       Function
#define Label[( Parameters)] Text  Define a Label that
                                  will be replaced
                                  with "Text" when
                                  it is found in the
                                  code. If
                                  "Parameters" are
                                  specified, then
                                  replace them in
                                  the code, similar
                                  to a macro.
#undefine Label                 Erase the defined
                                  Label and Text in
                                  Memory.
#include "File" | <File>        Load the Specified
                                  File in Line to
                                  the Text. When "<"
                                  ">" encloses the
                                  Filename, then the
                                  file is found
                                  using the
                                  "INCLUDE"
                                  Environment Path
                                  Variable. If """
                                  """ encloses the
                                  Filename, then the
```

	file in the current directory is searched before checking the "INCLUDE" Path.
#error Text	Force the Error listed in "Text"
#if Condition	If the "Condition" is True, then Compile the following code to "#elif", "#else" or "#endif". If the "Condition" is False, then ignore the following code to "#elif", "#else" or "#endif".
#ifdef Label	If the "#define" Label exists, then Compile the Following Code. "#elif", "#else" and "#endif" work as expected with "#if".
#ifndef Label	If the "#define" Label does NOT exist, then Compile the Following Code. "#elif", "#else" and "#endif" work as expected with "#if".
#elif Condition	This Directive works as an "#else #if" to avoid lengthy nested "#if"s. If the previous condition was False, checks the Condition.

`#else`	Placed after "#if" or "#elif" and toggles the Current Compile Condition. If the Current Compile Condition was False, after "#else", it will be True. If the Current Compile Condition was True, after "#else", it will be False.
`#endif`	Used to End an "#if", "#elif", "#else", "#ifdef" or "#ifndef" directive.
`#pragma String`	This is a Compiler dependent Directive with different "Strings" required for different cases.

The following words cannot be used in "C" applications as labels:

```
break
case
continue
default
do
else
for
goto
if
return
switch
while
```

"Backslash" characters

String	ASCII	Character
\r	0x00D	Carriage Return ("CR")
\n	0x00A	Line Feed ("LF")
\f	0x00C	Form Feed ("FF")
\b	0x008	Backspace ("BS")
\t	0x009	Horizontal Tab ("HT")
\v	0x00B	Vertical Tab ("VT")
\a	0x007	Bell ("BEL")
\'	0x027	Single Quote ("'")
\"	0x022	Double Quote (""")
\\	0x05C	Backslash ("\")
\ddd	N/A	Octal Number
\xddd	0x0dd	Hexadecimal Character

Common C functions

As Defined by Kernighan and Ritchie:

Function	Operation
int getchar(void)	Get one Character from "Standard Input" (the Keyboard). If no Character available, then wait for it.
int putchar(int)	Output one Character to the "Standard Output" (the Screen).
int printf(char *Const[, arg...])	Output the "Const" String Text. "Escape Sequence" Characters for Output are embedded in the "Const" String Text. Different Data Outputs are defined using the "Conversion Characters": %d, %i - Decimal Integer %o - Octal Integer %x, %X - Hex Integer (with upper or lower case values). No leading "0x" character String Output

%u - Unsigned Integer
%c - Single ASCII
Character
%s - ASCIIZ String
%f - Floating Point
%#e, %#E - Floating
Point with the precision
specified by "#"
%g, %G - Floating Point
%p - Pointer
%% - Print "%" Character
Different C
Implementations will
have different "printf"
parameters.

```
int scanf( char *Const, arg [, *arg...])
```

Provide Formatted Input
from the user. The
"Const" ASCIIZ String is
used as a "Prompt" for
the user. Note that the
input parameters are
always pointers.
"Conversion Characters"
are similar to "printf":
%d - Decimal Integer
%i - Integer. In Octal
if leading "0" or hex
if leading "0x" or "0X"
%o - Octal Integer
(Leading "0" Not
Required)
%x - Hex Integer
(Leading "0x" or "0X"
Not Required)
%c - Single Character
%s - ASCIIZ String of
Characters. When Saved,
a NULL character is put
at the end of the String
%e, %f, %g - Floating
Point Value with
optional sign, decimal
point and exponent
%% - Display "%"
character in prompt

```
handle fopen( char *FileName,     Open File and Return
  char *mode )                    Handle (or NULL for
                                  Error).
                                  "mode" is a String
                                  consisting of the
                                  optional characters:
                                  r - Open File for
                                      Reading
                                  w - Open File for
                                      Writing
                                  a - Open File for
                                      Appending to
                                      Existing Files
                                  Some systems handle
                                  "Text" and "Binary"
                                  files. A "Text" file has
                                  the CR/LF characters
                                  represented as a single
                                  CR. A "Binary" file does
                                  not delete any
                                  characters.
int fclose( handle )              Close the File.
int getc( handle )                Receive data from a file
                                  one character at a time.
                                  If at the end of an
                                  input file, then "EOF"
                                  is returned.
int putc( handle, char )          Output data to a file one
                                  character at a time.
                                  Error is indicated by
                                  "EOF" returned.
int fprintf( handle, char *Const[, arg...])
                                  Output String of
                                  Information to a File.
                                  The same "Conversion
                                  Characters" and
                                  arguments as "printf"
                                  are used.
int fscanf( handle, char *Const, arg[, arg...])
                                  Input and Process String
                                  of Information from a
                                  File. The same
                                  "Conversion Characters"
```

```
                              and arguments as "scanf"
                              are used.
int fgets( char *Line, int LineLength, handle )
                              Get the ASCIIZ String
                              from the file.
int fputs( char *line, handle )
                              Output an ASCIIZ String
                              to a file.
strcat( Old, Append )         Put ASCIIZ "Append"
                              String on the end of the
                              "Old" ASCIIZ String.
strncat( Old, Append, # )
                              Put "#" of characters
                              from "Append" on the end
                              of the "Old" ASCIIZ
                              String.
int strcmp( String1, String2 )
                              Compare two ASCIIZ
                              Strings. Zero is
                              returned for match,
                              negative for "String1" <
                              "String2" and positive
                              for "String1" >
                              "String2".
int strncmp( String1, String2, # )
                              Compare two ASCIIZ
                              Strings for "#"
                              characters. Zero is
                              returned for match,
                              negative for "String1"
                              < "String2" and positive
                              for "String1"
                              > "String2".
strcpy( String1, String2 )    Copy the Contents of
                              ASCIIZ "String2" into
                              "String1".
strncpy( String1, String2, # )
                              Copy "#" Characters from
                              "String2" into "String1".
strlen( String )              Return the length of
                              ASCIIZ Character
                              "String"
int strchr( String, char )    Return the Position of
                              the first "char" in the
                              ASCIIZ "String".
```

```
int strrchr( String, char )
```
Return the Position of the last "char" in the ASCIIZ "String".

```
system( String )
```
Executes the System Command "String".

```
*malloc( size )
```
Allocate the Specified Number of Bytes of Memory. If insufficient space available, return NUL.

```
*calloc( #, size )
```
Allocate Memory for the specified "#" of data elements of "size".

```
free( * )
```
Free the Memory.

```
float sin( angle )
```
Find the "Sine" of the "angle" (which in Radians).

```
float cos( angle )
```
Find the "Cosine" of the "angle" (which in Radians).

```
float atan2( y, x )
```
Find the "Arctangent" of the "X" and "Y" in Radians.

```
float exp( x )
```
Calculate the natural exponent.

```
float log( x )
```
Calculate the natural logarithm.

```
float log10( x )
```
Calculate the base 10 logarithm.

```
float pow( x, y )
```
Calculate "x" to the power "y".

```
float sqrt( x )
```
Calculate the Square Root of "x".

```
float fabs( x )
```
Calculate the Absolute Value of "x".

```
float frand()
```
Get a Random Number.

```
int isalpha( char )
```
Return Non-Zero if Character is "a"-"z" or "A"-"Z".

```
int isupper( char )
```
Return Non-Zero if Character is "A"-"Z".

```
int islower( char )
```
Return Non-Zero if Character is "a"-"z".

```
int isdigit( char )
```
Return Non-Zero if Character is "0"-"9".

```
int isalnum( char )            Return Non-Zero if
                               Character is "a"-"z",
                               "A"-"Z"  or "0"-"9".
int isspace( char )            Return Non-Zero if
                               Character is " ", HT,
                               LF, CR, FF or VT.
int toupper( char )            Convert the Character to
                               Upper Case.
int tolower( char )            Convert the Character to
                               Lower Case.
```

PICmicro® MCU enhancement functions

Useful Functions in PICmicro® MCU C implementations:

```
Function               Operation
inp, outp              Provide method for directly
                       accessing system registers.
SerIn, SerOut          NR2 Non-Return to Zero
12C                    12C Interface
PNM                    Measure/output PNM signals
```

Chapter

16

Constants and Data Tables

Mathematical and Physical Constants

```
Symbol     Value                      Description
AU         149.59787x(10^6) km        Astronomical Unit
           92,955,628 miles             (Distance from
                                         the Sun to the
                                         Earth)
c          2.99792458x(10^8) m/s      Speed of Light in
           186,282 miles/s              a Vacuum
e          2.7182818285
Epsilon-o  8.854187817x(10^-12) F/m
                                      Permittivity of
                                      Free Space
```

Ev	$1.60217733 \times (10^{-19})$ J	Electron Volt Value
g	32.174 ft/sec^2	Acceleration due to gravity
	9.807 m/sec^2	
h	$6.626 \times (10^{-34})$ Js	Planck Constant
k	$1.380658 \times (10^{-23})$ J/K	Boltzmann Entropy Constant
me	$9.1093897 \times (10^{-31})$ kg	Electron Rest Mass
mn	$1.67493 \times 10^{-27})$ kg	Neutron Rest Mass
mp	$1.67263 \times (10^{-27})$ kg	Proton Rest Mass
pc	$2.06246 \times (10^5)$ AU	Parsec
pi	3.1415926535898	Ratio of circumference to Diameter of a circle
R	8.314510 J/(K * mol)	Gas Constant
sigma	$5.67051 \times (10^{-8})$ W/(m^2 * K^4)	Stefan-Boltzmann Constant
u	$1.66054 \times (10^{-27})$ grams	Atomic Mass Unit
mu-o	$1.25664 \times (10^{-7})$ N/A^2	Permeability of Vacuum
None	331.45 m/s	Speed of Sound at Sea Level, in Dry Air at 20C
	1087.4 ft/s	
None	1480 m/s	Speed of Sound in Water at 20C
	4856 ft/s	

ASCII

The ASCII Definition uses the seven bits of each ASCII character.

3-0	6-4 -> 000	001	010	011	100	101	110	111
V	Control		Characters					
0000	NUL	DLE	Space	0	@	P	`	p
0001	SOH	DC1	!	1	A	Q	a	q
0010	STX	DC2	"	2	B	R	b	r
0011	ETX	DC3	#	3	C	S	c	s
0100	EOT	DC4	$	4	D	T	d	t
0101	ENQ	NAK	%	5	E	U	e	u

0110 |	ACK	SYN |	&	6	F	V	f	v
0111 |	BEL	ETB |	`	7	G	W	g	w
1000 |	BS	CAN |	(	8	H	X	h	x
1001 |	HT	EM |	)	9	I	Y	i	y
1010 |	LF	SUB |	*	:	J	Z	j	z
1011 |	VT	ESC |	+	;	K	[	k	{
1100 |	FF	FS |	,	<	L	\	l	|
1101 |	CR	GS |	–	=	M	]	m	}
1110 |	SO	RS |	.	>	N	^	n	~
1111 |	SI	US |	/	?	O	_	o	DEL

ASCII control characters

The ASCII Control Characters were specified as a means of allowing one computer to communicate and control another. These characters are actually commands and if the BIOS or MS-DOS display or communications APIs are used with them they will revert back to their original purpose. Writing these values (all less than 0x020) to the display will display graphics characters in the IBM PC.

Normally, only "Carriage Return"/"Line Feed" are used to indicate the start of a line. "Null" is used to indicate the end of an ASCIIZ string. "Backspace" will move the cursor back one column to the start of the line. The "Bell" character, when sent to MS-DOS will cause the PC's speaker to "beep". "Horizontal Tab" is used to move the cursor to the start of the next column that is evenly distributed by eight. "Form Feed" is used to clear the screen.

Hex	Mnemonic	Definition
00	NUL	"Null" – Used to indicate the end of a string
01	SOH	Message "Start of Header"
02	STX	Message "Start of Text"
03	ETX	Message "End of Text"

04	EOT	"End of Transmission"
05	ENQ	"Enquire" for Identification or Information
06	ACK	"Acknowledge" the previous transmission
07	BEL	Ring the "BELL"
08	BS	"Backspace" - Move the Cursor on column to the left
09	HT	"Horizontal Tab" - Move the Cursor to the Right to the next "Tab Stop" (Normally a column evenly divisible by eight)
0A	LF	"Line Feed" - Move the Cursor down one line
0B	VT	"Vertical Tab" - Move the Cursor down to the next "Tab Line"
0C	FF	"Form Feed" up to the start of the new page. For CRT displays, this is often used to clear the screen
0D	CR	"Carriage Return" - Move the Cursor to the leftmost column
0E	SO	Next Group of Characters do not follow ASCII Control conventions so they are "Shifted Out"
0F	SI	The following Characters do follow the ASCII Control conventions and are "Shifted In"
10	DLE	"Data Link Escape" - ASCII Control Character start of an Escape sequence. In most modern applications "Escape" (0x01B) is used for this function
11	DC1	Not defined - Normally application specific
12	DC2	Not defined - Normally application specific
13	DC3	Not defined - Normally application specific
14	DC4	Not defined - Normally application specific
15	NAK	"Negative Acknowledge" - the previous transmission was not properly received
16	SYN	"Synchronous Idle" - If the serial transmission uses a synchronous

		protocol, this character is sent to ensure the transmitter and receiver remain synched
17	ETB	"End of Transmission Block"
18	CAN	"Cancel" and disregard the previous transmission
19	EM	"End of Medium" – Indicates end of a file. For MS-DOS files, 0x01A is often used instead
1A	SUB	"Substitute" the following character with an incorrect one
1B	ESC	"Escape" – Used to temporarily halt execution or put an application into a mode to receive information
1C	FS	Marker for "File Separation" of data being sent
1D	GS	Marker for "Group Separation" of data being sent
1E	RS	Marker for "Record Separation" of data being sent
1F	US	Marker for "Unit Separation" of data being sent

ANSI display control sequences

From MS-DOS applications you can move the cursor or change the current display colors one of two ways. Normally I use the BIOS functions and direct writes to video RAM. The second way is to load the "ANSI.SYS" device driver in the "config.sys" using the statement:

```
device = [d:][path]ANSI.SYS
```

When the "Escape Sequences" listed below are output using the standard output device (using the MS-DOS APIs), the commands are executed.

This method is not often used for two reasons. The first is that it is much slower than using the BIOS APIs and writing directly to video RAM. For an application that

seems to change the screen in the blink of an eye, the
ANSI Display Control Sequences are not the way to do it.
The second is that "ANSI.SYS" takes away 10 KBytes of
memory that would normally be available for applications.

There are two advantages to using the ANSI Display
Control Sequences. The first is that it will make applications very portable. Passing the source to another system's just requires recompilation and linking. The
second advantage is that sending data serially to a receiver set up able to receive these sequences (set up as
an "ANSI" or "VT100 Compatible" Terminal), will provide simple graphic operations in an application.

In the table below, "ESC" is the ASCII "Escape"
Character 0x01B.

```
Sequence     Function
Esc[=#h      Set the PC's Display mode. This is not
             available in "true" ANSI compatible
             devices.
             #  =   0 - 40x25 Monochrome
             #  =   1 - 40x25 Color
             #  =   2 - 80x25 Monochrome
             #  =   3 - 80x25 Color
             #  =   4 - 320x200 Color Graphics
             #  =   5 - 320x200 Monochrome Graphics
             #  =   6 - 640x200 Monochrome Graphics
             #  =   7 - wrap to next line at line end
             #  =  14 - 640x200 Color Graphics
             #  =  15 - 640x350 Monochrome Graphics
             #  =  16 - 640x480 Color Graphics
             #  =  17 - 640x480 Color Graphics
             #  =  18 - 640x480 Color Graphics
             #  =  19 - 320x200 Color Graphics
Esc[=#l      Reset the PC's Display mode. This is not
             available in "true" ANSI compatible
             devices.
             #  =   0 - 40x25 Monochrome
             #  =   1 - 40x25 Color
             #  =   2 - 80x25 Monochrome
             #  =   3 - 80x25 Color
```

```
              # =  4 - 320x200 Color Graphics
              # =  5 - 320x200 Monochrome Graphics
              # =  6 - 640x200 Monochrome Graphics
              # =  7 - do not wrap at line end
Esc[#m        Set Character Attributes
              # =  0 - Normal (gray on black)
              # =  1 - Intensity Bit set for
                       Foreground Colors
              # =  4 - Underscore Characters in MDA
              # =  5 - Blink Characters in MDA
              # =  7 - Reverse the Character
                       Foreground Color with the
                       background
              # =  8 - Make MDA Characters Invisible
              # = 30 - Black Foreground
              # = 31 - Red Foreground
              # = 32 - Green Foreground
              # = 33 - Yellow Foreground
              # = 34 - Blue Foreground
              # = 35 - Magenta Foreground
              # = 36 - Cyan Foreground
              # = 37 - White Foreground
              # = 40 - Black Background
              # = 41 - Red Background
              # = 42 - Green Background
              # = 43 - Yellow Background
              # = 44 - Blue Background
              # = 45 - Magenta Background
              # = 46 - Cyan Background
              # = 47 - White Background
Esc[2j        Clear the Display
Esc[K         Erases from the Current Cursor Position
              to End of the Line
Esc[6n        Device Status Report - request the
              current position to be returned in the
              "Standard Input" Device
Esc[#;%R      This is the Current Cursor Row ("#") and
              Column ("%") loaded into the "Standard
              Input" after a "Device Status Report"
Esc[#;%f      Move Cursor to Row "#" and Column "%"
Esc[#;%F      Move Cursor to Row "#" and Column "%"
Esc[#;%H      Move Cursor to Row "#" and Column "%"
Esc[#A        Move the Cursor Up # Rows
Esc[#B        Move the Cursor Down # Rows
Esc[#C        Move the Cursor to the Right by #
              Columns
```

```
Esc[#D     Move the Cursor to the Left by # Columns
Esc[s      Saves the Current Cursor Position
Esc[u      Restores the Cursor Position to the
             saved position
Esc[F      Move the Cursor to the "Home" Position
             (Row = Column = 1)
Esc[H      Move the Cursor to the "Home" Position
             (Row = Column = 1)
Esc[#;%p   Reassign key "#" to "%"
Esc[#;STRp Reassign key "#" to String "STR"
```

IBM PC extended ASCII characters

The additional 128 characters shown in Fig. 16.2 can do a lot to enhance a character mode application without having to resort to using graphics. These enhancements include special characters for languages other than English, engineering symbols, and simple graphics characters. These simple graphics characters allow lines, and boxes in applications can be created (Figs. 16.1 and 16.2).

Windows ASCII characters

ASCII control characters do have meaning in Windows applications and do not have corresponding graphics characters for video RAM. The Windows character set starts with the "Blank" (ASCII 0x020) and only has the 232 upper characters defined. This character set is based on ASCII with the upper 128 characters defined for special functions and "National Languages" (Fig. 16.3).

EBCDIC

"Extended Binary-Coded Decimal Interchange Code". In the Table below, empty spaces do not have any characters. Note that EBCDIC is an 8-bit code.

Hex	0 x	1 x	2 x	3 x	4 x	5 x	6 x	7 x
x 0	_0_	▬ _16_	SP _32_	0 _48_	@ _64_	P _80_	` _96_	p _112_
x 1	☺ _1_	◄ _17_	! _33_	1 _49_	A _65_	Q _81_	a _97_	q _113_
x 2	● _2_	↕ _18_	" _34_	2 _50_	B _66_	R _82_	b _98_	r _114_
x 3	♥ _3_	‼ _19_	# _35_	3 _51_	C _67_	S _83_	c _99_	s _115_
x 4	♦ _4_	¶ _20_	$ _36_	4 _52_	D _68_	T _84_	d _100_	t _116_
x 5	♣ _5_	§ _21_	% _37_	5 _53_	E _69_	U _85_	e _101_	u _117_
x 6	♠ _6_	▬ _22_	& _38_	6 _54_	F _70_	V _86_	f _102_	v _118_
x 7	• _7_	↕ _23_	' _39_	7 _55_	G _71_	W _87_	g _103_	w _119_
x 8	◘ _8_	↑ _24_	(_40_	8 _56_	H _72_	X _88_	h _104_	x _120_
x 9	◙ _9_	↓ _25_	) _41_	9 _57_	I _73_	Y _89_	i _105_	y _121_
x A	◙ _10_	→ _26_	* _42_	: _58_	J _74_	Z _90_	j _106_	z _122_
x B	♂ _11_	← _27_	+ _43_	; _59_	K _75_	[_91_	k _107_	{ _123_
x C	♀ _12_	∟ _28_	, _44_	< _60_	L _76_	\ _92_	l _108_	¦ _124_
x D	♪ _13_	↔ _29_	- _45_	= _61_	M _77_	] _93_	m _109_	} _125_
x E	♫ _14_	▲ _30_	. _46_	> _62_	N _78_	^ _94_	n _110_	~ _126_
x F	☼ _15_	▼ _31_	/ _47_	? _63_	O _79_	_ _95_	o _111_	△ _127_

Figure 16.1 IBM PC "Extended ASCII" Set 0–0x07F

Hex	8x	9x	Ax	Bx	Cx	Dx	Ex	Fx
x0	ç 128	É 144	á 160	▓ 176	└ 192	╨ 208	α 224	≡ 240
x1	ü 129	æ 145	í 161	▓ 177	┴ 193	╤ 209	β 225	± 241
x2	é 130	Æ 146	ó 162	▓ 178	┬ 194	╥ 210	Γ 226	≥ 242
x3	â 131	ô 147	ú 163	│ 179	├ 195	╙ 211	π 227	≤ 243
x4	ä 132	ö 148	ñ 164	┤ 180	─ 196	╘ 212	Σ 228	⌠ 244
x5	à 133	ò 149	Ñ 165	╡ 181	┼ 197	╒ 213	σ 229	⌡ 245
x6	å 134	û 150	ª 166	╢ 182	╞ 198	╓ 214	µ 230	÷ 246
x7	ç 135	ù 151	º 167	╖ 183	╟ 199	╫ 215	τ 231	≈ 247
x8	ê 136	ÿ 152	¿ 168	╕ 184	╚ 200	╪ 216	Φ 232	° 248
x9	ë 137	Ö 153	⌐ 169	╣ 185	╔ 201	┘ 217	Θ 233	∙ 249
xA	è 138	Ü 154	¬ 170	║ 186	╩ 202	┌ 218	Ω 234	· 250
xB	ï 139	¢ 155	½ 171	╗ 187	╦ 203	█ 219	δ 235	√ 251
xC	î 140	£ 156	¼ 172	╝ 188	╠ 204	▄ 220	∞ 236	ⁿ 252
xD	ì 141	¥ 157	¡ 173	╜ 189	═ 205	▌ 221	φ 237	² 253
xE	Ä 142	₧ 158	« 174	╛ 190	╬ 206	▐ 222	∈ 238	■ 254
xF	Å 143	ƒ 159	» 175	┐ 191	╧ 207	▀ 223	∩ 239	255

Figure 16.2 IBM PC "Extended ASCII" Set 0x080-0x0FF

	!	"	#	$	%	&	'	(	)	*	+	,	-	.	/	0	1	2	3	4	5	6	7	8	9	:	;	<	=	>	?
@	A	B	C	D	E	F	G	H	I	J	K	L	M	N	O	P	Q	R	S	T	U	V	W	X	Y	Z	[	\	]	^	_
`	a	b	c	d	e	f	g	h	i	j	k	l	m	n	o	p	q	r	s	t	u	v	w	x	y	z	{	|	}	~	□
€	□	,	ƒ	„	…	†	‡	^	‰	Š	‹	Œ	□	□	□	□	'	'	"	"	•	–	—	~	™	š	›	œ	□	□	Ÿ
	¡	¢	£	¤	¥	¦	§	¨	©	ª	«	¬	-	®	¯	°	±	²	³	´	µ	¶	·	¸	¹	º	»	¼	½	¾	¿
À	Á	Â	Ã	Ä	Å	Æ	Ç	È	É	Ê	Ë	Ì	Í	Î	Ï	Ð	Ñ	Ò	Ó	Ô	Õ	Ö	×	Ø	Ù	Ú	Û	Ü	Ý	Þ	ß
à	á	â	ã	ä	å	æ	ç	è	é	ê	ë	ì	í	î	ï	ð	ñ	ò	ó	ô	õ	ö	÷	ø	ù	ú	û	ü	ý	þ	ÿ

Figure 16.3 Microsoft Windows "Arial" Font

3-0 7-4>	0	1	2	3	4	5	6	7	8	9	A	B	C	D	E	F
V																
0					SP	&	-									0
1						/			a	j		A	J			1
2									b	k	s	B	K	S	2
3									c	l	t	C	L	T	3
4									d	m	u	D	M	U	4
5				LF					e	n	v	E	N	V	5
6									f	o	w	F	O	W	6
7									g	p	x	G	P	X	7
8									h	q	y	H	Q	Y	8
9									i	r	z	I	R	Z	9
A				CT	!		:									
B				.	$	,	#									
C				<	*	%	@									
D				(	)	_										
E				+	;	>	=									
F				|	?	"										

"SP" is "Space" and "CT" is a "Cents" ("¢") character.

Audio Notes

Notes around Middle "C". Note that an Octave above is twice the note frequency and an Octave below is one-half the note frequency.

Note	Frequency
G	392 Hz

G#	415.3 Hz
A	440 Hz
A#	466.2 Hz
B	493.9 Hz
C	523.3 Hz
C#	554.4 Hz
D	587.3 Hz
D#	622.3 Hz
E	659.3 Hz
F	698.5 Hz
F#	740.0 Hz
G	784.0 Hz
G#	830.6 Hz
A	880.0 Hz
A#	932.3 Hz
B	987.8 Hz

"Touch-Tone" Telephone Frequencies

Frequency	1209 Hz	1336 Hz	1477 Hz
697 Hz	1	2	3
770 Hz	4	5	6
852 Hz	7	8	9
941 Hz	*	0	#

Modem "AT" Commands

"AT" refers to the command "prefix" that is sent before each command to the modem. All Commands (except for "A/") must start with the ASCII Characters "AT" and end with an ASCII Carriage Return (0x00D).

Command "A/" will cause the modem to repeat the last command. The command will repeat upon receipt of the "/" character.

Command "+++" will force the modem from "on-line" state to local ("AT Command Set") state. Do not pass data to the modem for one second before and one second after this command.

Command	Operation	Expected Reply
AT	Sent without a prefix, then Modem is tested	
A	If "AT" "ATA" forces the modem to take the line "off hook". Before executing this command, make sure the string "RING" has been received by the modem	"OK" "OK"
B#	Set the Communications Preference # = 0, CCITT Mode # = 1, Bell 103/212A Default - V.21/V.22 (High Speed)	"OK"
DP	"ATDP ######" Dial the Specified Number using "Pulse Dialing". "," in digit string causes a delay. "W" in digit string causes the modem to wait for a dial tone before continuing. ":" in digit string causes a wait for calling card tone. "@" in digit string causes a wait for quiet period "!" in digit string causes the modem to go on hook and off hook momentarily. "R" at the end of the digit string causes the modem to go on hook and into "auto answer" mode after dialing. ";" at the end of the digit string causes the modem to go into the local command state after connecting. "S#" dials the number stored in location "#"	NO DIALTONE NO ANSWER NO CARRIER BUSY CONNECT 300 CONNECT 300/REL CONNECT 1200 CONNECT 1200/REL CONNECT 2400 CONNECT 2400/REL CONNECT 4800 CONNECT 4800/REL CONNECT 7200 CONNECT 7200/REL CONNECT 9600 CONNECT 9600/REL CONNECT 12000 CONNECT 12000/REL CONNECT 14400 CONNECT 14400/REL

617

Command	Operation	Expected Reply
DT	"ATDT ######" Dial the Specified Number using "Tone Dialing". "," in digit string causes a delay. "W" in digit string causes the modem to wait for a dial tone before continuing. ":" in digit string causes a wait for calling card tone. "@" in digit string causes a wait for quiet period. "!" in digit string causes the modem to go on hook and off hook momentarily. "R" at the end of the digit string causes the modem to go on hook and into "auto answer" mode after dialing. ";" at the end of the digit string causes the modem to go into the local command state after connecting. "S#" dials the number stored in location "#"	NO DIALTONE NO ANSWER NO CARRIER BUSY CONNECT 300 CONNECT 300/REL CONNECT 1200 CONNECT 1200/REL CONNECT 2400 CONNECT 2400/REL CONNECT 4800 CONNECT 4800/REL CONNECT 7200 CONNECT 7200/REL CONNECT 9600 CONNECT 9600/REL CONNECT 12000 CONNECT 12000/REL CONNECT 14400 CONNECT 14400/REL
E#	Turn on or off the AT Command "Echo" State. # = 0, Turn off "Echo" Mode # = 1, Turn on "Echo" Mode (Default)	OK
H[#]	First Enter "+++" Command and then send "ATH#". # = 0, Put modem on hook # = 1, Put modem off hook	OK

Command	Description	Response
I#	Request Modem Information	# = 0, Product ID
		# = 1, Modem Code
		# = 2, "OK"
		# = 3, Country Code
		# = 4, Return Features
L#	Speaker Code (0 Soft, 9 Loud)	
M#	Control Speaker	OK
	# = 0, Turn off Speaker	OK
	# = 1, Turn on Speaker until Carrier Established (Default)	
	# = 2, Leave Speaker on Continuously	
	# = 3, Speaker on except when dialing	
N#	Specify Communication Preference	OK
	# = 0, Use S37 for Speed Selection. If S37 = 0, then connect at Highest Speed Possible	
	# = 1, Connect at Speed Set in S37	
O#	Return to on line state	OK
	# = 0, Return to on line state after using "+++" Command Sequence	
	# = 1, Return to on line state after Carrier "retrain"	
P	Enable Pulse Dialing	
Q#	Specify modem reply returned. See "V#"	OK
	# = 0, Send Result Codes (Default)	
	# = 1, Turn off Messages	
	# = 2, Send Result Codes when Originating call	

Command	Operation	Expected Reply Register Contents
S#?	Return the Contents of the Register "#"	OK
S#=Constant	Set the Register "#" to "Constant"	
T	Enable Tone Dialing	
V#	Verbalize Commands. See "Q#"	OK
	# = 0, Displays Response Numbers	
	# = 1, Displays Response Reply (Default)	
W#	Process Result Codes	OK
	# = 0, Do not display "Carrier" Information (Default)	
	# = 1, Display "Carrier" Information	
X#	Output Active Result Codes	OK
	# = 0, Return only Error and "CONNECT" Replies	
	# = 1, Return only Error and Initial "CONNECT" Replies	
	# = 2, Return only Error and Initial "CONNECT" Replies	
	# = 3, Return all Error and Initial "CONNECT" Replies	
	# = 4, Return all Replies (Default)	
Y#	Indicate "Break" interval	OK
	# = 0, Breaks are Ignored (Default)	
	# = 1, Hang up when Break Received	
	# = 2, Return to Command State but do NOT hang up when Break Received	
Z#	Modem Reset	OK
	# = n, Load Profile n	

Command	Description	Response
&F	Recall Default Profile	OK
&G#	Specify Guard Tone Transmission	OK
	# = 0, No Guard Tone (Default)	
	# = 1, Output 1.8 KHz Guard Tone	
&L#	Specify Leased Line for Signal Lock	OK
	# = 0, Dial up line (Default)	
	# = 1, Conditioned Leased Line	
&Q#	Select Connection Mode	OK
	# = 0, Asynchronous, No Error Control	
	# = 5, Fastest Connection Possible made, Fallback if Problems	
&V	Display Current and Saved Profiles	Active Profile
		Saved Profile 0
		Saved Profile 1
		Saved Telephone Numbers
&W#	Save Current Profile	OK
	# = 0, Save Current in Profile 0	
	# = 1, Save Current in Profile 1	
&Y#	Specify Start up Profile "#"	OK
&Z#=###...	Save Specified Telephone Number. Note Digit String can have the parameters Listed in "DP" and "DT"	OK

Modem registers

All registers are 8 bits in size and take the range 0x000 to 0x0FF unless otherwise noted. Registers handle numeric data as decimal rather than Hex. Below are Hex Values shown for compatibility with this chapter.

Register	Description	Default
S0	Number of Rings before Auto-Answer	0
S1	Ring Counter	N/A
S2	Escape Character (7 Bit ASCII)	0x02B
("+")		
S3	Line End Character	0x00D (CR)
S4	Line Feed Character	0x00A (LF)
S5	Backspace Character	0x008 (BS)
S6	Initial Dialing Wait (in Seconds)	2
S7	Carrier Wait (Seconds)	50
S8	Pause Time (Seconds)	2
S9	Carrier Detect Response Time	6
	(1/10 Seconds)	
S10	Disconnect Time (1/10 Seconds)	14
S11	Tone Dialing Spacing (msecs)	95
S12	Escape Code Guard Time (1/50 Seconds)	50
S18	Self Test Duration (Seconds)	0
S36	Negotiation Failure Response (Settings)	5
	# = 0, Attempt V.42	
	# = 3, Attempt V.42/Attempt MNP	
	# = 4, Attempt V.42/Attempt MNP	
	# = 5, Attempt V.42/Attempt MNP/Attempt	
	Asynchronous Connection	
S37	Desired Connection Speed	0
	# = 0, Connect at Highest Possible	
	Speed (Default)	
	# = 3, 300 bps	
	# = 5, 1200 bps	
	# = 6, 2400 bps	
	# = 7, 4800 bps	

Register	Description	Default
	# = 8, 7200 bps	
	# = 9, 9600 bps	
	# = 10, 12000 bps	
	# = 11, 14400 bps	
S38	Delay before hang up (Seconds)	0
S46	V.42 bis Data Compression Settings	138
	# = 136, V.42 only	
	# = 138, V.42 with V.42 bis compression	
	(Default)	
S48	Feature Negotiation	7
	# = 0, Negotiation Disabled	
	# = 3, Negotiation without Detection	
	Phase	
	# = 7, Negotiation with Detection Phase	
	(Default)	
S95	Error Control Negotiation Messaging	32
	# = 1, Not Used	
	# = 4, Enables Carrier Messages Only	
	# = 8, Enables Carrier and Protocol	
	Messages Only	
	# = 32, Enables Carrier, Protocol and	
	Compression Messages (Default)	

Morse Code

```
".″ - Dot
"-″ - Dash
Character        Code
A                .-
B                -...
C                -.-.
D                -..
E                .
F                ..-.
G                --.
H                ....
I                ..
J                .---
K                -.-
L                .-..
M                --
N                -.
O                ---
P                .--.
Q                --.-
R                .-.
S                ...
T                -
U                ..-
V                ...-
W                .-
X                -..-
Y                -.--
Z                --..
1                .----
2                ..---
3                ...--
4                ....-
5                .....
6                -....
7                --...
8                ---..
9                ----.
0                -----
Period           .-.-.-
,                --..--
:                ---...
Dash             -....-
```

```
/                    - . . - .
?                    . . - - . .
Error                . . . . . . . .
End Trans            . - . - .
Inv Trans            - . -
```

Phonetic Alphabets

Letter	Engineering	Aviation
A	Able	Alpha
B	Baker	Bravo
C	Charlie	Charlie
D	Dog	Delta
E	Easy	Echo
F	Fox	Foxtrot
G	George	Gulf
H	Harry	Hotel
I	Izzy	India
J	Joe	Juliet
K	Kitten	Kilo
L	Larry	Lima
M	Mike	Mike
N	Nancy	November
O	Oscar	Oscar
P	Peter	Papa
Q	Quincy	Quebec
R	Robert	Romeo
S	Sam	Sierra
T	Tom	Tango
U	Under	Uniform
V	Vic	Victor
W	Walter	Whiskey
X	X-Ray	X-Ray
Y	Young	Yankee
Z	Zebra	Zulu

"Ten" Radio Codes

Code	Message
10-1	Receiving Poorly, Bad Signal
10-2	Receiving OK, Strong Signal
10-3	Stop Transmitting
10-4	Message Received

10-5	Relay Message
10-6	Busy, Please Stand By
10-7	Out of Service
10-8	In Service
10-9	Repeat Message
10-10	Finished, Standing By
10-11	Talk Slower
10-12	Visitors Present
10-13	Need Weather/Road Conditions
10-16	Pickup Needed at _____
10-17	Urgent Business
10-18	Is there anything for us
10-19	Nothing for you, Return to Base
10-20	My Location is _____
10-21	Use a Telephone
10-22	Report in Person to _____
10-23	Stand By
10-24	Finished Last Assignment
10-25	Contact _____
10-26	Disregard Last Information
10-27	I'm Changing to Channel _____
10-28	Identify your Station
10-29	You Time is up for Contact
10-30	Does not Conform to FCC Rules
10-32	I'll Give you a Radio Check
10-33	Emergency Traffic at this Station
10-34	Help Needed at this Station
10-35	Confidential Information
10-36	The Correct Time is _____
10-37	Wrecker needed at _____
10-38	Ambulance needed at _____
10-39	Your Message has been Delivered
10-41	Please Change to Channel _____
10-42	Traffic Accident at _____
10-43	Traffic Congestion at _____
10-44	I have a Message for _____
10-45	All Units Within Range Please Report In
10-50	Break Channel
10-60	What is the Next Message Number
10-62	Unable to Copy, Please call on Telephone
10-63	Net Directed to _____
10-64	Net Clear
10-65	Standing By, Awaiting Your Next Message
10-67	All Units Comply
10-70	Fire at _____

10-71	Proceed with Transmission in Sequence
10-73	Speed Trap at _____
10-75	Your Transmission is Causing Interference
10-77	Negative Contact
10-81	Reserve Hotel Room for _____
10-82	Reserve Room for _____
10-84	My Telephone Number is _____
10-85	My Address is _____
10-89	Radio Repairman is Needed at _____
10-90	I have TVI
10-91	Talk Closer to the Microphone
10-92	Your Transmitter Needs Adjustment
10-93	Check my Frequency on this Channel
10-94	Please give me a Long Count
10-95	Transmit Dead Carrier for 5 Seconds
10-99	Mission Completed, All Units Secure
10-200	Police Needed at _____

Miscellaneous Electronics

Resistor Color Coding

Color Coding on resistors is based on the "Bands" around the device (Fig. 17.1).

The Actual Value is determined as:

```
Resistance = ( First Digit * 10 + Second Digit ) *
Multiplier
```

Number	Color	Band1	Band2	Band3	Band4			Optional Band5
0	Black	N/A	0	0	10	**	0	N/A
1	Brown	1	1	1	10	**	1	1% Tolerance
2	Red	2	2	2	10	**	2	2% Tolerance

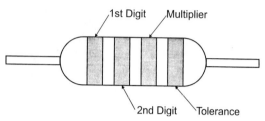

Figure 17.1 Resistor Bands

		1st Digit	2nd Digit		Multiplier	Tolerance
3	Orange	3	3	3	10 ** 3	N/A
4	Yellow	4	4	4	10 ** 4	N/A
5	Green	5	5	5	10 ** 5	0.5% Tolerance
6	Blue	6	6	6	10 ** 6	0.25% Tolerance
7	Violet	7	7	7	10 ** 7	0.1% Tolerance
8	Gray	8	8	8	10 ** 8	0.05% Tolerance
9	White	9	9	9	10 ** 9	N/A
N/A	Gold	N/A	N/A	N/A	10 ** -1	5% Tolerance
N/A	Silver	N/A	N/A	N/A	10 ** -2	10% Tolerance

Electromagnetic Spectrum

Frequency	Use
0 Hz	Direct Current (DC)
15-20,000 Hz	Audio Frequencies
30-15,000 Hz	Human Hearing
16-4186.01 Hz	Musical Scales
10 KHz - 16 KHz	"Ultrasonics"
3-30 KHz	Very Low Frequency
3-30 Hz	Extremely Low Frequency Radio Transmissions
30-300 Hz	Ultra Low Frequency Radio Transmissions
30 KHz - 300 MHz	Radio Frequencies
53.5 KHz - 170.5 KHz	AM Broadcast Bands

3.5 MHz - 4 MHz	80 Meter Amateur Band
7 MHz - 7.3 MHz	40 Meter Amateur Band
10.100 MHz - 10.150 MHz	30 Meter Amateur Band
14.10 MHz - 14.35 MHz	20 Meter Amateur Band
18.068 MHz - 18.168 MHz	17 Meter Amateur Band
21.00 MHz - 21.45 MHz	15 Meter Amateur Band
24.890 MHz - 24.990 MHz	12 Meter Amateur Band
26.965 MHz - 27.405 MHz	Citizens Band ("CB")
26.95 MHz - 27.54 MHz	Industrial, Scientific, Medical Use
28.00 MHz - 29.70 MHz	10 Meter Amateur Band
30 MHz - 300 MHz	Very High Frequencies
30 MHz - 50 MHz	Police, Fire, Forest, Highway, Railroad
50 MHz - 54 MHz	6 Meter Amateur Band
54 MHz - 72 MHz	TV Channels 2 to 4
72 MHz - 76 MHz	Government
76 MHz - 88 MHz	TV Channels 5 and 6
88 MHz - 108 MHz	FM Broadcast Band
108 MHz - 118 MHz	Aeronautical Navigation
118 MHz - 136 MHz	Civil Communication Band
148 MHz - 174 MHz	Government
144 MHz - 148 MHz	2 Meter Amateur Band
174 MHz - 216 MHz	TV Channels 7 through 13
216 MHz - 470 MHz	Miscellaneous Communication
220 MHz - 225 MHz	1 ¼ Meter Amateur Band
225 MHz - 400 MHz	Military
420 MHz - 450 MHz	0.7 Meter Amateur Band
462.55 MHz - 563.20 MHz	Citizens Band
300 MHz - 3000 MHz	Ultra High Frequencies/Radar
470 MHz - 806 MHz	TV Channels 14 through 69
806 MHz - 890 MHz	Cellular Telephone
890 MHz - 3000 MHz	Miscellaneous Communication
3 GHz - 30 GHz	Miscellaneous Communication/Radar
30 GHz - 300 GHz	Super High Frequencies/Radar
Wavelength	Radiation Type
30 um - 0.76 um	Infrared Light and Heat

```
0.76 um - 0.39 um          Visible Light
6470 - 7000 Angstroms      Red Light
5850 - 6470 Angstroms      Orange Light
5750 - 5850 Angstroms      Yellow Light
5560 - 5750 Angstroms      Maximum Visibility Light
4912 - 5560 Angstroms      Green Light
4240 - 4912 Angstroms      Blue Light
4000 - 4240 Angstroms      Violet Light

320 - 4000 Angstroms       Ultraviolet Light

0.032 - 0.00001 um         X-Rays

0.00001 - 0.0000006 um     Gamma Rays

< 0.0005 Angstroms         Cosmic Rays
```

Radar bands

```
Frequency                  Band
390 - 1,550 MHz            L
1,550 - 5,200 MHz         S
5,200 - 10,900 MHz        X
10,900 - 36,000 MHz       K
36,000 - 46,000 MHz       Q
46,000 - 56,000 MHz       V
56,000 - 100,000 MHz      W
```

Digital Logic

The Output/Threshold Levels for +5V Logic is:

```
Technology  Input       Output "Low"  Output "High"
            Threshold
TTL         1.4 Volts   0.3 Volts     3.3 Volts
HC          2.4 Volts   0.1 Volts     4.9 Volts
HCT         1.4 Volts   0.1 Volts     4.9 Volts
CMOS        2.5 Volts   0.1 Volts     4.9 Volts
```

Gates

The six most common Logic Gates are:

Type	Symbol	State Table		

NOT   A —▷o— Output

"A"	Output
0	1
1	0

AND   A, B —D— Output

"A"	"B"	Output
0	0	0
0	1	0
1	0	0
2	1	1

OR   A, B —D— Output

"A"	"B"	Output
0	0	0
0	1	1
1	0	1
1	1	1

XOR   A, B —))D— Output

"A"	"B"	Output
0	0	0
0	1	1
1	0	1
1	1	0

NAND   A, B —Do— Output

"A"	"B"	Output
0	0	1
0	1	1
1	0	1
1	1	0

NOR   A, B —Do— Output

"A"	"B"	Output
0	0	1
0	1	0
1	0	0
1	1	0

Flip flops

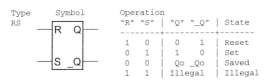

Type	Symbol	Operation			

RS

"R"	"S"	"Q"	"_Q"	State
1	0	0	1	Reset
0	1	1	0	Set
0	0	Qo	_Qo	Saved
1	1	Illegal		Illegal

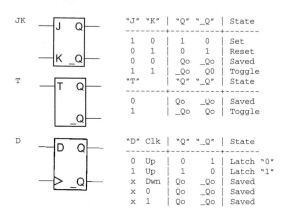

```
JK                         "J" "K" | "Q" "_Q" | State
                           --------+----------+-------
                            1   0  |  1    0   | Set
                            0   1  |  0    1   | Reset
                            0   0  | Qo   _Qo  | Saved
                            1   1  | _Qo   Q0  | Toggle
T                           "T"    | "Q" "_Q" | State
                           --------+----------+-------
                            0      | Qo   _Uo  | Saved
                            1      | _Qo   Qo  | Toggle

D                         "D" Clk  | "Q" "_Q" | State
                          ---------+----------+-------
                           0   Up  |  0    1  | Latch "0"
                           1   Up  |  1    0  | Latch "1"
                           x   Dwn | Qo   _Qo | Saved
                           x   0   | Qo   _Qo | Saved
                           x   1   | Qo   _Qo | Saved
```

Formulas

DC Electronics Formulas

Ohm's Law:

$$V = IR$$

Power:

$$P = VI$$

Series Resistance:

$$Rt = R1 + R2 \ldots$$

Parallel Resistance:

```
Rt = 1 / (( 1/R1) + (1/R2) ... )
```

Two Resistors in Parallel:

```
Rt = ( R1 * R2 ) / ( R1 + R2 )
```

Series Capacitance:

```
Ct = 1 / (( 1/C1) + (1/C2) ... )
```

Parallel Capacitance:

```
Ct = C1 + C2 . . .
```

Wheatstone Bridge:

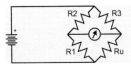

Ru = R1 * R3 / R2

When NoCurrent Flow
in the Meter

AC Electronics Formulas

Resonance:

```
frequency = 1 / ( 2 * pi * SQRT( L * C ))
```

RC Time Constant:

```
Tau = R * C
```

RL Time Constant:

```
Tau = L / R
```

RC Charging:

```
V(t) = Vf * ( 1 - e^(-t/Tau))

i(t) = if * ( 1 - e^(-t/Tau))
```

RC Discharging:

```
V(t) = Vi * e ^ ( -t/Tau )

i(t) = ii * e ^ ( -t/Tau )
```

Coil Inductance Formulas:
• Coil around Linear Form:

```
Inductance = Permeability of Form *
             ( Number of Turns ^ 2 )
             * Form Area / Coil Length
```

• Coil Around Toroid with a square cross-section:

```
Inductance = ln( Outer Diameter / Inner
             Diameter ) * Permeability
             of Form * ( Number of
             Turns ^ 2 ) * Thickness of
             Toroid / ( 2 * pi )
```

Transformer Current/Voltage:
• Turns Ratio = Number of Turns on Primary ("p") Side/Number of Turns on Secondary ("s") Side

```
Turns Ratio = Vs / Vp = Ip / Is
```

• Transmission Line Characteristic Impedance:

```
Zo = SQRT( L / C )
```

Mathematical Formulas

```
Frequency = Speed / Wavelength
```

For Electromagnetic Waves:

```
Frequency = c / Wavelength
```

Perfect Gas Law:

```
PV = nRT
```

Boolean Arithmetic

Identify Functions:

```
A AND 1 = A
A OR 0 = A
```

Output Set/Reset:

```
A AND 0 = 0
A OR 1 = 1
```

Identity Law:

```
A = A
```

Double Negation Law:

```
NOT( NOT( A )) = A
```

Complementary Law:

```
A AND NOT( A ) = 0
A OR NOT( A ) = 1
```

Idempotent Law:

$$A \ AND \ A = A$$
$$A \ OR \ A = A$$

Commutative Law:

$$A \ AND \ B = B \ AND \ A$$
$$A \ OR \ B = B \ OR \ A$$

Associative Law:

$$(A \ AND \ B) \ AND \ C = A \ AND \ (B \ AND \ C)$$
$$= A \ AND \ B \ AND \ C$$
$$(A \ OR \ B) \ OR \ C = A \ OR \ (B \ OR \ C)$$
$$= A \ OR \ B \ OR \ C$$

Distributive Law:

$$A \ AND \ (B \ OR \ C) = (A \ AND \ B) \ OR \ (A \ AND \ C)$$
$$A \ OR \ (B \ AND \ C) = (A \ OR \ B) \ AND \ (A \ OR \ C)$$

De Morgan's Theorem:

$$NOT(A \ OR \ B) = NOT(A) \ AND \ NOT(B)$$
$$NOT(A \ AND \ B) = NOT(A) \ OR \ NOT \ (B)$$

Note:

AND is often represented as multiplication, nothing between terms, "." or "*".

OR is often represented as addition with "+" between terms.

Conversions

1 Inch = 2.54 Centimeters

1 Mile = 1.609 Kilometers

1 Ounce = 29.57 Grams

1 Gallon = 3.78 Liters

1 Atmosphere = 29.9213 Inches of Mercury

            = 14.6960 Pounds per Square Inch

            = 101.325 kiloPascals

10,000,000,000 Angstroms = 1 Meter

1,000,000 Microns = 1 Meter

Tera = 1,000 Giga

Giga = 1,000 Mega

Mega = 1,000 Kilo

Kilo = 1,000 Units

Unit = 100 Centi

Unit = 1,000 Milli

1 Hour = 3,600 Seconds

1 Year = 8,760 Hours

Microchip

Microchip's corporate headquarters is

Microchip Technology, Inc.
2355 W. Chandler Blvd.
Chandler, AZ 85224
Phone: (480) 786-7200
Fax: (480) 917-4150

Their Web site ("Planet Microchip") is at **http://www.microchip.com** and contains a complete set of data sheets in .pdf format for download as well as the latest versions of MPLAB. Also on the website is the link to **http://buy.microchip.com,** which is Microchip's on-line ordering system for parts and development tools.

Microchip puts on a series of seminars throughout the world every year. Information on these events can be found on the microchip Web page.

PICmicro® MCU Books

Note that Microchip has excellent datasheets available for download from their Web site as well as available on CD-ROM.

Programming and Customizing the PICmicro® MCU Eight Bit Microcontroller—Second Edition

Author: M. Predko
ISBN: 0-07-136175-8

Design with PIC Microcontrollers
Author: J. B. Peatman
ISBN: 0-13-759259-0

PICTUTOR
Author: J. Becker
URL:

http://www.matrixmultimedia.co.uk/picprods.htm

PIC'n Techniques
Author: D. Benson
ISBN: 0-9654162-3-2

PIC'n Up the Pace
Author: D. Benson
ISBN: 0-9654162-1-6

Serial PIC'n
Author: D. Benson
ISBN: 0-9654162-2-4

Easy PIC'n
Author: D. Benson
ISBN: 0-9654162-0-8

*The Microcontroller Beginner's Handbook—
2nd Edition*
Author: L. Duarte
ISBN: 0-79061-153-8

An Introduction to PIC Microcontrollers
Author: R. A. Penfold
ISBN: 0-85934-394-4

Practical PIC Microcontroller Projects
Author: R. A. Penfold
ISBN: 0-85934-444-4

*A Beginners Guide to the Microchip PIC—2nd
Edition*
Author: N. Gardner
ISBN: 1-899013-01-6

PIC Cookbook
Author: N. Gardner
ISBN: 1-899013-02-4

Useful Books

Here are a collection of books that are useful for developing electronics and software for applications. Some of these are hard to find, but definitely worth the effort in finding them in a used bookstore.

The Art of Electronics–1989

Horowitz and Hill's definitive book on electronics—a complete electrical engineering course wrapped up in 1125 pages. Some people may find it to be a bit too complex, but just about any analog electronics question you could have will be answered in this book. The digital information in this book is less complete.

ISBN: 0-521-37095-7

Bebop to the Boolean Boogie–1995

Somewhat deeper in digital electronics (and less serious) than *The Art of Electronics*, Clive Maxwell's introduction to electronics stands out with clear and insightful explanations of how things work and why things are done the way they are. It distinguishes itself from other books by explaining Printed Wiring Assembly technology (PCB Boards, Components, and Soldering). This book complements *The Art of Electronics* very nicely.

ISBN: 1-878707-22-1

The Encyclopedia of Electronic Circuits–Volume 1 to 7

Rudolf Graf's Encyclopedia series of Electronic Circuits is an excellent resource of circuits and ideas that have been cataloged according to circuit type. Each book contains thousands of circuits and can really make your life easier when you are trying to figure out how to do something. Each volume contains an index listing circuits for the current volume and the previous ones.

Volume 1, ISBN: 0-8306-1938-0; Volume 2, ISBN: 0-8306-3138-0; Volume 3, ISBN: 0-8306-3348-0; Volume 4, ISBN: 0-8306-3895-4; Volume 5, ISBN: 0-07-011077-8; Volume 6, ISBN: 0-07-011276-2; Volume 7, ISBN: 0-07-015116-4.

CMOS Cookbook–Revised 1988

In *CMOS Cookbook*, Don Lancaster introduces the reader to basic digital electronic theory, while also explaining the operation of CMOS gates, providing hints on soldering and prototyping, listing common CMOS parts (along with TTL pinout equivalents) and providing a number of example circuits (including a good basic definition of how NTSC video works). The update by Howard Berlin has made sure the chips presented in the book are still available. In the 1970s, Don Lancaster also wrote the *TTL Cookbook*, *555 Timer Cookbook*, and *Active Filter Cookbook*.

ISBN: 0-7506-9943-4

The TTL Data Book for Design Engineers–Texas Instruments

I have a couple of 1981 printed copies of the second edition of this book and they are all falling apart from overuse. The Texas Instruments TTL data books have been used for years by hundreds of thousands of engineers to develop their circuits. Each datasheet is complete with pinouts, operating characteristics, and internal circuit diagrams. While the data books are not complete for the latest "HC" parts, they will give you just about everything you want to know about the operation of small scale digital logic.

ISBN: N/A

PC PhD–1999

This book/CD-ROM package was written to give a clear introduction to the PC, from a "bottoms up" hardware perspective as well as an explanation of how code works in the PC. Along with explaining the architecture, there are also over twenty applications that will help the reader understand exactly how MS-DOS and Windows code executes in the PC and how hardware is accessed using the various interfaces available within the PC.

ISBN: 0-07-134186-2

PC Interfacing Pocket Reference–1999

This book is designed as an easy to use pocket reference for programmers and engineers working on the PC. Along with detailing the PC's architecture, the Intel 8086 and later microprocessors are described. The instruction sets used in the processor are listed along with addressing and value information. The information is useful for all PCs from the first 8088s to the most modern multi-Pentium III systems.

ISBN: 0-07-135525-1

The Programmer's PC Source Book–2nd Edition, 1991

Thom Hogan's 850 page book is just about the best and most complete reference that you can find anywhere on the PC. This book basically ends at the 386 (no 486, Pentiums of any flavor, PCI, Northbridge, Southbridge or SuperIO, or any ASICs of any type), but is the most complete PC reference that explains BIOS, all the "Standard" I/O, DOS and Windows 3.x Interfaces you can find.

ISBN: 1-55615-118-7

The Embedded PC's ISA Bus: Firmware, Gadgets and Practical Tricks–1997

Ed Nisley's book is an almost complete opposite to the previous two books and *The Programmer's PC Source Book.* Where the others' books focus is on documenting the innards of the PC, Nisley's shows you how to practically interface to the PC's "Industry Standard Architecture" ("ISA") bus and if you follow through the book you will end up with an LCD graphic display. Theory, register addresses, and programming information is available in this book, but it is presented as Ed works through the projects. This book is a resource that you can go back to and look at actual scope photographs of bus accesses or discussions on how to decode bus signals. There are a lot of great tricks in this book that can be used for many different PC interfacing applications.

ISBN: 1-5739-8017-X

Handbook of Microcontrollers–1998

Introduction and complete reference package for modern 8-bit embedded microcontrollers. As well as providing technical and processor programming information on the: Intel 8051, Motorola 68HC05, Microchip PICmicro® MCU, Atmel AVR and Parallax Basic Stamp, datasheets, development tools and sample applications are included on the included CD-ROM. To help with your future applications, interfacing to RS-232, I2C, LCD and other devices is explored and a fair amount of space is devoted to such advanced topics as Fuzzy Logic, Compilers, Real Time Operating Systems (I have included a sample one for the 68HC05), and Network Communications.

ISBN: 0-07-913716-4

IBM PC and Assembly Language and Programming–
4th Edition, 1997

This is an excellent introduction to assembly language programming with a fairly low level approach concentrating on Microsoft's "MASM" and Borland's "TASM". "Debug.com" is used exclusively as a debug tool, which makes this book reasonably inexpensive to get involved with.

ISBN: 1-1375-6610-7

*The C Programming Language–*2nd Edition, 1988

Brian W. Kernighan, Dennis M. Ritchie's classic text explaining the "C" programming language has not lost any of its usefulness since its first introduction. This book has probably been used by more students in more colleges and universities than any other. Despite the fact that the book was written originally for a programming course, the writing is crisp, sharp, and easily understandable.

ISBN: 0-13110-362-8

PICList Internet List Server

These guidelines should be used and followed for any list server or news group. After the guidelines, there are instructions for subscribing to the PICList.

1. Don't subscribe to a list and then immediately start sending questions to the list. Instead, wait a day or so to get the hang of how messages are sent and replied to on the list and get a "feel" for the best way of asking questions.

2. Some lists send an email sent to them back to the author (while others do not). If you receive a copy of your first email, don't automatically assume that it is a "bounce" (wrong address) and resend it. In this case, you might want to wait a day or so to see if any replies show up before trying to resend it. Once you have been on the list for a while, you should get an idea of how long it takes to show up on the list and how long it takes to get a reply.

3. If you don't get a reply to a request, don't get angry or frustrated and send off a reply demanding help. There is a good chance that nobody on the list knows exactly how to solve your problem. In this case, try to break down the problem and ask the question a different way.

4. Do not count on getting replies to questions within minutes. Nobody on the PICList is paid to reply to your questions. The majority of people who reply are doing so to help others. Please respect that and do not badger, and help out in anyway that you can.

5. If you are changing the "Subject" line of a post, please reference the previous topic (i.e., put in "was: '...'"). This will help others keep track of the conversation.

6. When replying to a previous post, try to minimize how much of the previous note is copied in your note and maximize the relevance to your reply. This is not to say that none of the message should be copied or referenced. There is a very fine balance between having too much and too little. The sender of the note you are replying to should be referenced (with their name or ID).

My rule of thumb is, if the original question is less than ten lines, I copy it all. If it is longer, then I cut it down (identifying what was cut out with a "SNIP" Message), leaving just the question and any relevant information as quoted. Most mail programs will mark the quoted text with a ">" character, please use this convention to make it easier for others to follow your reply.

7. If you have an application that doesn't work, please don't copy the entire source code into an email and post it to a list. As soon as I see an email like this I just delete it and go on to the next one (and I suspect that I'm not the only one). Also, some lists may have a message size limit (anything above this limit is thrown out) and you will not receive any kind of confirmation.

If you are going to post source code: keep it short. People on the list are more than happy and willing to answer specific questions, but simply copying the complete source code in the note and asking a question like "Why won't the LCD display anything" really isn't useful for anybody. Instead, try to isolate the failing code and describe what is actually happening along with what you want to happen. If you do this, chances are you will get a helpful answer quickly.

A good thing to remember when asking why something won't work, make sure you discuss the hardware that you are using. If you are asking about support hardware (i.e., a programmer or emulator), make sure you describe your PC (or workstation) setup. If your application isn't working as expected, describe the hardware that you are using and what you have observed (i.e., if the clock lines are wiggling, or the application works normally when you put a scope probe on a pin).

8. You may find a totally awesome and appropriate Web page and want to share it with the list. Please make it easier on the people in the list to cut and paste the URL by putting it on a line all by itself in the format:

```
http://www.awesome-pic-page.com
```

9. If you have a new application, graphic, or whatever, that takes up more than 1K which you would like to share with everyone on the list, please don't send it as an attachment in a note to the list. Instead, either indicate that you have this amazing piece of work and tell people that you have it and where to request it (either to you directly or to a Web server address). If a large file is received many list servers may automatically delete it (thrown into the "bit bucket") and you may or may not get a message telling you what happened.

   If you don't have a Web page of your own or one you can access, requesting that somebody put it on their Web page or ftp server is acceptable.

10. Many of these List Servers are made available, maintained, and/or moderated by a device manufacturer. Keep this in mind if you are going to advertise your own product and understand what the company's policy is on this before sending out an advertisement.

   The PICList is quite tolerant of advertisements of *relevant* products. If you are boarding puppies or

have something equally non-PICmicro® MCU re-
lated, find somewhere else to advertise it.

11. Putting job postings or employment requests *may*
be appropriate for a list (like the previous point,
check with the list's maintainer). However, I don't
recommend that the rate of pay or conditions of employment should be included in the note (unless
you want to be characterized as cheap, greedy, un-
reasonable, or exploitive).

12. "Spams" are sent to every list server occasionally.
Please do not "reply" to the note even if the message
says that to get off the spammer's mailing list just
"reply". This will send a message to everyone in the
list. If you must send a note detailing your disgust,
send it to the spam originator (although to their ISP
will probably get better results).

NOTE: There are a number of companies sending
out bogus spams to collect the originating addresses
of replying messages and sell them to other companies or distributors of addresses on CD-ROM. When
receiving a spam, see if it has been sent to you per-
sonally or the list before replying—but beware if
you are replying to the spam, you may be just sending your e-mail address for some company to resell
to real spammers.

13. Following up with the previous message, if you are going to put in pointers to a list server, just put a hyper-
link to the list server request email address, NOT TO
THE LIST SERVER ITSELF. If you provide the address to the list server, spammers can pull the link
from your page and use it as an address to send spams
to. By not doing this, you will be minimizing the op-
portunity for spammers to send notes to the list.

14. By sending off-topic messages, while it is tolerated, you will probably bring lots of abuse upon yourself, especially if you are belligerent about it. An occasional notice about something interesting or a joke is fine as long as it is unusual and not likely to attract a lot of replies.

    If you feel it is appropriate to send an off-topic message; some lists request that you put "[OT]" in the subject line, some members of the list use mail filters and this will allow them to ignore the off-topic posts automatically.

    Eventually a discussion (this usually happens with off-topic discussions) will get so strung out that there are only two people left arguing with each other. At this point stop the discussion entirely or go 'private'. You can obtain the other person's e-mail address from the header of the message—send your message to him or her and not to the entire list. Everyone else on the list would have lost interest a long time ago and probably would like the discussion to just go away (so oblige them).

15. Posts referencing Pirate sites and sources for "cracked" or "hacked" software are not appropriate in any case and may be illegal. If you are not sure if it is okay to post the latest software you've found on the Web, then DON'T until you have checked with the owners of the software and gotten their permission. It would also be a good idea to indicate in your post that you have the owner's permission to distribute cracked software.

    A variety of different microcontrollers are used in "Smart Cards" (such as used with Cable and Satellite scrambling) or video game machines and asking how they work will probably result in abusive replies at worst or having your questions ignored at best. If you have a legitimate reason for asking about smart cards, make sure you state it in your email to the list.

16. When you first subscribe to a list, you will get a reply telling you how to unsubscribe from the list. DON'T LOSE THIS NOTE. In the past in some lists, people having trouble unsubscribing have sent questions to the list asking how and sometimes getting angry when their requests go unheeded. If you are trying to unsubscribe from a list and need help from others on the list, explain what you are trying to do and how you've tried to accomplish it.

17. When working with a list server, do *not* have automated replies sent. If they are enabled, then all messages sent by the server to you will be replied to back to the list server. This is annoying for other list members and should be avoided.

18. Lastly, please try to be courteous to all on the list. Others may not have *your* knowledge and experience or they may be sensitive about different issues. There is a very high level of professionalism on all the lists presented below, please help maintain it. Being insulting or rude will only get you the same attitude back and probably will lead to your posts and legitimate questions being ignored in the future by others on the list who don't want to have anything to do with you.

To put this succinctly: **"Don't be offensive or easily offended."**

To subscribe to the PICList, send an email to

```
listserv@mitvma.mit.edu
```

with the message:

```
subscribe piclist <I>your name</I>
```

in the body of your email.

Save the confirmation message; this will give you the instructions for signing off the list as well as instructions on how to access more advanced PICList List Server functions.

To sign off the list, send a note to the same address **(listserv@mitvma.mit.edu)** with the message:

```
signoff piclist
```

When signing off the PICList make sure that you are doing it from the ID that you used to sign on to the list.

Once you have subscribed to the PICLIST, you will begin receiving mail from

```
piclist@mitvma.mit.edu
```

Emails can be sent to this address directly or can be replied to directly from your mailer. The list archive is available at:

```
http://www.iversoft.com/piclist/
```

and it has a searchable summary of the emails that have been sent to the PICList.

Recommended PICmicro® MCU Web Sites

At the time of writing, there is somewhere in the neighborhood of one thousand Web pages devoted to the PICmicro® MCU with different applications, code snippets, code development tools, programmers, and other miscellaneous information on the PICmicro® MCU and other microcontrollers. The following sites are excellent places to start and work through.

The author's Web page has the latest PICmicro® MCU information as well as errata for this book and sample PICmicro® MCU projects.

```
http://www.myke.com
```

Alexy Vladimirov's outstanding list of PICmicro® MCU resource pages. Over 700 listed as of February 2000.

```
http://www.geocities.com/SiliconValley/Way/5807/
```

Bob Blick's Web site. Some interesting PICmicro® MCU projects that are quite a bit different than the run of the mill.

```
http://www.bobblick.com/
```

Scott Dattalo's highly optimized PICmicro® MCU math algorithms. The best place to go if you are looking to calculate Trigonometric Sines in a PICmicro® MCU.

```
http://www.dattalo.com/technical/software/software.html
```

Along with the very fast PICmicro® MCU routines, Scott has also been working on some GNU General Purpose License Tools designed to run under Linux.

The tools can be downloaded from:

```
http://www.dattalo.com/gnupic/gpsim.html
```

```
http://www.dattalo.com/gnupic/gpasm.html
```

Marco Di Leo's "PIC Corner". Some interesting applications including information on networking PICmicro® MCUs and using them for cryptography.

```
http://members.tripod.com/~mdileo/
```

Dontronics Home Page. Don McKenzie has a wealth of information on the PICmicro® MCU as well as other electronic products. There are lots of useful links to other sites and it is the home of the SimmStick.

```
http://www.dontronics.com/
```

Fast Forward Engineering. Andrew Warren's page of PICmicro® MCU information and highly useful question/answer page.

```
http://home.netcom.com/~fastfwd/
```

Steve Lawther's list of PICmicro® MCU Projects. Interesting PICmicro® MCU (and other microcontroller) projects.

```
http://ourworld.compuserve.com/homepages/steve_lawther/
ucindex.htm
```

Eric Smith's PIC Page. Some interesting projects and code examples to work through.

```
http://www.brouhaha.com/~eric/pic/
```

Rickard's PIC-Wall. Good site with a design for PICmicro® MCU-based composite video game generator.

`http://www.efd.lth.se/~e96rg/pic.html`

PicPoint—Lots of good projects to choose from including 5 MB free to anyone that wants to start their own PICmicro® MCU Web page.

`http://www.picpoint.com/`

MicroTronics—Programmers and Application reviews.

`http://www.eedevl.com/index.html`

Tony Nixon's "Pic 'n Poke" Development Systems Home page to the "Pic 'n Poke" development system. This system includes an animated simulator that is an excellent tool for learning how data flows and instructions execute in the PICmicro® MCU microcontroller.

`http://www.picnpoke.com/`

Periodicals

Here are a number of magazines that do give a lot of information and projects on PICmicro® MCUs. Every month, each magazine has a better than 50% chance of presenting a PICmicro® MCU application.

Circuit Cellar Ink
Subscriptions:
   P.O. Box 698
   Holmes, PA 19043-9613
   1(800)269-6301
   Web Site: **http://www.circellar.com/**
   BBS: (860)871-1988

Poptronics
Subscriptions:
  Subscription Department
  P.O. Box 55115
  Boulder, CO
  1(800)999-7139
  Web Site: **http://www.gernsback.com**

Microcontroller Journal
  Web Site: **http://www.mcjournal.com/**
This is published on the Web.

Nuts & Volts
Subscriptions:
  430 Princeland Court
  Corona, CA 91719
  1(800)-783-4624
  Web Site: **http://www.nutsvolts.com**

Everyday Practical Electronics
Subscriptions:
  EPE Subscriptions Dept.
  Allen House, East Borough,
  Wimborne, Dorset,
  BH21 1PF
  United Kingdom
  +44 (0)1202 881749
  Web Site: **http://www.epemag.wimborne.co.uk**

Useful Web Sites

While none of these are PICmicro® MCU specific, they
are a good source of ideas, information, and products
that will make your life a bit more interesting and
maybe give you some ideas for projects for the
PICmicro® MCU.

Seattle Robotics Society

http://www.hhhh.org/srs/

The Seattle Robotics Society has lots of information on interfacing digital devices to such "real world" devices as motors, sensors, and servos. They also do a lot of exciting things in the automation arena. Most of the applications use the Motorola 68IIC11.

List Of Stamp Applications (L.O.S.A)

http://www.hth.com/losa.htm

The List of Parallax Basic Stamp Applications will give you an idea of what can be done with the Basic Stamp (and other microcontrollers, such as the PICmicro® MCU). The list contains projects ranging from using a Basic Stamp to giving a cat medication to providing a simple telemetry system for model rockets.

Adobe PDF Viewers

http://www.adobe.com

Adobe .pdf file format is used for virtually all vendor datasheets, including the devices presented in this book (and their datasheets on the CD-ROM).

"PKZip" and "PKUnZip"

http://www.pkware.com

PKWare's "zip" file compression format is a "Standard" for combining and compressing files for transfer.

Hardware FAQs

**http:paranoia.com/~filipg/HTML/LINK/LINK_IN
.html**

A set of FAQs (Frequently Asked Questions) about the PC and other hardware platforms that will come in useful when interfacing a microcontroller to a Host PC.

http://www.innovatus.com

Innovatus has made available "PICBots", an interesting PICmicro® MCU simulator which allows programs to be written for virtual robots which will fight amongst themselves.

Part Suppliers

The following companies supplied components that are used in this book. I am listing them because they all provide excellent customer service and are able to ship parts anywhere you need them.

Digi-Key

Digi-Key is an excellent source for a wide range of electronic parts. They are reasonably priced and most orders will be delivered the next day. They are real lifesavers when you're on a deadline.

Digi-Key Corporation
701 Brooks Avenue South
P.O. Box 677
Thief River Falls, MN 56701-0677

Phone: 1(800)344-4539 [1(800)DIGI-KEY]
Fax: (218)681-3380

http://www.digi-key.com/

AP Circuits

AP Circuits will build prototype bare boards from your "Gerber" files. Boards are available within three days. I have been a customer of theirs for several years and they have always produced excellent quality and been helpful in providing direction to learning how to develop my own bare boards. Their Web site contains the "EasyTrax" and "GCPrevue" MS-DOS tools necessary to develop your own Gerber files.

Alberta Printed Circuits Ltd.
#3, 1112-40th Avenue N.E.
Calgary, Alberta T2E 5T8

Phone: (403)250-3406
BBS: (403)291-9342
Email: staff@apcircuits.com

http://www.apcircuits.com/

Wirz Electronics

Wirz Electronics is a full service Microcontroller component and development system supplier. Wirz Electronics is the main distributor for projects contained in this book and will sell assembled and tested kits of the projects. Wirz Electronics also carries the "SimmStick" prototyping systems as well as their own line of motor and robot controllers.

Wirz Electronics
P.O. Box 457
Littleton, MA 01460-0457

Toll Free in the USA & Canada: 1(888)289-9479
[1(888)BUY-WIRZ]
Email: sales@wirz.com

http://www.wirz.com/

Tower Hobbies

Excellent source for Servos and R/C parts useful in homebuilt robots.

Tower Hobbies
P.O. Box 9078
Champaign, IL 61826-9078

Toll Free Ordering in the USA & Canada: 1(800)637-4989
Toll Free Fax in the USA & Canada: 1(800)637-7303
Toll Free Support in the USA & Canada: 1(800)637-6050
Phone: (217)398-3636
Fax: (217)356-6608
Email: orders@towerhobbies.com

http://www.towerhobbies.com/

Jameco

Components, PC Parts/Accessories, and hard to find connectors.

Jameco
1355 Shoreway Road
Belmont, CA 94002-4100

Toll Free in the USA & Canada: 1(800)831-4242

http://www.jameco.com/

JDR

Components, PC Parts/Accessories, and hard to find connectors.

JDR Microdevices
1850 South 10th St.
San Jose, CA 95112-4108

Toll Free in the USA & Canada: 1(800)538-5000
Toll Free Fax in the USA & Canada: 1(800)538-5005
Phone: (408)494-1400
Email: techsupport@jdr.com
BDS: (408)494-1430
Compuserve: 70007,1561

http://www.jdr.com/JDR

Newark

Components—Including the Dallas Line of Semi-
conductors (the DS87C520 and DS275 is used for
RS-232 Level Conversion in this book).

Toll Free in the USA & Canada: 1(800)463-9275
[1(800)4-NEWARK]

http://www.newark.com/

Marshall Industries

Marshall is a full-service distributor of Philips microcontrollers as well as other parts.

Marshall Industries
9320 Telstar Avenue
El Monte, CA 91731
1(800)833-9910

http://www.marshall.com

Mouser Electronics

Mouser is the distributor for the Seiko S7600A TCP/IP Stack Chips.

> Mouser Electronics, Inc.
> 958 North Main Street
> Mansfield, Texas 76063
> Sales: (800) 346-6873
> Sales: (817) 483-6888
> Fax: (817) 483-6899
> Email: sales@mouser.com

> **http://www.mouser.com**

Mondo-tronics Robotics Store

Self-proclaimed as "The World's Biggest Collection of Miniature Robots and Supplies" and I have to agree with them. This is a great source for Servos, Tracked Vehicles, and Robot Arms.

> Order Desk
> Mondo-tronics Inc.
> 524 San Anselmo Ave #107-13
> San Anselmo, CA 94960

> Toll Free in the USA & Canada: 1(800)374-5764
> Fax: (415)455-9333

> **http://www.robotstore.com/**

Index

Note: **Boldface** numbers indicate illustrations.